From Buddy To Boss
EFFECTIVE FIRE SERVICE LEADERSHIP

From Buddy To Boss
EFFECTIVE FIRE SERVICE LEADERSHIP

Chase Sargent

Fire Engineering

> **Disclaimer.** The recommendations, advice, descriptions, and methods in this book are presented solely for educational purposes. The author and publisher assume no liability whatsoever for any loss or damage that results from the use of any of the material in this book. Use of the material in this book is solely at the risk of the user.

Copyright© 2006 by
PennWell Corporation
1421 South Sheridan Road
Tulsa, Oklahoma 74112-6600 USA

"The Paradoxical Commandments" are reprinted by permission of the author.
Copyright© Kent M. Keith 1968, renewed 2001

Cover photos by Chase Sargent and Martin C. Grube.

800.752.9764
+1.918.831.9421
sales@pennwell.com
www.pennwellbooks.com
www.pennwell.com

Director: Mary McGee
Managing Editor: Marla Patterson
Production / Operations Manager: Traci Huntsman
Production Editor: Tony Quinn
Book Designer: Susan E. Ormston Thompson
Cover Designer: Beth Caissie

Library of Congress Cataloging-in-Publication Data

Sargent, Chase, 1955-
  From buddy to boss : effective fire service leadership / Chase Sargent.
      p. cm.
  ISBN-13: 978-1-59370-075-1
  ISBN-10: 1-59370-075-X
  1. Fire departments--Management. 2. Leadership. I. Title.
  TH9158.S27 2006
  363.37068'4--dc22
                                                          2006028554

All rights reserved. No part of this book may be reproduced, stored in a retrieval system, or transcribed in any form or by any means, electronic or mechanical, including photocopying and recording, without the prior written permission of the publisher.

Printed in the United States of America
9 10 11 12 13 17 16 15 14 13

To Kathy and Brandon Sargent, for all the years you put up with me being gone for my career at your expense, and still supporting and encouraging me; for always being there regardless of the situation and continually showing me unconditional love; and for helping me with the transition from buddy to boss as I got older.

To my parents, Jack and Gloria Sargent, for sticking by me and teaching me right from wrong and believing in me enough to allow me to succeed in life and for being there even when it looked like I was done. For your love, respect, and encouragement, I can never repay you.

To my old friends Mike Brown, Dean Paderick, Rich Alfes, Mike and Danielle Purdy, Chris and Lisa Scott, and of course Tim Gallagher, who have been driving forces in my life, giving me advice and sticking with me regardless of the situation—I love all of you dearly—and to my first fire chief, Harry Diezel, for giving me a shot at lots of things and believing in me; for always being there to listen, advise, bail me out, and help; and just for being my lifelong friends. The list is short, but I know I can always count on you guys for whatever I need, wherever I am.

To the fire service, my other family, for giving me a career that I can be proud of, that I love, and that will always hold a special place in my heart and soul. I am proud to say I was a firefighter!

Contents

Preface: Leadership Lost xv
    Ray Downey xv
    Jim Page xvii

Introduction xix

1   **The Organizational Foundation for Leadership** 1
    The Governance Process 2
    Time-Management Model 6
    Mission, Vision, and Core Values 10
    How the Organization Sets the Foundation 11
    Values 14
    Organizational versus Individual Leadership 16
        Organizational leadership 18
        Individual leadership... 21
    Why Senior Leaders Must Lead 25
    Who Packed Your Parachute Today? 27
    Notes 30

2   **Knowing Yourself and Others** 31
    Understanding How People Are Made 31
    Why Know Yourself? 35
    Using Styles and Tendencies to Strengthen Your Team 39
    Leading Your B-Team Players 44
    Cain and Abel in the Workplace 47
        Knowledge is power 48
        Liars 49
        Familiarity 49
        Manage it—don't prevent it 49
        Master of deniability 50
    Notes 50

3   **Universal Rules for Survival** 53
    Vision: The Essence of Leadership 54
    Why We Don't Listen and Hear 56
        On the frontlines 57
        Middle management 58
        It's not what you say, but how you say it 58
        Your undivided attention 59

      Two-way communication. 59
      In closing. 60
  Values-Based Decision Making 61
  The Moment of Truth 64
  Of Officers and Riverboat Captains. 65
      What does it all mean?. 65
      Super models 66
      Growing a good crop 67
      On a mission 67
      The task at hand. 68
      My message to you 68
  Understanding Your Duty as a Leader 69
  The Universal Three 77
  The ABCs of Leadership 77
      A. Trust your subordinates 77
      B. Develop a vision 78
      C. Keep your cool 79
      D. Encourage risk. 79
      E. Be an expert. 80
      F. Invite dissent 81
      G. Simplify. 81
  Methods to Improve Your Leadership 82
  The Best Leaders I Ever Had: Characteristics to Embrace 84
      Caring for the job 85
      Technical competence. 85
      Understanding people 86
      Expecting the best. 87
      Taking a risk. 87
      A final word. 88

4 Keeping Humanity in Your Leadership 89
  People-Based Organizations 91
  People before Plans. 92
  Keeping in Touch. 95
  The Second Chance. 97
  Some Final Words about People. 98
  Notes 98

Contents

5 Being Tested as a New Leader ... 99
How Personnel Test a New Leader ... 99
- First day on the job ... 99
- Well guess what? ... 100
- The ballers ... 101
- The routine call ... 103
- Actions speak louder... ... 104

Keeping the Kittens in the Box ... 105
Giving Orders ... 106
- Intent ... 107
- End state ... 107
- Main effort ... 107
- Rules of engagement ... 108

The Bank Account ... 108
- Go first ... 109
- Stay in touch ... 110
- Make meaning ... 111
- Teach through stories ... 111
- Reflect on moments of learning ... 113

Big Stuff, Little Stuff—You Would Be Surprised ... 113
Notes ... 117

6 Maintaining Technical Competence ... 119
Education—What's It Worth? ... 119
Credibility—Earning It and Keeping It ... 120
Trust, Respect, and Other Concepts ... 120
Job Maturity ... 121
- Getting into trouble ... 123
- Liability—how we create it or avoid it ... 125

"People Are My Most Important Asset" ... 129
- At the company level ... 131
- At the organizational level ... 132

Sharpening the Saw ... 134
- Physical competence ... 134
- Mental competence ... 134
- Spiritual competence ... 135
- Social and emotional competence ... 135

A Final Word on Training ... 135
Notes ... 137

7 Understanding and Enforcing Policy ... 139
The Culture of Policy and Procedure Making ... 139
- Your name on this policy ... 140
- I'm the king or queen ... 140
- Good sound policy ... 141
- Statutes, laws, ordinances, and consensus standards ... 141
- Crisis policy ... 142
- Political policy direction ... 142
- Influence ... 143

Methods to Expand Your Influence ... 144
- Understanding the difference ... 145

The Consequences of Nonenforcement ... 147
- Argue at the right time and place ... 147
- Don't talk bad about policy in front of the troops ... 147
- You don't have to like it; you just have to do it ... 148
- Know the process and the policy ... 148
- Don't pick and choose the policies you want to enforce ... 148
- Don't allow or have a "look over the fence" mentality ... 149

Retooling ... 149
- Rules versus values systems ... 152

Inheriting Policy Violations ... 154
- Relief for Firefighter Jones ... 155

Conclusion ... 158
Notes ... 158

8 Evaluating and Compensating People ... 159
The Culture of Forming Evaluations ... 159
The Myth of Merit Raises ... 160
- The bar ... 163

What Do We Know about Evaluations? ... 164
Beginning the Process ... 165
Using the Door and the Desk ... 167
Positive Influence for Positive Employees ... 168
Myths and Truths about Discipline and Employees ... 169
Fringe Employees ... 171
- Rules for dealing with fringe employees ... 172

Documentation and Evaluations ... 175
- A word about personnel files ... 176
- Some learning points on documentation and attitude ... 178
- The Hoover ... 179

Notes ... 182

9 Prejudice, Diversity, and Sexual Harassment 183
It Is What It Is 183
The Changing Face of the Fire Service 185
Diversity in the Workplace 187
Equal Employment Opportunity: A Primer 189
Sexual Harassment: What It Is and Is Not 191
    What the courts have said 193
    Supervisory actions 194
Pornography in the Firehouse 195
Showstoppers in the Firehouse 198
Notes 198

10 Anger and Violence in the Workplace 199
What Is Violence in the Workplace? 199
Managing Anger and Disappointment 202
Recognizing the Potential for Violence 204
Evaluating Threats and Taking Action 208
Defusing Violent Situations 209
Conclusion 211
Notes 211

11 Decision Making—It's Not Tarot Cards 213
Neurology and Decision Making 213
The Pitfalls of the Decision-Making Process 214
Colin Powell's Decision-Making Guidelines 216
Gut Decision Making—and Why It Is Important 218
Completed Staff Work 219
Making Choices 221
Fire Ground Decision Making and Choices 224
People Choices 225
    Emotional decisions 226
    Past experience 226
    Values 227
    Loyalty 227
    Politics 227
    Fiscal considerations 227
Choices about Moving On 228
Notes 229

12 Accountability and Responsibility........................... 231
Firehouse to Fire Ground 231
Why People Follow You: Understanding Trust 234
    Relationship-based trust... 235
    If you talk the talk, you had better walk the walk 238
    Communications—the key to long-term commitment 239
    Don't fly if you don't know how 240
The Look-over-the-Fence Mentality...................................... 243
The Virus; or, My Stomach Hurts, I Don't Feel Well! 244
Identifying the Disease... 245
Making Accountability the Normal Operating Procedure 247
Inoculations for Success 248
Maintaining Accountability 251
Making Mistakes and Taking Your Lumps 253
Some Leadership Traits for Accountability and Responsibility......... 255
Notes 257

13 Battlefield Firefighting 259
Ruining Your Career in 10 Minutes or Less 259
It's All Yours! 260
Prioritization: Start with the End in Mind 260
Identify the Main Problem 261
Communications, Command, and Control—First-Due Perspective 262
Why Fire or Evolving Rescue Grounds Are Your Enemy................ 263
The Enemy at Work 265
    Friction 265
    Self-induced friction 266
    Uncertainty.. 267
    Fluidity 270
    Disorder... 271
    The human factor... 271
RIT—Doing It Right 272
Hypercool—How to Act When Things Look Bad 273
Final Considerations on Being in the Soup 276
Notes 277

14 Transitional Team Life Cycles 279
Constants When Evaluating the Team 280
    Teams enhance the organization's capabilities 280
    through high-performance service 280
    Small teams accomplish 90% of everything we do.................. 280

	Teams of any kind have a predictable life cycle 281

       Teams of any kind have a predictable life cycle 281
       To reach peak performance, teams require leadership and vision 282
       Teams are living entities.. 283
   Team Life Cycles 284
   Understanding Team Life Cycles—Managing the Big Picture 286
       Phase I: Development 287
       Phase II: Buildup 287
       Phase III: Fine-tuning... 289
       Phase IV. Drawdown 291
       Phase V. Buildup.. 292

15 Team Decision Training... 295
   The Concept of Team Decision Making................................. 295
   Where Teams Reside... 297
   What Works and What Doesn't... 298
   Developing the Model 301
   Benchmarking Behaviors... 303
       Team identity 305
       Team conceptual level... 305
       Team self-monitoring.. 307
   Putting It All Together 307
   Notes.. 308

16 Planning and Implementation 309
   Planning or Decision Making—the Consequences of Doing Neither 309
   Planning at the Company Level.. 311
   Planning to Pay Your Dues 315
   Paralysis by Analysis.. 316
   Functional Planning Tools 317

17 Managing Change.. 319
   Change—the Perilous Journey 319
   The Reality of Change... 321
   The Model of Placebo Change 322
   What Makes Change Successful?....................................... 323
       Lessons from 2001 325
       Lessons from 2006 326
   Change as an Educational Leadership Tool............................. 328
   Winning the Stakeholders—Key Decisions for Success 331
   Organizational Big Picture Mistakes 334
   Notes.. 336

18   The Business of the Business 337
   The Essence of Being a Firefighter 337
   The 5% Rule 339
   9/11 Retrospective 340
   Why You Are Dangerous 342
   Leading or Doing: Failing Either Will Alter the Moment of Truth 344

19   When Leadership Fails 347
   Why Leadership Fails 347
   Setting the Stage for Organizational and Political Failures 350
      Case study: How trust and respect affect ability 352
   Failure to Educate 354
      Promotion of noncombatants 355
   Individual Failure 357
      Walking the walk and talking the talk 359
   Failure to Exploit Small Teams 361
   How to Survive Failing Leadership 361
      Leading up 364
      Bunker mentality 366
      Political or managerial action 367
      Moving on 368
   What Next? 371
   Notes 371

Conclusion—Changing the Culture One Person at a Time 373

Epilogue 374

Appendix A: Sargent's Critical Commandments of Leadership 377

About the Author 379

Index 381

Preface
Leadership Lost

This book is about leadership, and it would not be complete if I did not recognize two mentors, fire service leaders and friends I thought would be around forever. I found out how wrong I could be when both were taken from my life much too early. Deputy Chief Ray Downey and James O. Page were friends, confidants, mentors, and teachers and laid the foundation for my leadership development.

Ray Downey

This is in remembrance of my dear friend, partner, and mentor, the late Deputy Chief Raymond Downey of the New York City Fire Department (FDNY) and in honor of his wife, Rose; his sons Chuck and Joe, who are both on the job with FDNY, and Ray Jr.; and his daughters, Marie and Lisa. Ray gave his life to protect others on September 11, 2001. Ray epitomized leadership throughout his entire career, even up to the last moments of his life. As did many others on that day, Ray knew full well the consequences of going south while he sent everyone else north after the first tower fell. We should all be as blessed as Chief Downey to have chosen our moment, with our men, doing what we loved, when God calls us home.

As a firefighter in New York City, Ray rose through the ranks to become one of the world's leading authorities on specialized rescue and was well known throughout the special operations community. Ray was a true leader who thought of his command, his personnel, and his family above all else, even up to the last moment.

Those of us who knew Ray Downey well saw another, more eloquent side to him besides his commitment to rescue, and that was his commitment to leadership, mentoring, and teaching. Ray exhibited the unique quality of being incredibly humble yet incredibly knowledgeable. Chief Downey never missed an opportunity to pass on what he knew to the younger members of the service and exercised a magnificent understanding of human character and soul. Ray gave younger members of the community a model that might have changed even New York City's culture had he lived long enough! Ray's charisma was grounded in his education, faith in God, integrity, loyalty, knowledge, moral and physical courage, and undying commitment to excellence.

Ray demonstrated leadership—candidly and subtly—each and every time I saw him. Whether on our morning run in a blizzard before a Federal

Emergency Management Agency (FEMA) Urban Search and Rescue (US&R) meeting in Washington, D.C., or working the podium of the Task Force Leaders meeting, or testifying before congress, Ray had the stuff that leaders dream of. I spent many hours listening and talking with Ray, trying desperately to understand what made him a natural born leader. Perhaps it was his faith in God, his love of family, his loyalty and love of FDNY and the fire service, or his great moral and physical courage; or perhaps it was a quality that very few in history ever possess: I tend to believe that Ray simply had a God-given gift that allowed him to do what he did in the most magnificent manner.

Regardless of what mystical quality Ray possessed, Ray's presence on this earth was a blessing for those of us who were fortunate enough to call him our friend. He was a teacher, a mentor, a friend, and above all, a firefighter's firefighter.

May his family and friends find some comfort in the fact that because of Deputy Chief Raymond Downey, all of us are better leaders, better people, and better firefighters. In ancient times, legends were told around the campfires after great battles. The legends of people like Ray Downey will be told around the new campfires of today, the firehouse, after each battle we wage. The legacy of Ray Downey will be passed down from generation to generation, to the new generation of leaders being created for the fire service. May we each take some of Ray's knowledge, skills, abilities, humbleness, courage, integrity, and honor with us on our leadership quest. He is without a doubt watching us and smiling as he begins to create a special operations command in heaven for those of us worthy enough to follow him.

Ray demonstrated every characteristic of good and competent leaders that I speak about in this book. We must remember to hold the ideals that Ray espoused close as we begin our journey. We must remember that in every new recruit's eyes, in every team member's soul, lies a potential Ray Downey just waiting to reach out and lead. What we seek are everyday leaders like Ray. Ray was able to lead men and women under great stress, through great battles, because he was an everyday leader. On that fateful day in September 2001, he simply did what he did every single day he came to work. He led from the front, created a vision, and showed honor, strength, courage, and loyalty. It is the everyday leader who succeeds both daily and in the largest of crises. It is the everyday leader who solves the small problems for you each day who is most likely to succeed in the long run.

It is with tears in my eyes and a very heavy heart that I write, read, rewrite, and reread this dedication. Despite the words committed to paper, I find it hard to believe that my good friend is gone. I wish above all that I had had more time with Ray—time to say things I never had the chance to say, time to say just how much his friendship meant to me, and time to get to know him more as an individual. As you read this book and digest these words, remember this very first lesson in leadership that Ray often spoke of: in our business, "sometimes goodbye really means goodbye." God rest his soul.

Jim Page

None of us who knew Jim will ever forget the day he died while doing his morning swim. When John Sinclair called me and told me to sit down, I knew something horrible had happened. Jim was my friend and mentor, and I clearly remember back 20 years to when he gave me one of my first breaks—lecturing at EMS Today as a snot-nosed rookie instructor who would have traveled anywhere on his own dime to be able to teach at such an event.

I truly appreciate the unique opportunity to have spent time with a man of such character and wisdom that one wonders how God managed to put so much into a single human being. The love the community expressed for Jim after his much-too-early departure was a testament to his leadership during his time on earth.

Jim always has and always will hold a special place in my heart and soul. Jim took me under his wing when I was a young man and gave to me something that I would never be able to repay in a dozen lifetimes: his friendship. Jim gave me opportunity, he took time to impart wisdom and knowledge, and most of all, he gave me a friendship that you just knew came directly from his heart. It was such a privilege to be able to spend time with a man of such character, loyalty, and integrity that I often felt uncomfortable, hoping that I did not disappoint him by my actions or words. I am so glad to have been given the blessing of calling him my friend.

I am reminded of the story of the prodigal son. While the parable is open to interpretation, I would aver that when we celebrate Jim's life, it should be about our loss. In our lives, we take for granted what we have and the people in our lives whom we love until they are gone. Too often, it is only then that we realize what a true blessing it was to have them in our lives. Jim was one of those unique blessings from God who was bestowed on those who had the privilege to call him friend, mentor, teacher, boss, confidant, or partner. We should never forget what Jim has meant to us and to the community to which we belong. We should also remember what Jim will mean, because of his work, dedication, commitment, and love of humanity, to the next generation that is preparing to fill our shoes; although they will never have the opportunity that we were given, and some may not even know the name Jim Page, rest assured that their ability to serve their fellow human beings was deeply influenced by Jim's time on this earth.

At this very moment, Jim is being asked to serve on a dozen advisory boards in heaven, or working on a project long awaited, or perhaps repainting his wings fire engine red! Just as he was on earth, he will always be looking over us, working to protect us, teach us, and place us on the right path as we continue our journey in the wake of the love, knowledge, and the wealth of friendship he shared with us during his lifetime.

Jim was a very dear friend whom we loved very much, and we will travel with a void in our hearts as we continue our journey. The love he showed us can never be replaced; it is a debt that cannot be repaid. On this day, let us resolve to remember Jim by doing what he would have wanted us to do, to take care of and love each other and take care of our fellow human beings. Let us take no friendship, no love, no handshake, no commitment for granted—so that we will not look back years from now wishing we had realized what we had at the moment it was given to us.

We must remember the examples of these two men and of the many others who blazed the leadership trail for us to follow. Relationships, friendship, integrity, loyalty, faith, honor, duty, desire, commitment, drive, and love are all words that we use too carelessly. Leadership is a lifelong commitment. Remember that the spirits of Ray and Jim are watching, waiting, and smiling on those of you who take to the path to leadership.

Introduction

> *Individual accountability creates organizational greatness.*
> —*Michael Abshroff*

As a young officer, I was ready for the challenge—or so I thought. The department had spent 26 weeks training me as a firefighter, and I had spent at least a year studying and preparing for the promotional examination. I had scored well on the written, done well on the in-basket, and managed to provide all the right answers to the board. So I found myself at number one on the Lieutenant's List and ready to go.

At the promotion ceremony, there were no warnings about the pitfalls and land mines that lay ahead. While pinning the badge on me, my wife smiled, as did Chief Diezel (though inside he was surely saying, "My God what have I done!"). The crowd applauded, and I was told to go forth and prosper. That was it: overnight I had gone from riding backward to riding forward, and the only preparation I had was my knowledge as a firefighter/paramedic and a nice pat on the back from the chief.

I knew I could fight fire; I had been doing that for 10 years and had gained a reputation as a young sled dog capable of holding my own, pulling my own weight, dropping tubes down bloody tracheas, and stirring up (as well as receiving) my share of firehouse mischief. I could make the long hallways, do the rescue, and provide the emergency medical services (EMS), and I had the burn scars to prove it. Your reputation as a firefighter is made on the basis of how you fare in the heat of battle: Are you an operator, do you hang tough when things go bad, can you take the heat, and are you aggressive? I thought to myself that the departmental authorities who had promoted me must know what they were doing—since it had been that way forever, I guessed I would be okay.

On my first shift at my new station, I sat at my desk and began to think about what to do. I opened the drawer, and inside the drawer was a letter with a matchbook attached and one match in the book. It was a note from the previous lieutenant, a veteran firefighter of 24 years, and the note said, "Rookie, use this, set the world on fire, when it goes out, STOP!"

Thus began my journey as a company officer, which eventually led me to where I am today, as a division chief. Never did a little voice tell me to beware of the human resources issues about to confront me. Never did the department show me what was expected of me as an officer. Never were the key aspects of initially managing the fire ground spelled out. Never had I been schooled on evaluations, violence in the workplace, sexual harassment, teams, planning, or any of the other tools I would need to be successful. For all these things, I was left to my own devices, the experience passed down from more-seasoned officers, my experience with the knuckleheads I had already encountered, and my run-ins with policy and policy makers during my years as a firefighter/paramedic. Common sense and experience were all that I had in my toolbox as a new officer, and considering what I was about to face, it was a sparse set of tools.

Each and every item in this book is born of my own experience or the experience of others I have known or worked with. I wish I could tell you honestly that I did not step on any of the land mines described in this book, but that would be, well, economical with the truth. I assure you that I have received my fair share of "bullet holes" in the foot—and I have tasted my foot often enough. I wish I could tell you that every officer I ever worked with was a model of leadership, integrity, values, and knowledge, but that would be a dreadful lie.

What I have discovered is that just like any family, members who live, work, and eat together 56 hours a week interact. With interaction comes an entire set of consequences, both good and bad, which the company officer and sometimes the battalion officer must account for. Coupled with these interactions is the brutal fact that at any given moment you and your team might be called to make the ultimate sacrifice for someone you don't even know. You will see the best and the worst that society has to offer, yet none of that will be as difficult as the human resources issues you will deal with daily. Nothing will be as difficult as leading effectively, both in times of calm and when surrounded by nightmarish scenery.

I have also discovered that just because you reach the top of the organizational ladder, this does not make you a good leader. It does not make you any more of a leader than you were before you got there, and in some circumstances, it may make you a worse leader. Reaching the pinnacle does not entitle you to trust or respect. Those are earned by action, not by the weight of gold on your collar!

You will discover that when people call you, they literally release their life to you, trusting everything you do. They will allow you to walk into their six-year-old daughter's room at 0200, tearing their house up, and they will thank you for it, taking your directions as if they were gospel. Yet when you return to the firehouse, you may not be able to deal effectively with an employee conflict that is making the workplace miserable.

I have spent the past seven years doing one- and two-day Buddy to Boss seminars nationally and internationally. To tell the truth, when I first thought up the idea, I was skeptical. Much of the credit for the program goes to my friend Dean Paderick, who kept hearing me complain about things and challenged me to develop a process to educate others. Dean also provided the name of the program after I named it something along the lines of "All the Idiots I Have Known, Including Myself." Dean did not think that was a politically correct title and suggested in his fatherly way that I change it. The rest, as they say, is history.

I thought that other fire departments surely must have been providing this kind of training for their officers. What I found was shocking! It turned out that the vast majority of the fire service was providing nothing of the sort alongside the process of promotion. Sure, some departments had human resources departments that offered a class here and a class there. There are leadership and management courses taught at the national fire academy and at most major fire and rescue academies, but most of them, in my experience as a student, used outdated texts and outdated concepts and were too broad in their scope of application. It became apparent very quickly that the leadership and human resources training being provided was as scarce as water in a desert and that the fire service was thirsty.

In the past, I had wondered what it took to survive in the fire service leadership game. It's not a game at all, but I have been astounded by the people I have seen in major departments and in supposed positions of leadership at home and abroad. They must have really played the game to get where they are today! It was apparent that the fire service had become accustomed to manager-politicians, instead of leaders. We promote people on the basis of their ability to write journalism, with no consideration to their operational ability or knowledge, skills, and abilities (KSAs) on either the fire ground or the human ground. The fire service has forgotten that it's in the people business when it fills the top spots in the organization.

It has always been said of the fire service that we are a family. Just like a family, we raise our kids (new firefighters) surrounded by ideals, values, and expectations, and we teach them until they become (or at least are designated) adults (officers). If the current status of leadership in the fire service is representative of our dynamic as a family, we better call social services, because we have a dysfunctional family.

From this description, you might assume that I see the fire service as a leaderless, valueless, political organization. While there is more to that than I would like to admit, there are also wonderful examples of leadership thoughout the fire service. It's unfortunate that we lost true leaders like Ray Downey, Terry Hatton, and Patty Brown, to name only a few, at the World Trade Center attack. Nevertheless, we should be happy that the Freddy Edrikats, Bill Lokeys, Alan Brunacinis, Geoff Williams, and Ronnie Colemans of the world are still around. We mourn the loss of firefighters such as Jim Page, who was still alive when I began writing this. Rest assured that others will rise to take their places and that we still have some outstanding examples of leadership at the chief of department level and below. I am sure you can name many, just as I can.

The point is that we must begin to make leadership the norm, not the exception. We must begin to change our organizational culture one person at a time, starting with ourselves. We must begin to teach, mentor, and provide the training for our up-and-coming officers, new officers, and current officers. We must balance the need for formal education with the real-world education gained from years on the job, and we must begin to base promotion on this balance, rather than on seniority or the ability to write journalistically and with proper grammar. Some of our best leaders are being left behind because they don't educate themselves and we don't educate them. Moreover, in the promotional process, we forget entirely that the job of firefighting still entails and will always entail dirty, hot, dangerous, skilled labor and split-second decision-making work.

The Chinese say that "A journey begins with a single step." I know that we will not change the fire service overnight. It will take years, maybe even a decade, to stop this downward spiral and to once again value and espouse the leadership, supported by management skills, that will be required in order to usher the fire service into the decades ahead. This book is about taking the first step. Many of us, myself included, have had to go back and walk the same path many times to get where we are today.

In each class of new recruits, I see the same excitement, the same drive, and the same ambition that I had the first day I walked into rookie school—and that I still have today. I see it in their eyes the first day they make the floor of the house and as they sit around for their first meal with the shift, and I hear the excitement in their voices after the first real job. I know the energy and commitment are there to create tomorrow's fire service leaders, because I feel it, see it, and taste it every day I go to work.

I am in the twilight of my fire service career. I know my time is short, and I see a whole world full of future officers, chiefs, and leaders in the faces of today's firefighters. Every individual member of the service can make a difference. Don't think that one person can't change an organization or a culture; on the contrary, I have seen it happen. This book is about change, tradition, and being humble and great all at the same time; about leadership supported by management skills; and ultimately, about the future of the fire service.

Chapter 1
The Organizational Foundation for Leadership

> *Brains every day become more precious than blood. You must give men new ideas, you must teach them new words, you must modify their manners, you must change their laws, you must root out prejudices, subvert convictions, if you wish to be great.*
>
> *Beware of endeavoring to become a great man in a hurry. One such attempt in ten thousand may succeed. These are fearful odds.*
>
> —Benjamin Disraeli

When I teach, I look around the room at new officers or officers in waiting, but what I actually see is a new organization. I see future fire chiefs, who are the future leaders of their organizations and of organizations that they have not even heard of yet. I see the opportunity, the desire to change our current trend of promoting and espousing manager/politicians, and the opportunity to create leaders. This chapter has nothing to do with your ability, but rather explores how certain things occur in organizations long before they reach people like us. As Stephen Covey says, "Seek first to understand, then to be understood." In conjunction with that is the issue of creating a fundamental understanding of some basic premises and ideas on which we will build throughout this text, those that relate to the abstract and contradictory world of politics.

There is an age-old question: "Are leaders made or born?" I say both theories are bogus. Most leaders become leaders because they are uncommon and they know it. The truth is that even though leaders are made, most make themselves. They do this because they believe they are leaders, and in fact many have a little bit of prima donna in them. So don't confuse rank with leadership. You can have all the gold in the world, promote well because you do well on the assessment center, written and practical and still never be able to lead effectively, because you cannot make yourself.

To make oneself, a foundational understanding must be formed of just what creates an environment in which leaders thrive, fight, live, create, and ultimately succeed. This chapter is about understanding the environment and prospects that define and shape leadership principles.

Most of us work for a department that, while relatively large in size, represents just one of many departments within the larger animal we call the city, state, or federal government or the corporation. For us to understand how things work in our little corner of the world—our fire department, our government structure, our corporation—we have to understand the details of how events unfold.

The questions that come to mind—and that demand to be understood when unanswered—are global in nature but fundamental in structure. First, how does the fuzzy political directive find its way into the river of the organization and work its way downstream to become something of substance? Where do you and I, based on our position in an organization, fit into the grand scheme of things, and where do we fit in the big model of the entire organization we belong too? Where should we be spending our time now and as we are promoted through the ranks? What do a strategic business plan and core values have to do with leadership, and how do they affect what we do? All of these questions provide some understanding about what and what not to do. In some cases, they provide firm foundational information on which we build, and in other circumstances, they provide us simply with an "aha"—that is, information that is nice to know, and that somehow relates to our leadership roles.

The Governance Process

Large organizations are extremely complicated, and they do not always represent what is best in people. Competing political agendas, poor or skewed core values, working outside the time-management model, and a range of issues lead us to believe that politics and leadership are in fact two distinct entities.

Like it or not, all of us still have to work for and report to—and are influenced by (good or bad)—the political system under which we work. If we must live with a dragon, we had better understand how the dragon works and how to avoid making it angry. To understand the dragon, I am going to suggest a governmental model that seeks to provide information on why and how things happen. Your governmental model may be council/manager, mayor/manager, county board of directors, or any one of a dozen models. Regardless of the model, the general responsibilities of each level remain the same.

The model I am presenting in figure 1–1 is not mine. This model, which represents a manager/council form of government, was presented to me by

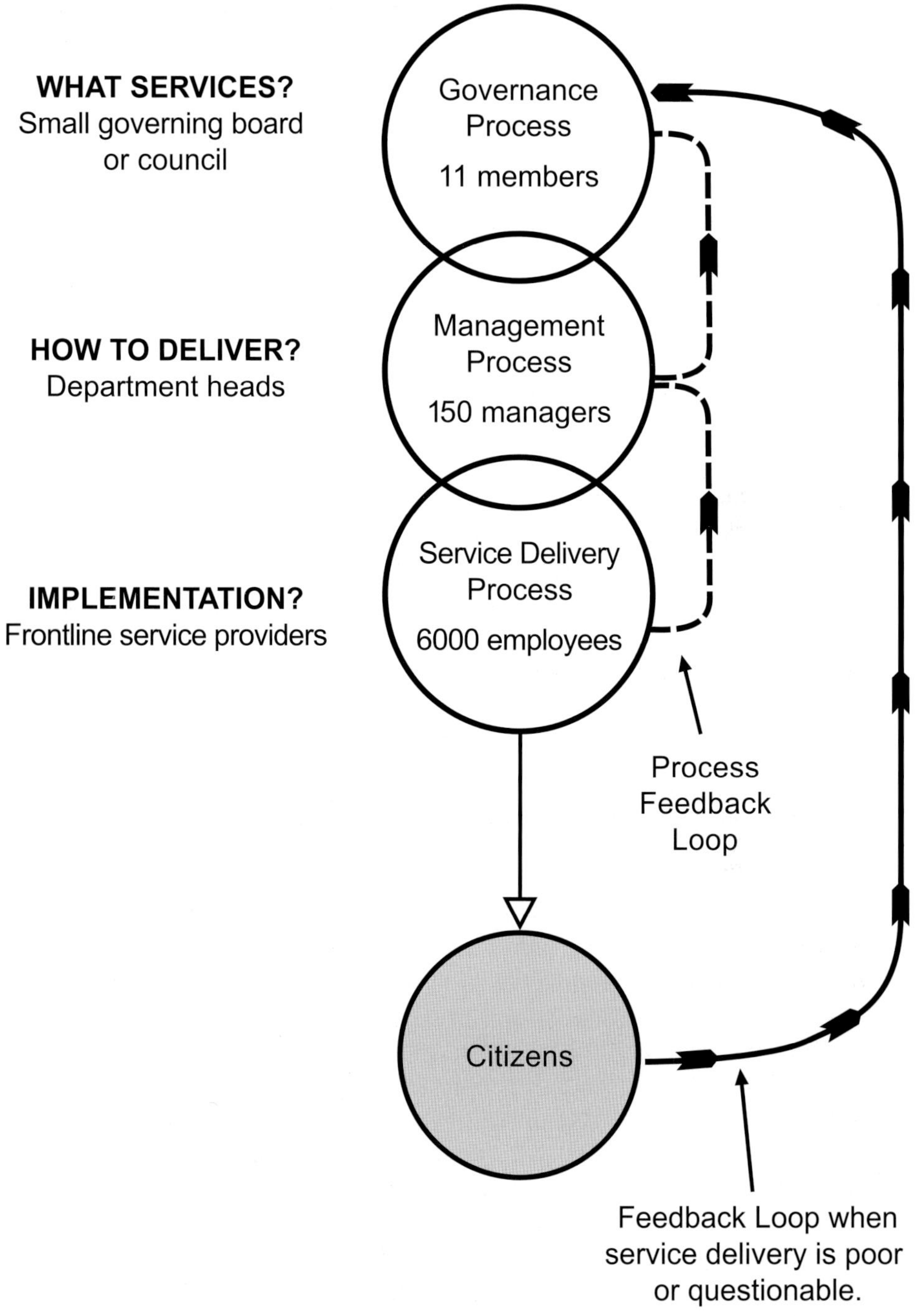

WHAT SERVICES?
Small governing board or council

HOW TO DELIVER?
Department heads

IMPLEMENTATION?
Frontline service providers

Governance Process
11 members

Management Process
150 managers

Service Delivery Process
6000 employees

Citizens

Process Feedback Loop

Feedback Loop when service delivery is poor or questionable.

Figure 1–1 Governance model

my city manager James Spore, from the City of Virginia Beach, during a leadership course. This model is his method of showing us how the big animal of government makes decisions at a macro level. I thought the model was significant enough to share with all of you. Don't get hung up on the fact that it's a manager/council form of government; you can plug in whatever form of government you work under. What's important is that you understand that regardless of the form of government, the flow and the process are the same.

There are five specific areas of interest. The first part of the model is called the *governance process*. This is composed of and controlled by elected officials (politicians), fire board, or other entity and represents the smallest number of stakeholders, in my system representing a mayor and 10 council members. The governance portion of government is primarily concerned with answering the question of *what*. What services do the citizens want? The whats of the governance system can in fact be quite fuzzy. The whats of government may indicate that citizens want quality schools; want to feel safe anywhere, anytime; want a community for a lifetime; and so on. Politicians that answer the question of *what* don't necessarily have the answer as to *how*.

The second portion of the model is the management process. This section if filled by appointed officials such as department heads (fire chief, police chief, superintendent of schools) and the city manager. In my city, this represents 150 department heads and one city manager. The management portion of the model is primarily concerned with answering the question of *how*. How do we take the fuzzy whats from the governance portion and convert them into programs that can meet the mandates from the politicians? What do we do to make citizens feel safe anywhere and anytime? What do we do to assure quality school services, for instance? The *hows* are most commonly addressed by a strategic business plan created by an individual department. Importantly, they are reflected in our mission statement and in the core values under which we operate. Yes, the mission statement; go find that obscure piece of paper hanging in the hallway and wipe the dust off it, because you are going to need it!

The third portion of the model is the *service delivery section*, and that's where the little people like you and me find ourselves. This section is composed of the employees and supervisors who actually deliver the service. The delivery system is primarily concerned with the implementation or day-to-day delivery of the *how*. The delivery system must take the how and convert it into meaningful tasks, or actually create the vision that is supposed to develop from the *how*, generated by department heads. Are you beginning to see how this ties into leadership? No? Okay, then let's try harder.

One notable feature of the model is that the core areas overlap. This reflects the need for meaningful discussion, feedback, and dialogue between governance and management and between management and delivery. In other words, there should be open feedback between the sections of government; when this fails or is not exercised, the first crack in the foundation forms. In fact, the connection between management (the *hows*) and delivery (the *dos*) reflects the need for frontline supervisors to constantly provide feedback and input on systems and service delivery, which we will explore further in the time-management model.

In this model there are two feedback loops. The first loop travels from the delivery section to management and the governance process. This feedback loop is critical to ensure that system problems and systems success are communicated, so that the system of what-how-do can be improved. The method by which this feedback occurs varies from system to system, but it must occur. If it does not occur, then we will never have systems alignment or systems improvement, but rather a constant flow of whats, subsequently translated to hows that may or may not get done! The bottom line is that people must implement plans, simply by writing them down; moreover, failing to engage the plan with human energy results in a plan that never gets realized.

The second feedback loop is what I call the *poop loop*. Very seldom do we get told we do a good job. However, provide poor service in the eyes of the citizen and they will completely bypass the delivery and management portions of government and go directly to the politicians. And as they say in leadership school, when the toilet flushes on high, it rolls downhill quickly! When the toilet flushes, it inevitably creates work upon work for those involved. The time spent answering the questions created by the toilet flush can usually be spent in a much more productive manner.

Avoid actions that create toilet flush. This imposes additional work on everyone in the chain of command. This is energy that could be spent for much more productive things!

The real challenge for you as a supervisor is to find a way to convert this fuzzy stuff into everyday action. More instruction on this is given later, but being capable of this conversion is critical to your success as a leader in any organization, especially as a frontline supervisor or middle manager.

Time-Management Model

To be effective, organizations must have a division of labor. In functional organizations, that is represented by job descriptions, span of control, chain of command, and rank structures. An important part of the division of labor is understanding where we, as employees, first-line supervisors, division heads, or department heads should be spending our time.

As you progress in rank through the department, your job will most certainly change. More important than your changing job is the focus or the aspect of the job or rank you reach. In other words, the chief of the department should not be doing or worrying about the same things as a firefighter; a division chief should not be doing or worrying about the same things as a captain or lieutenant. It is the nature of the beast that the further up the chain of command you move, the further you are from actually operating the system. Like it or not, you cannot, if you choose to promote up through the ranks, expect to pull hose, fight fire, or manage a company all your career. As you promote, you must learn to leave certain aspects of your job behind, and move "from the engine room to the bridge of the ship."

So what are the expectations of certain levels of the organization, and why is it important that we understand them? First, you must understand what is expected of you with regards to implementing a service delivery model at any given place in the organization. Not everyone can be out there on the fire truck or medic, or rescue. Likewise, not everyone can be working the budget or planning for the future. The organization requires certain tasks at certain levels, and everyone has to know what those are.

We do three specific things in the fire service: we operate the system (provide service); we improve the system (evaluate and improve our operations, policy, and people, etc.); and we create the future (look at the long-range goals, evaluate immediate and midrange threats, engage the management and governance portions of the model, and create, simplify, and communicate vision). Each of these functional areas provides a critical level of service that, when engaged properly, results in the execution of your organizational strategic business plan. Depending on your current location in the system, you must spend different amounts of time on each of these three things (fig. 1–2). This is what we call the *time-management model*, or where we should spend our time on organizational duties.

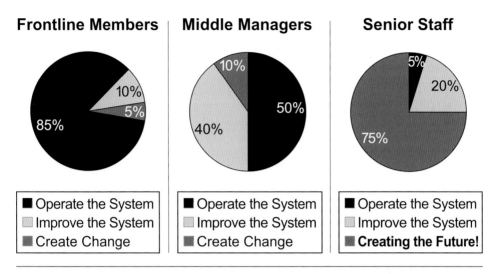

Fig. 1–2   Time management model

For definition purposes, we will refer to firefighters and company officers (lieutenant and captain) as frontline members; battalion officers as middle managers; and division chief through fire chief, in addition to administrative positions in staff held by captains and battalion officers, as senior staff. You can place your rank structure to fit here as you see fit.

Now that you have the model, don't get too hung up on the percentages; they represent a time frame, which is not locked into stone. Here's the bottom line. If you are a frontline member (service delivery), get out and see the customer (the company level), the vast majority of your time should be spent operating the system. A smaller but very important portion of your time is spent improving the system, by giving feedback through a variety of methods on how to improve customer service and customer delivery. Very little time is spent on the big global picture of creating change. At this level, there is a phenomenon called the *inverse change rule*. This means that even though officially you should spend very little of your actual workday "improving the overall system," the organization should understand that frontline suggestions for improvement should be taken very seriously for the following reasons:

- Frontline members are closest to the customer. If we have a customer service process, emergency response process, or anything that affects the business of the business, the frontline personnel will see it first.
- Frontline personnel are always right until proven otherwise. Personnel sitting behind a desk, removed

from the customer, will know a lot less than personnel who provide the service day in and day out.
- Change is always driven from the bottom up—never from the top down. Plans without people are worthless.

Personnel on the front line have an inordinate ability and power to improve the organization if you take time to listen and implement ideas. No other member of the service has this unique position and view of the world of fire and rescue services in either the emergency or the nonemergency setting.

If you are a battalion chief, you still operate the system but you go on certain types of incidents only, such as structure fires and technical rescues. In other words, you spend less of your time operating the system and more of your time on staffing, reporting, training, and all those other fun tasks that come with promotion. You spend a much larger time in improving the system by training and mentoring your officers, establishing training and expectations for your battalion, carrying the message of improved customer service methods up the chain, and acting as a buffer protecting your personnel from stuff that flows downhill. As a battalion officer, you will spend some time creating the future by communications with your chain of command, sitting on committees, or performing other activities that expand your circle of influence.

One of the greatest skills for a battalion-level officer is the ability to communicate effectively, frequently, and honestly. Battalion officers are often responsible for providing accurate and timely information to frontline members, which comes from senior staff, staff meetings, and other venues. At the same time, the organization expects you to be capturing the needs, expectations, and suggestions for improvements and getting them moved up the chain of command. Failure to do either of these well results in gray areas, half-truths, rumors, and frustration among everyone in the organization. Your inability to capture information and effectively pass it on in a timely manner will affect your ability to lead by affecting your credibility.

At the senior staff level, you should spend very little time operating the system. Otherwise, it's called micromanagement! As a division chief, I go only if it's a second alarm or greater, a significant rescue event, injury or death, or when something sounds like it's heading south quick (that's a Southern phrase for going bad). Of course, I can go out and take a look at the troops from time to time, but not to take command just to see how the guys are operating and to evaluate my battalion officers. My dear friend Alan Brunacini calls this a *sighting*. If you're the fire chief or deputy, you can come anytime you want, but really don't need to unless all hell is breaking loose; furthermore, since you accept all the liability at that level, it's your right to come and see your career going up in smoke! You spend some time improving the system, by taking the suggestions provided by service providers and making sure they are implemented or looking at systems-wide problems and figuring out a way to correct or improve them. The vast majority of your time at this level is spent on global issues: creating the future and, in some instances, surviving the present. Budget, meetings, policy, politics, creating influence, large-scale projects, and oversight of your division are what you should be doing. In other words, it is your duty to create a *vision*, something that is so important we spend some time talking about it.

What you should not be doing is going to the company level and counting lockers that were ordered to make sure they really need them. This includes camping out on a senior chief's desk because you want to know about discipline to a firefighter, requiring that all decisions affecting the organization come to your desk, or otherwise micromanaging areas where you should have trusted and competent officers functioning. I have discovered that officers who exhibit these types of behavior do so because they are very uncomfortable and even dysfunctional in their present assignment. It's easier for them to do what they feel comfortable with (working layers below where they should be), rather than doing what their advanced rank requires, because they really can't do the latter job very well.

As you can see, your role changes as you climb the organizational ladder. Some people have a hard time adjusting and end up working outside their time-management areas, becoming known as micromanagers (the first sign of lack of trust and the first sign that they have been promoted to a level beyond their capability), or simply lack the competence to operate at the current level and revert to the previous comfort zone. The bottom line is that if you plan on promoting, understand where your responsibilities will lie and expect change. Don't try to do someone else's job for them.

Mission, Vision, and Core Values

If your organization does not have a mission statement, a vision, and printed core values, don't feel too bad; mine didn't until about five years ago. I never understood how important these items were until I realized that our former leadership guidelines were either nonexistent or simply drawn from tradition. What we called leadership was nothing more than the cultural abstract of behavior passed down from one supervisor to another.

I have read book after book that speaks to the culture of an organization—how CEOs (read fire chiefs) must create a vision for people to follow and how a mission statement is critical to success. In many cases, the first and easiest thing for a fire department to create is a mission statement; however, this is often just a set of words tacked up on a firehouse wall or in some chief's office, rather than something to live by. Whenever I see a mission statement that reads something like "To save lives and property," I laugh because I know it took all of three seconds to create, has no value, and means nothing to anyone.

The reality is that if you really want to lead, want those around you to follow, and want to learn how to lead, then you have to lay a foundation, to be engaged in something larger than yourself while at work. The challenge for leaders is to take the fuzzy stuff that is reflected in a mission statement, in a vision statement, and in core values and convert it into something firm that people can sink their teeth into. It's easy to print words like loyalty, strength, honor, and courage and then to stand on your soapbox and say, "These are what we believe in." It's entirely different to be able to take those same words, communicate their meaning and an associated group of actions to people, and then show people how to use them every day by your own actions. In essence, words become powerful only when they mean something to people deep down in their soul, in their hearts, and when they understand what it means to express those words through solid action every day. For those of you who have ever raised children, think about when you tried to talk to them. Most children don't really listen to what you say; in fact, they may expend real effort in trying not to listen to what you say. However, what they do extremely well is watch what you do. What this boils down to, with the people you lead or with your kids is that it's what you do that's important, not what you say.

Before you can create a vision, which is less of a destination and more of an organizational lifestyle, you must have the right people. You must get the right people on the bus, the wrong people off the bus, and the right people in the right seats on the bus.

How the Organization Sets the Foundation

Mission or vision statements should always be created after much thought, some facilitation, some heated dialogue, and lots of drawing on charts and should ultimately reflect what you, the members of the organization, believe your organization really stands for—the reason the organization even exists at all. Despite all that work or many months, maybe years (though not if you do it right), your mission statement or vision statement is still, when it's all said and done, only words. What brings the mission or vision statement to life is the action of the members of the organization. How we act, what we value, where we spend our time, and how we treat our members and our customers are what the mission statement is all about.

Why do I drag this out in such detail? It's simple: you, as a leader, as well as the other leaders of the organization, will have to effectively communicate the mission and vision to the members of the organization. I'm not talking about going to the fire station, gathering the men and women around, and saying, "Okay, everyone, listen up, the following is our mission statement: To Save Lives and Property in an Efficient Manner." (That kind of mission statement is just as hollow and meaningless as it seems on reading it, but nonetheless is frequently encountered.) "Does everyone understand our mission? Any questions? Good, now go out and carry out our mission." Once, during a senior staff meeting at which the discussion centered on why we were not getting the message out about our strategic plan, I watched a senior chief exclaim, "Hell, just send the plan out and make them read it." This reminded me of the timeless quote: "Never assume malicious intent when ignorance is sufficient to explain it." Its amazing how ignorant people can be when you really give them the chance to shine!

To demonstrate what I mean, let's examine a mission statement by dissecting it in terms of how, as leaders, we begin to turn this into a method of organizational living—for our members, for ourselves, for our customers, and ultimately for the people who will inherit the leadership positions whom we are mentoring or preparing to follow us. You need to take your mission statement and do the same type of surgery on it so that you can understand how to implement it.

> *The purpose of the Virginia Beach Fire Department is to be a customer service organization partnering with communities, members, citizens, and visitors to foster a feeling of safety anyplace, at anytime, through planning, mitigation, response, and restoration.*

What we have here are simply words that have been molded into a purpose as a mission statement. However, without substantial effort by leadership and our members, they remain meaningless words. No business plan ever put out a fire, made a rescue, installed a car seat, completed an inspection, or delivered a baby. Our challenge is to find methods, actions, and processes that allow us to convert these words to everyday actions by members of our organization.

What we can extract first from this is that members of the organization place high value on customers, as expressed in a number of ways in the mission statement itself. What we should recognize next is that we want all of our customers to have a feeling of being safe. Notice that it does not mention anything about extinguishing fires, saving property, or any of the clichés that commonly find their way into mission statements. Certainly those specific functions and actions are important, as are the KSAs necessary to accomplish those outcomes. The key is interpreting those words to mean that the organization values every definable method to make people feel safe. Finally we see that there are four core areas of our "business": planning, mitigation, response, and restoration.

The challenge now begins as senior staff and members of the organization must find concrete examples and convert them into action. The mission statement should express our reasons for existing—our reasons for coming to work each day. We could no doubt spend countless hours and pages defining these and discussing the philosophy behind that concept. Instead, let's examine some specific examples that fall outside delivery of normal fire and rescue services tasks. Whenever I took command of a battalion or division, I wanted to find such examples because, if I could give my officers just a few examples of what I expected, they would consistently come up with dozens of other methods, which they implemented every single day.

A simple example, which I constantly used, was the old story of Mrs. Smith broken down at 0200 on a dark road. I told my officers, "If you are coming back from an incident and you find Mr. or Mrs. Smith on the side of the road you had better stop, change the tire, start the car, get a wrecker, or if you have to, put Mrs. Smith in the truck and return her to her home." Regardless of what it takes, it is our responsibility to restore Mrs. Smith's feeling of safety. Although this example demonstrates a simple concept, it indicates clearly that our actions speak louder than our words.

Another example is as follows. Organizationally, we were pretty good at mitigation and response but were miserable in restoration, even at the lowest level of our service delivery model, the everyday structure fire. We did a great job responding, rescuing the small child, and extinguishing a fire, but all too often, we rode off into the sunset leaving the family with a burned-out home and nowhere to relocate to, no clothes, no food, and a feeling of being alone and abandoned. In other words, we did not do a very good job of restoring their feeling of safety.

To this end, we developed a system to ensure that whenever there was a displaced family, we could find them food, shelter, and clothing. This system might be a simple as having them stay with neighbors or family, or it might involve getting into our database and locating a hotel, the Red Cross, or any one of a dozen other agencies that could provide these services. Also, we developed an "After the Fire" brochure that we could provide to the family at the conclusion of our involvement. This brochure provided checklists and directions on everything from how not to increase the destruction to your personal property (by not opening your safe right away, etc.), to how to secure your property, to how to contact your insurance agent.

Car seat installations, fire prevention and education activities, pre-incident planning, and apparatus and station maintenance all represent programs and specific actions that on a daily basis meet our mission statements. The real challenge to company officers and their superiors is to convert these words into everyday action, for everyday leaders.

Let me leave you with one last look at what I mean. If you walk into any fire station in the world and find a firefighter sweeping or mopping the floor and asked them why they are doing it, they would probably answer, "Because the lieutenant told me to." Not so, I say, not so at all.

I ask you this, do you like cleaning the toilets at your house, or painting the house, or cutting the grass, or vacuuming the house, or doing laundry? Of course you don't, but you do it anyway, and the question is Why? Rather than because anyone told you to (except perhaps your spouse, who may be the domestic equivalent of a lieutenant and chief all rolled into one), you do it because it meets your unwritten values and family vision of having a house that is safe and healthy and has some equity. It keeps your house free of disease, keeps your family safe, and makes you feel good for any number of other reasons. In other words, it meets your family values and requirements for accomplishing that part of supporting a family.

The bottom line with the firefighter mopping the floor is that he or she is not doing it because the lieutenant said to but because at the very basic, visceral level, it meets the organizations goals of having a safe and healthy work site for employees and visitors. Recognizing and communicating that notion is the first step in understanding what values and organizational mission are all about.

Officers must take fuzzy words and philosophies such as mission statements and core values and convert them into everyday actions. This is what everyday leaders do: They see, and they simplify without anyone ever realizing it.

Values

Everyone has values in their personal life, things that they believe in and that guide the way they live. We use values to make decisions about how to act, how to treat other people, and what limits to place on our actions; in addition, many times we make decisions based on our values, rather than on what might be popular. These values are in turn passed on to our children, as a model of what the family believes in, with respect to how we should live our lives, treat others, and communicate. Certainly, there are families and individuals who have values that you and I would not consider to be moral and that to some might even seem evil; names like Adolf Hitler,

Joseph Stalin, and Charles Manson conjure up visions of values gone astray! In general, values are learned through exposure and by witnessing actions. No child learns to hate or to be a racist; these are learned by example from someone else who espouses the values of hatred and racism. It is the same with noble values: either personal or organizational, they are taught and lived until they become part of the individual—and finally part of the larger body we call family or organization. Ultimately, values become the filter through which we make decisions.

If we have values in our personal lives, then why should organizations be any different? Why shouldn't organizations have values that we express and that we expect our members to use in making those same types of decisions that we must make as leaders in our organizational life? The answer is that organizations must have values if they are to create a method of decision making that is fair and honest, creates trust, and shows a level of respect and accountability for all members of the organization, from the chief on down to the newest rookie.

Organizational values are based on the belief that with the use of these values, we will create a better organization and provide a better quality of life for our customers and for our members. To be truly effective, values are those things that the members of the department hold dear and are unwilling to compromise. Values should be written down in your strategic plan and, along with the mission and vision statement, should be used as one method to evaluate any employee's behavior in the annual appraisal report. Values include

- Caring
- Integrity
- Trust
- Honesty

Still, like mission statements and vision, these words have little worth or use if they are not defined and demonstrated through everyday action. Without members of the organization acting on these words, they remain simply words.

The challenge at this point is to incorporate these values into the way we make organizational decisions, how we treat our members and customers, and more important, how we lead. As leaders, even when we don't completely agree with a decision, we must let organizational values drive how we do business. Failing this, we violate the most basic tenet of leadership, which dictates that true leaders always lead not for themselves but for something greater than themselves.

> *The fire service is a people game: Win people—win the game; lose people—lose the game. I am not talking about not holding members accountable for their actions or kissing anyone's rear end. Instead I am suggesting that everything we do in the organization is with and about people. We live, eat, train, respond, and even die with people in our organization. In addition, we don't make widgets; we serve people. Every action we take is intended to prepare for or actually deliver service to people who may be facing the worst days of their lives.*

Organizational versus Individual Leadership

Too often leadership is presented as an abstract undertaking—a matter of vision and values, rather than practical detail. Leadership is only possible when the ground has been prepared in advance, and preparing that ground falls squarely on the shoulders of the organization. The organization functions much as the parents in a family unit; dysfunctional parents are more likely to have a set of dysfunctional kids, unless they are smart enough to recognize it and rise above it. It may also be that the kids seek mentorship outside the family, from an individual who is more functional and grounded.

Why this discussion about family, you ask? Well, organizations and the men and women that command them are directly responsible for the organizational family. They are responsible for how the family acts, thinks, and responds, and they are responsible for preparing the ground for future leaders.

The further up the organizational ladder you move, the less you get to operate the system, or at least that's the way it's supposed to work. Those who cannot make that transition are micromanagers, not actual leaders. Had they prepared the ground before they found themselves at the top of the heap, they would be able to trust the leadership they have developed below them to operate the system and make decisions. The top job involves looking globally to create the future of the organization. To do so, one must take fuzzy political directives, mission statements, visions, and values and communicate them to the troops in real-world language and subsequently actions. In other words, you must be able to give the members of your organization a path to follow that is based on tangible and understandable directions associated with real-world tasks.

Because much of leadership is presented in the abstract, it is critical to understand how to convert the abstract fuzzy words into something of use. That new product must then be communicated into something that our personnel, our organization, can accomplish. We will look at some very specific conversions from fuzzy to functional later in this chapter, but right now let's just set the expectations for leadership.

Throughout this book run some continual themes that you may well get tired of hearing. Many of those themes have been extracted from fuzzy directions, mission statements, vision statements, and core values and converted into task-level, understandable, achievable actions that true leaders implement each day. The world needs everyday leaders: normal men and women who come to work each day to lead—not supermen or superwoman, but everyday mortals who lead from the front and set examples that can be passed down from generation to generation.

To understand the essence of leadership, we must understand that there are two forms of leadership that are at work here: (1) organizational leadership; and (2) individual leadership. Before we move on, let's try to understand the difference between the two and discern how one affects the other for better or worse. Let's take the fuzzy words and try to convert them into concrete examples, so that we all recognize the animal when we see it. It's one thing to say platypus, but it's quite another thing to recognize one.

Leaders are shaped by the environment in which they are raised. Not only are they shaped by the environment but they also take active roles in shaping the environment in which they live and work. In essence, true leaders create organizations that support, educate, and exercise the cultivation of leadership. This is accomplished when leaders, using a variety of means, achieve systematic organizational development.

The work of leadership is both individual and organizational. The bad news is that leadership is a draining and difficult business. Members that practice individual leadership are required to walk the walk, not just talk the talk. The great news is that leaders can be made, and anyone who wants to work hard enough can become a leader. Some members will always have the edge on leadership, the organization succeeds only with everyday leaders at all levels.

Organizational leadership

Organizational leadership involves the cultural aspects of leadership, which the organization espouses and preaches. It is the undisputable responsibility of the chief fire executive and the deputies and senior staff to establish this culture and to communicate it, nurture it, change it when necessary, and manage it from day to day. How this foundational aspect of leadership is established affects individual leadership and the entire organization. If this is done effectively, individual leaders will prosper and develop the ability to communicate and implement a plan much larger than themselves. Poor individual leadership at the executive level will result in the failure of organizational leadership. When that occurs, what is left are pockets of leadership in the field undertaking bunker mentality in order to survive! In essence, organizational leadership is the glue that holds the pieces together. If leadership were an aircraft carrier, organizational leadership would be the entire ship and the preparation for a mission, the carrying out of the mission and the safe return to the homeport. The foundational aspects of organizational leadership include the following:

Converting words into action. Every ship has a captain and a group of senior officers who must organize and direct the operations of the ship. It is no different in an organization. We call this chain of command. Above this chain of command are a group of politicians who have politically driven agendas or who are looking at an entire city, state, or government of which the fire department is simply one division. In this context, these people include the city council, city managers, town councils, boards, and so forth.

It is ultimately the chief executive's responsibility to develop the vision with input from the stakeholders from all levels of the organization. This person must find and develop a method using the energy of surrounding personnel to create educational opportunities and a learning environment, so that the organization and its officers can understand the vision and convert it into actions.

Creating a sound executive leadership team. Too often executives surround themselves with personnel who cannot do the job effectively. In other words, the executives must surround themselves with great people to create and bring to life a great vision. Unfortunately, this is a double-edged sword: if executives fail to recognize that the senior staff members are unable to function at their level or if senior executives surround themselves with like people, then they are bound to fail.

Failure here results from three specific errors on the part of the senior executive:

1. The chief executive is a poor visionary, unable to communicate effectively with personnel that work directly for him or her or with frontline personnel. The result is poor people skills in a people business.
2. The chief executive is a competent leader but fails to recognize that those who serve directly under him or her are unable or unwilling to effectively lead at the executive level.
3. The chief executive fails to make the hard choice of replacing deputy- or division-level personnel when they are unable to function effectively.

Recognizing that staff support operations. The function of staff is to support the operations and customer service delivery side of the organization, not the other way around. A real dichotomy exists in executive leadership. Once you reach the top of the ladder, if you choose to be effective, you must become the biggest servant of the organization. To successfully accomplish that very difficult balancing act, you must ensure that the following occur at the organizational leadership level:

- You must listen and hear what frontline service providers are saying. Remember that the personnel seeing the customer every day are the first ones who are able to tell you if a policy does not work, if there is a safety issue, and if the customer is happy, as well as myriad other aspects that deal directly with the customer.
- You must accept that frontline providers are always right until proven otherwise. There is a saying that "A desk is a dangerous place from which to judge the world." Members of your organization are in contact with your customers every day; also, they are working under and applying departmental policy and procedure, undertaking programs, evaluating effectiveness, and providing feedback. To have a living organization, you must leave your ego at home and listen to and evaluate and implement ideas from frontline members.

Developing and communicating a plan. Organizational leadership is the fabric that holds the organization together and navigates it through both calm seas and horrendous storms. To ensure that this is possible, the organization must develop a plan of action—either a formal business plan or any other method to communicate expectations and to track progress. This plan includes the development of a mission statement and core values and, most important, requires that the organization live by those values, constantly evaluate the plan, and drive aggressively forward by using team power and human initiatives.

This requires a great deal of trust from senior executives. It means sharing decision making at levels well below the crystal palace, listening and responding to suggestions for change, and taking risks by letting go of the reins and trusting in personnel. Executive leaders who micromanage, assume all responsibility, fail to take risks, don't include senior members in decision making, make or review all decisions, and constantly work outside of the time-management model do so because of several factors. First, they may be unable to effectively work in their positions, because they lack the KSAs or education to function effectively. Second, they cannot let go of the perceived power of their position and therefore cannot trust anyone to make a decision other than themselves. These are classic Peter principle behaviors. These actions in turn create a valueless environment for them and their subordinates. Whatever the factor, it is a clear sign that organizational leadership either has failed to communicate the plan effectively or simply does not have the KSAs to communicate and implement the plan.

Developing a life-long leadership legacy. The final and perhaps the most important aspect of organizational leadership is the ability of the organization to create a learning environment. In human terms, this is akin to procreation, or having children. If the organization is going to survive long term, the next generation must be prepared to take up the cause.

This is a multitiered process that involves education, listening, and changing. First, organizations must develop a method to train and educate new leaders in all aspects of organizational life. Without this ability, organizational leadership will fail in the long run, as individual leadership fails. Next, organizations must be listening and hearing organizations. If we are going to implement what is best for our personnel and our customers, we must be willing to continually listen and hear and implement new ideas. In other words, once the organization has procreated, it must be able to grow up. Finally, we must anticipate and prepare for change, both good and bad. Change is driven by a wide range of internal and external factors, ranging from budget to personnel turnovers. Members and leaders must receive education in change as the organization matures.

Leadership is successful only when the ground has been prepared in advance. This process is beyond the capabilities of a lone individual in an organization. If the organization is not pulling for you, you are probably hobbled from the start.

Individual leadership

If organizational leadership is the entire ship, then what is individual leadership? Well, on any ship there are multiple departments, which must run effectively for the bigger ship to function. These departments must interact, communicate, share, work together, and understand the big picture while running from day to day with the small picture. Much like the parts of the body, which depend on each other for health, individual leadership combines to create a healthy learning organization that is progressive and able to change without the wailing and gnashing of teeth.

The organization creates the ground on which individual leaders are grown. The individual leaders are not simply an empty, passive container into which the organization pours its intentions. There are many traits that members and officers must demonstrate to become effective individual leaders.

Charisma and presence. No individual leader ever forgets the first time that their command presence was put to the test. I can remember moments that have tested my leadership skills from dozens of significant fires and a number of technical rescues, as well as Oklahoma City and the collapse of the World Trade Center, but one particular incident reminds me time and again of the defining moment when I was put to the test.

Virginia Beach Fire and Rescue (VBF&R) had responded to assist Chesapeake with a collapse in an auto parts store in which two firefighters, Frank and Johnny, were trapped and missing. On arrival I was assigned as the rapid intervention team (RIT) chief and gathered two crews from the technical rescue team and additional engine companies to begin the search. It was clear by this time that it would have been impossible for the two men to make it out alive. Adding to the tragedy, all of us had known

Frank and Johnny, having worked, trained, and socialized with them. There was no doubt that the search was going to be a physically demanding task and that when someone found evidence of the firefighters, the emotions would run deep.

I gathered the team together and tried to muster the correct words, which ran along these lines:

> *Gentleman, we are about to undertake a task which no firefighter ever wants to undertake: we have been tasked with the search for our two fallen brothers. Look around you and see the faces of the men and women here today—look at their expectations and hope. Stay focused, stay safe, and remember why we are here! We are responsible for this mission, and we will successfully complete this. I want you to put your emotions aside and focus on what we are required to do here today.*

The team acknowledged the task, and we moved into the structure. I hung back, taking a look at the big picture and watching the team's search patterns and observing where they were going. This process seemed to take forever, until Master Firefighter Duane Krohn indicated he had located a self-contained breathing apparatus (SCBA) bottle. As the crews moved forward and began to remove debris, it became obvious that we had located the two fallen firefighters, wrapped in each other's arms, as a team, neither leaving the other. The team began to falter, so I quickly spoke over their shoulders, "Folks, stay focused on what we have to do here. This is your job, your task. Members of the department are depending on you to get this done. Now let's do what we have to do!"

As the bodies were removed and the team's mission was finished, I sequestered the team to an adjoining building in one of the exposures. We were alone with our thoughts, and I said, "You have done a difficult thing today, but you did it professionally and with great skill. I am proud of all of you, and the families of the men you removed today are grateful." Tears flowed, hugs went around, and the team moved on to the next task. At the time, moving on to rehab seemed appropriate based on our task assignment. No one had to listen or follow, but they did. I had no idea how significant the words and actions were until the funerals and several months later, when members of the team started calling to tell me how much it meant to have someone focus them and lead them.

Expertise. For individual leadership to be effective, people must believe you know what you are talking about and that you are doing the right thing. This is all related to the concepts of trust, integrity, and credibility, along with other traits directly attributable to being an expert. These concepts will be discussed in greater detail later in this book.

Expertise also means learning more than one job. Most of us end up with expertise in a specific functional area, and we get very comfortable there. We believe that our entire organizational life will and should be spent in operations. To be a true expert, you must practice the following:

1. *Diversify your knowledge base.* Take the opportunity to spend some time in other organizational locations and 40-hour staff jobs. Accept that some assignments are not exactly where you want to be placed, and use them as a learning tool.
2. *Educate yourself.* Get your formal education, and do it now! Even though I have a master of public administration, I am still just an educated idiot. To this day, firefighters I work with tell me, "Chief, that last e-mail you sent, I don't think that —— is really a word, and perhaps you should use spell check!"
3. *Learn from outsiders.* Get out of your organization to conferences, to other departments, to the National Fire Academy (low bid) and see how others live and work. Of course, the organization also has to make this a priority, as will be discussed in the section on technical competence. It is also important to bring people into the organization to teach with a different perspective. Never isolate yourself—organizationally or individually. Anyone who reaches the rank of deputy chief or above and has isolated himself from his own organization over his career is a loser!
4. *Learn from insiders.* There are great people, as well as not-so-great people, in any organization. Find the great ones and learn from them. Put your ego aside and take the time to learn. Greatness and the ability to teach and mentor have nothing to do with rank. Some of the most important lessons I have learned were taught to me by firefighters and company officers.
5. *Listen, hear, and respect.* The ability to lead is based not on people who want to follow you but on who follows what you are able to create: something bigger than you and them, something that everyone has a stake in. If you are going to do that, then you must listen, hear, and respect the ideas of others. In combat, the frontline soldier is closest to the shooting, the patrolling, and the dying. If something is not working, they can tell you before anyone else. In the fire service, the frontline members are closest to the customer, our policy impact on operations, and our overall capability. If you want to create an environment that establishes

fertile ground for leadership—current and future—then spend more time listening than you do implementing your own ideas.

6. *Know yourself.* This is so important that there is an entire chapter in this book on knowing yourself and others. But consider these questions as a teaser: What's your expertise? What are your strengths—and just as important, what are your weaknesses? To sustain leadership, you must continually self-examine.

7. *Have empathy.* I learned this from my former fire chief, now Councilman Harry Diezel. I will always remember what he once said: "Don't make personnel a green shade operation," meaning that accounts and bean counters look only at money and not at human beings. I call this keeping humanity in your leadership, and it too is the topic of an entire chapter in this book. As long as we are in the people business, we are must hold our members accountable and responsible for their actions. It also means that we have to look deeper than the exterior, into the motive and into the very human being involved in the action.

8. *Have fun.* Some people make life way too serious. The business of the business that we are in is serious enough. For goodness' sake, have some fun at work and let your members have some fun. I always get the strangest looks when, during a meeting with senior members, I find something funny about the nature of our discussion. I remember buying instant-scratch lottery tickets and giving them away during drills for correct answers. One of my firefighters actually won some money! I remember Don Moss putting dog poop on the hydrants after he would grease and flow them as we followed him to paint them. Lieutenant White used to get so mad, but inside we were all busting out laughing. It was good fun. I remember the day that Gary Reich slipped while painting hydrants and covered himself in white and red paint. It took us nearly two hours to clean him up with Varsol! There is plenty of time to be serious in this business, so take some time for your soul now and then.

There has never been a decent crop grown by a farmer if the field was not prepared in advance. Without proper attention to the needs of the soil, the weather, and the specifics of the crop, nothing grows. Creating a foundation for leadership is like preparing a field. If you want great crops you had better mix the soil right and throw in a little luck with regards to rain and sun.

Leadership is a group effort. It requires organizational support and foundational development, and it takes individual leadership to create collective excellence.

Why Senior Leaders Must Lead

> *The best executive is the one who has the sense enough to pick good men to do what he wants done, and the self-restraint enough to keep from meddling with them while they do it.*
>
> —Theodore Roosevelt

    Those of you who have raised children know that they listen to everything you say. Many of you probably fell out of your chairs in disbelief of that statement. Those of us who raised children know that, in reality, they spend an inordinate amount of time figuring out how not to listen to you. However, the one thing that does not escape them is your actions.

    Children, adolescents, teenagers, and young adults—and the members of your organization—watch everything you do. This should be no great surprise, since your mother used to tell you all the time "Show me, don't tell me." How many of you have been called on your actions or inactions by your children? Dad, I thought you said we shouldn't be doing that? Dad, I thought you said we should help out the poor, so how come you wouldn't give that guy a dollar? Dad, how come you yell at me about my language and then use those words? The bottom line is you cannot go down the road in your car with a drink in one hand and a cigarette in the other and say to your children, "You should never drink and drive, and don't smoke!"

So it is with leadership at all levels, but especially at the senior level. No leader operates in isolation. Nor do they actually command in the literal sense of the word, issuing one-way orders and expecting complete compliance all the time. Leadership always involves cooperation—collaborations that can occur only when senior leadership shows the way and sets the example. This requires *centralized control*—where control is a relative term, since we often don't control much of anything without cooperation from and others who buy in to our ideas—with *decentralized execution*, meaning people have to engage and believe to get things done.

Senior leadership must surround itself with educated, competent, and committed members who have the expertise necessary to fulfill the jobs at hand, so that delegation becomes a matter of trust and respect. There can be no more damning action than to ignore what others say on a continual basis and implement only one's own ideas. If you surround yourself with knuckleheads, you are going to get knucklehead solutions, and you are going to wonder why, four or five years (or sooner) down the road, no one believes in you or will follow you.

Senior staff must develop a method to effectively communicate on a continual basis, seeking most importantly to listen to information from frontline providers. Avoid having a yes-man or yes-woman on your staff. As Colin Powell once said, "If I have a yes man or a yes woman on my staff, one of us is redundant!"[1] In the article, "Leadership in a Combat Zone," retired army General William Pagonis coined the phrase "Staff grows, paperwork flows, and no one knows."[2] Information flow is one thing, but getting the right information is quite another.

Senior leadership is responsible for shaping and communicating the vision. Simple is always better, since the ability to do your job and trust that the delegated task is being accomplished is based entirely on shared understandings of organizational goals. Without this simplification, subordinates—who are ultimately responsible for defining those objectives and moving them forward, toward a desired outcome—will struggle to succeed. Objectives must be specific and quantifiable, thereby giving members the ability to both act and assess the impact of their actions: "Let's put this fire out" is a statement of vision; by contrast, "Let's get a line to the second floor and pull ceilings" is an objective articulated to advance the vision.

Senior leadership must educate and be educated. Personnel entering the organization must get a clear sense of what you are about, what your style is, what your expectations are, and what the vision and shared objectives for the organization are—and they must be simplified. You cannot hand someone a 50-page business plan and say "this is our vision, now make it a reality." Senior staff must continually learn from the organization, remembering that frontline members are always right until proven otherwise and that the closest people we have to the customer are the lowest on the food chain.

Part of education is holding meetings and skull sessions. Nevertheless, I have partaken in an inordinate number of meetings that duplicate information, frustrate people, and accomplish nothing. Let's face it, if it were not for meetings, I would not have a job; however, much of our time is wasted and in turn disrespects the value of the time that chiefs and others commit to meetings. A key aspect of senior leadership is to make each meeting productive through identified outcomes and to limit the number of meetings to the level necessary to accomplish the task and communicate effectively. If you delegate and trust your personnel, all of those departments on that big ship should be meeting, so you don't have to pull everyone together all the time.

Communications in senior leadership takes on two forms: informal skull sessions and formal communications structures designed to complement the chain of command. In either case, feedback is critical to movement up and down the chain. An open and regular method of informal communications can be gathered through "leadership by walking around"; if you want straight talk, nothing compares to a dinner in the fire station, time spent on the fire ground (after the incident is mitigated) talking, or an informal talk in the middle of the hall.

The stark reality is that every day that senior leaders show up, they must practice and show leadership, not simply speak about it. People will watch what you do every day and will judge the leadership activity and success on action, not words.

Finally, leadership at all levels of the organization is only possible if the ground has been prepared in advance. Only though constant self-vigilance and leadership action can senior staff prepare the ground for our members today and those coming tomorrow.

Who Packed Your Parachute Today?

One of the great downfalls of leaders at all levels of the organization is that they often take themselves much too seriously. We should remember that we lead only because our followers allow us to lead and that we will experience as many failures as other leaders in the organization.

My point here is that in the fire service, there is no single job that is any more important than any other job. We are all linked together by our success or our failure, at all levels of the organization. As I see it, we are the largest legal gang in America. Take a look at any fire department and you will see "colors" worn proudly both on the job and off work. For example, if a group of firefighters go to a bar and some drunk gives one a little grief, he or she will have the entire clan to deal with. We are very proud of who we are and normally wear our colors with pride.

However, there are occasionally times when you put the colors away for a while, as when something happens in the organization, at either the administrative or the operational end of the firehouse, that makes the paper. Have you burnt a house down by accident or because of poor tactics? Or was someone caught dipping into the till or not following the purchasing guidelines? Because we are responsible for each other and we are all interlinked, it reflects poorly on the chief whenever the firefighters and company officers do poorly on the street. Similarly, if the chief or his staff do poorly with the council or in other matters, it reflects poorly and affects the ability of personnel to do their jobs.

This resonates in a story sent to me many years ago:

> *Charles Plumb, a U.S. Naval Academy graduate, was a jet pilot in Vietnam. After 75 combat missions, his plane was destroyed by a surface-to-air missile. Plumb ejected and parachuted into enemy hands. He was captured and spent six years in a Communist Vietnamese prison.*
>
> *He survived the ordeal and now lectures on lessons learned from that experience.*
>
> *One day, when Plumb and his wife were sitting in a restaurant, a man at the other table came up and said, "You're Plumb! You flew jet fighters in Vietnam from the aircraft carrier Kitty Hawk. You were shot down!"*
>
> *"How in the world did you know that?" asked Plumb*
>
> *"I packed your parachute," the man replied. Plumb gasped in surprise and gratitude. The man pumped his hand and said, "I guess it worked!"*
>
> *Plumb assured him, "It sure did. If your chute hadn't worked, I wouldn't be here today."*
>
> *Plumb couldn't sleep that night, thinking about the man. Plumb said, "I kept wondering what he might have looked like in a Navy uniform: a white hat, a bib in the back, and bell-bottom trousers. I wonder how many times I might have seen him and not even said,*

> *"Good morning, how are you?" Or anything because, you see, I was a fighter pilot and he was just a sailor.*
>
> *Plumb thought of the many hours the sailor had spent on a long wooden table in the bowels of the ship, carefully weaving the shrouds and folding the silks of each chute, holding in his hands each time the fate of someone he didn't know.*
>
> *Now, when Plumb lectures, he asks, "Who's packing your parachute?" Everyone has someone who provides what he or she needs to make it through the day. Plumb also points out that he needed many kinds of parachutes when his plane was shot down over enemy territory—he needed his physical parachute, his mental parachute, his emotional parachute, and his spiritual parachute. He called on all of these supports before reaching safety.*

Sometimes in our organizations, we miss what is really important. We fail to say hello, to say thank you, to congratulate someone for a job well done, to give a compliment, or just to do something nice for no reason at all.

The question is, "Who is packing your parachute today?" Every day you go to work, regardless of whether you are the chief of the department or the newest probationary firefighter, someone is packing your parachute. Our success or failure as leaders—and ultimately as an organization—is inextricably linked. Each of us should remember this when we begin to take our positions too seriously. As you go through your organizational life, recognize who packs your parachute.

And so the first step of our journey is finished, yet we still have many more miles to go on our leadership quest. These basic guiding principles provide the foundation of our leadership house. Since all great structures start with a solid foundation, make yours solid through your everyday leadership actions.

Never underestimate the power of dedicated people willing to do what is necessary to succeed. At the same time, never underestimate the viral infection caused by incompetent people floundering in leadership positions. Fail to surround yourself with great people or fail to remove those that are unable to function, and you will fail.

Notes

1. Harari, Oren. *The Leadership Secrets of Colin Powell*. New York: McGraw-Hill, 2002.
2. Pagonis, William. "Leadership in a Combat Zone." *Harvard Business Review*, December 2001.

Chapter 2
Knowing Yourself and Others

Small people, casual remarks, and little things very often shape our lives more powerfully than the deliberate solemn advice of great people at critical moments.

—*Winston Churchill*

Understanding How People Are Made

In a business dominated by people, it seems logical that one must understand how people are made and what makes them tick. Taken at face value, this is a dangerous proposition, since understanding something as complex as a human being, as well as human emotions, is a daunting task. We innately know what makes people tick, what motivates them, what undermines their morale, and where they draw their talents from because each of us lives with one each day—that is to say, we live with ourselves! Often we spend so much time analyzing others that we really don't take the time to reflect on who we are and what makes us change and grow.

Leadership requires the work of both individuals and organizations in order to create an environment where people and their ideas can thrive. Thus, the question that needs to be answered is, *how much can people be changed?* Surely if they are going to buy in to the grand plan, the vision, the direction, they must change their view. But is that really possible—is that the most effective method for getting people to lead or follow? If you hate the bright lights, can you be taught the passion for public speaking? If you have grave difficulty with confrontation, can you learn to sharply debate and defend? If you are shy and withdrawn, can you become outgoing and the life of the party? There is an old saying, "You can't teach an old dog new tricks," onto which has been tacked, "but you can beat him into doing what you want."

In many organizations, managers—notice that I did not say leaders—believe that the most productive method of getting members of an organization on board with a plan or a new course of action is to dedicate themselves to ensuring that the members learn new behaviors. They espouse the idea that everyone in the organization has the same potential: we spend time sending our members to countless programs to teach them everything from empathy to assertiveness. Yet in the end, these courses seem to create very few changes in the core person who has attended the course. As I write this chapter, a new study from Harvard Business School has just been released, in which many Fortune 500 companies, as well as many successful CEOs, were asked three questions:

1. *When you hire someone, how long does it take for you to realize they are not right for the job?*
Amazingly, the answer was 10 days.

2. *When you realize that they are not right for the job, how long does it take the organization to fix it?*
You guessed it, 10 years!

3. *If you had to hire for technical skills or human interaction skills, which would you put priority on?*
The answer: human interaction skills. Why? Because you can teach someone the technical aspects of the job, but you cannot make him or her interact with, understand, motivate, and communicate with people.

There is do doubt that providing education is a critical element in assisting members and new leaders. If I send you to a public-speaking class for two weeks and you are a horrible public speaker, there is no doubt in my mind that you will come back better educated and better prepared to speak publicly after two weeks. However, the fundamental question remains, Will you be capable of producing much better speeches now that you graduated from a two-week school? Will your actual performance under the lights be so drastic that you will be a different person in the public-speaking arena? Quite frankly, the answer is no.

In 1999, Marcus Buckingham and Curt Coffman published a book entitled First, *Break All the Rules*,[1] containing the findings of a Gallup Organization research study of over 80,000 managers in over 400 companies, the largest study of its kind ever undertaken. Buckingham and Coffman found that the best leaders and managers don't believe that people can change very much at all; the following coinage echoes the Gallup findings:

People don't change that much.
Don't waste time trying to put in what was left out.
Try to draw out what was left in.
That is hard enough.[2]

Interestingly, the Gallup study found that a person's talents—their mental filter—is actually what has been left in. If that is true, then it would be folly to expect any amount of charm school or public-speaking school to change fundamental talents or, in this case, non-talents.

The bottom line is that the so-called self-help books are wrong. An individual's mental filter is as unique as one's fingerprints or retina. Neuroscience has confirmed what the Gallup poll found and what true leaders and managers have known all along, soundly refuting the classic self-help texts.

In early 1990, Congress began funding research that assisted the scientific community in working to untangle the mysteries of the human brain and its workings. Money, along with new technology such as magnetic resonance imaging and positron-emission tomography, has allowed scientists and doctors to observe the brain at work. As a result of these breakthroughs, we can now diagnose mental illness by seeking areas of the brain that are malfunctioning or damaged. We know that chemicals in the brain play an important part in how we act and react as human beings, and we understand, as paramedics have for years, that dopamine slows and calms us whereas serotonin creates stimulus and action.

We have learned that men use both sides of their brain during everyday activity, but that in women, the left and right brain actually communicate with each other more effectively than a man's. We now know that the brain grows and that memories are stored throughout the brain, not in a singular location.

When a child is born, the brain contains hundreds of billions of neurons, which regularly grow and die throughout this human's life. However, these neurons do not comprise what we would term a mind. They are simply areas where information is taken in and stored. The mind as we know it—a giant master computer completing unlimited tasks, calculations, emotions, sensor actions, and other activities—is actually composed of connections between the cells at synapses. Scientists now know that during the first 15 years of life, the connections between neurons and synapses create a system that allows the individual to begin to carve out who and what he or she is; it is here, in the connections, that the traits that will eventually characterize this unique human being take shape.

From the day a child is born, every neuron in the brain acts as a giant transmitter, sending out thousands and thousands of signals in an attempt to talk with other parts of the brain. According to scientists, by the time a child reaches the age of three years, the number of successful connections is astronomical, resulting in about 15 thousand synaptic connections for each hundred billion neurons in the brain. These connections begin to create relationships between pieces of information, categorizing them into a neural database from which are drawn inferences, information, talents, and decisions.

This incredible connectivity creates a brain overload, and the level of information that the human being is attempting to decipher is so great that the brain begins to attempt to make sense of it all. For the next decade, this human being's brain will begin to refine the connections. In *First, Break All the Rules*, Buckingham and Coffman quote Dr. Harry Chugani, professor of neurology at Wayne State University, who equates the categorization of data and streamlining of information with a highway. Dr. Chugani states, "Roads with the most traffic get widened. The ones that are rarely used fall into disrepair."[3]

Subsequently, our internal neurological composition is designed and ingrained long before we show up for our first job or at the fire station. Quite simply, it boils down to the physiological and neurological fact that we are each different from everyone else. Deep within the recesses of our brains are established our own methods of seeing and deciphering the world and of decision making and the things that motivate and excite us, as well as the things that we are uninterested in or in which we display a nontalent.

Different is not wrong. It's just different.

Let me offer you an example of how this affects us when dealing with employees. Suppose that, as my supervisor, you decide that I am not performing at the desired level in a certain area. As a leader, you make the decision to sit me down and explain the issue and your expectations. In reality, what you want is for me to improve my performance; or in terms of a performance scale, you want me to move from point A to point B.

After we have our little talk you lay out how you want me to get from point A to point B. But stop and think about this for a moment. When you lay out specific steps that you expect me to take, how are you describing the process of getting from A to B? That's right, you are outlining the process that you yourself would use to get from point A to point B.

Now, if what we said earlier holds true, then perhaps I get to point B differently than you do and thus need a different path, different motivators, and different methods. After all, what's more important to you: that I get to point B and improve or that I get there the way you would get there? Understanding just this one idiosyncrasy about people could save you lots of time and grief.

> *When dealing with people, find out what motivates, demotivates, excites, energizes, and drives them. Then, when you find yourself in this position you can work together to put a plan together to get the members from point A to point B using their talents.*

Why Know Yourself?

It may sound funny, but you really can't expect to lead anyone if you don't know yourself. To this, you may say, "Hey, I know exactly who I am and how I react." Although that may well be true, my point is that each of us has an individual style that is readily apparent to peers, subordinates, and ourselves. There is a wide range of instruments available including the DiSC profile, the leadership profile, and the Myers-Briggs Type Indicator to help figure this out.

Consider why it's important to know yourself. Our leadership reactions are often shaped by the environment in which we find ourselves, as well as perceptions regarding relative strengths and weaknesses within that environment. Moreover, most of us have a dominant style that serves us well most of the time. However, sometimes that style leads to trouble; when the environment and players change, we need to understand our tendencies and make the necessary adjustments for continued success. Following are a few suggestions on how to accomplish this.

All of us have tendencies, or those characteristics that determine how we make decisions, how we interact with others, and what our strengths and weaknesses are. If we recognize those tendencies (by using an instrument to help identify them), we can then identify what it will take to balance our styles. For example, when I apply the DiSC instrument to myself, I find that

I am interpreted as a dominance, or high D. From a behavioral standpoint, my leadership life is characterized by the following tendencies:[4]

- Getting immediate results
- Causing action
- Accepting challenges
- Making quick decisions
- Questioning the status quo
- Taking authority
- Managing trouble
- Solving problems

If these are really representative of the process that I use as an individual to make decisions that affect my leadership, then there are several valuable lessons here. The first lesson is that these strengths at times can become weaknesses. If you surround yourself with only one style, in a vacuum, then in the wrong environment, you may no longer be able to use that style. The second lesson is that if you are really going to work in a people business, constructed almost entirely of teams, then you had better learn to balance what you do and to seek help from others. This might be referred to as an action plan, in response to the question, What type of different behaviors do I need to surround myself with to get the most out of my style and ultimately make the right decisions? My own action plan—what I need from others on the basis of the tendencies listed previously—might look like this:[5]

- Someone who weighs the pros and cons of a decision
- Someone who calculates risk
- Someone who uses caution and advises me when to do so
- Someone (a boss) who can structure a predictable environment
- Someone who is better than me at researching facts and data
- Someone who is deliberate before deciding

In addition to what I need from others, I must force myself to recognize the needs of others.

These needs may all seem admirable to some and overbearing to others. To be truly effective and work with people with other styles and other tendencies, I have to answer certain questions. The principle is, What kind of environment do I desire to work in? This establishes where, based on my style, I feel most comfortable and most successful. That's all well and good, but manipulating my work environment over my entire career to ensure that these factors always fit is another matter. However, as bosses, finances, and other aspects of the organization and job change

so does the environment. My ideal environment, in which I would thrive and do my best work, would include the following:[6]
- Someone with power and authority
- Prestige and challenge
- Opportunities that provide individual accomplishment
- A wide scope of operations
- Direct answers
- Freedom from control or supervision
- New and varied activities

So what is the lesson about the environment? As the Rolling Stones said, "You can't always get what you want, but if you try sometimes, you just might find, you get what you need." Sometimes the environment is not perfect for you to thrive. Nevertheless, from a leadership perspective, if my boss is smart and wants the best from me, he will do his best to structure my work environment to meet these needs. Similarly, as a supervisor/leader, I would do well to realize that there are also certain environments in which my subordinates thrive and work toward meeting those needs.

What, you say, structure an environment for everyone—you're nuts! Well, it's not about structuring the environment; rather, it's about giving the individual member what they need with regards to workload, types of work, feedback, and other aspects of organizational life. Again using myself as an example, to be my most effective, I need the following:[7]
- To receive challenging assignments
- To understand that I need other people in order to be successful (see "Tendencies")
- To base my process on techniques and practical experience
- To receive an occasional shock (reality check)
- To identify with a group (be gregarious)
- To verbalize reasons for conclusion (since I fall short on data and research)
- To be aware of existing sanctions (because I push the boundaries)
- To pace myself and relax more (not confuse my job with my life)

The moral is if you are smart as a leader, then you will realize that everyone's behavioral dimensions are different, even as they fall into several classifications. When you realize this about yourself and others, it prevents you from making the critical mistake of attempting to make everyone like you. Understanding yourself allows you to understand others.

There are myriad other aspects about knowing yourself that affect your ability to lead. We could spend hours on them, and when I do Buddy to Boss seminars, I spend time with everyone, creating the behavioral profile and discussing its implications. Other aspects of knowing yourself that I will hint at now and discuss further throughout the book are

- *Emotions.* What demeanor do you typically exhibit in dealing with people, problems, politics, success, and failures?
- *Goals.* When you deal with people, what is your goal? Is it to control, or is it to be cautious? Do you like victory with flair, or do you seek popularity?
- *How you judge others.* Do you judge people based on their competence, their ability to tolerate others, their personal standards, or other imposed criteria?
- *How you influence others.* Do you attempt to find solutions to others problems, create personal relationships, accountability for you own work, or other aspects?
- *Your value to the organization.* Do you accomplish goals with a team, or do you avoid passing the buck? Do you initiate design changes, or do you remain stable and predictable?
- *How you overuse your style.* Do you overuse your authority or ingenuity? Do you try to control people or situations? Are you overenthusiastic? Do you demand too much data or create too many fail-safe controls?
- *How you react under pressure.* Do you become indecisive, become tactful and diplomatic, internalize your conflict, or become manipulative and quarrelsome?
- *What you fear.* Do you fear weak behavior, loss of social status, inferior work standards, failure, or loss?
- *How you would increase your effectiveness.* Would individual follow-through help? Would patience and empathy help? Would attention to realistic deadlines help?

When you work with people and deal with human behavior all day and all night, creating a framework from which you can observe your own behavior and increase your knowledge of other behavioral patterns is critical to success. Unfortunately, this is not emphasized when we teach our new leaders, nor is it typically considered important in leadership.

Using Styles and Tendencies to Strengthen Your Team

You never know till you try to reach them how accessible men are; but you must approach each man by the right door.

—Henry Ward Beecher

The real question is, How does that apply to you as a leader? See, I know you have these organizational shenanigans figured out. Organizationally, you know what the major problems associated with any fire or rescue service are right now, and the key problem is that people are not more like you. If everyone was more like you, this entire organizational shambles would fix itself! Yeah, right—and I have some property at the foot of Mount St. Helens I want to sell to you!

I am always amazed when I hear an officer or someone in the fire service mutter, "I'm just not a people person." I shudder to think about working for them, when our entire industry is about people. You see, the lesson here is that it's a people game: Win people—win the game; lose people—lose the game. This is fundamental to the business that you and I have chosen as our profession. The fire service is all about people. It's about working with people, living with people, and functioning as a team—it's a mini-family. Even when we go to work, the business of the business is about taking care of people, communicating with people, and solving people's problems.

At the executive level of the fire service, fire chiefs make their biggest mistake by accepting what they are given as far as senior staff. In the book *Good to Great*, Jim Collins affirms organizational leadership principles that we have known for a very long time. It's not initially about a new direction or a new vision; instead it's about people. First, get the right people on the bus and the wrong people off the bus; get the right people in the right seats, and then decide where to drive the bus. Collins states,

> *The good-to-great leaders understood three simple truths. First, if you begin with "who," rather than "what," you can more easily adapt to a changing world. If people join the bus primarily because of where it is going, what happens when you get ten miles down the road and you need to change direction? You've got a problem. But if the people are on the bus because of who else is on the bus, then it's much easier to change direction. . . . Second, if you have the right people on the bus, the problem of how to motivate and manage*

> people largely goes away. The right people don't need to be tightly managed or fired up; they will self-motivate by the inner drive to produce the best results and to be part of creating something great. Third, if you have the wrong people, it doesn't matter whether you discover the right direction; you still won't have a great company. Great vision without great people is irrelevant.[8]

The bottom line is that sometimes when you can't change people, you have to change people! As it turns out, the old adage "People are our most important asset" is wrong, because success is about getting the *right* people.

Because this job is all about people, it makes our jobs as supervisors (bosses) extremely complicated. The human being is the most dynamic, complicated, high-maintenance machine every constructed. The human brain is without a doubt the most complicated computer that will ever be built, and amazingly enough, 90% of everything we know about the brain will be learned in the next 10 years! Emotions, logic, input, and output are ongoing, and there is a constant interaction with the environment in which the human being works. From a neuroscience perspective, what we are is mostly in place by the time we are in our late teens. Talents and nontalents, approach to the world, how one filters information, whether one is outgoing or self-effacing, data driven or instinct driven, is all laid down in the brain via information superhighways. In essence, we are who we are, and attempting to make us something that we are not is a lot harder than using who we are. In *First, Break All the Rules*, Buckingham and Coffman have suggested that "Instead of spending time trying to stuff in what was left out, spend time trying to draw out what was left in."[9] It's critical that, as leaders, we realize that the people who come to the job are different and unique. While they are all expected to have the KSAs required of firefighter or paramedic or both, when it comes to what drives them, what motivates or demotivates them, what excites them, and what specific talents they bring to the job, they are all different.

When you deal with people, there are three pillars of success through which everything we do should be filtered. The ultimate win is that when we do something, it benefits the triad of customer, organization, and members (either the team or the individual). The real secret is the ability to effectively identify and apply individual talents for the good of the team.

The first rule about being a supervisor in a people business is that if you are going to be successful, you have to be a talent scout. You must identify how each member of your shift, or your battalion, division, or organization, can best contribute to the overall outcome. It's really about getting the right people in the right seats on the bus, and to do that, you have to recognize the talents individuals bring to your team, so that they can be channeled and used correctly to benefit the team and the individual. Why is this so important, you ask, because given the opportunity, the right people will always self-motivate? If you can give them the opportunity to apply a talent that they really enjoy, that really sparks them, then you have a member who is working hard and having fun and will go to the ends of the earth to be successful for you and the organization. The right people want to be part of something great and will work to achieve it.

At the beginning of this section, I tried to sell you on the concept that if you were more like me, everything would be all right. The reality, though, is that if you were more like me and we worked together, we would either kill each other or end up in jail. So the next lesson is that "Different ain't wrong; it's just different." This is the other side of being a talent scout. If you recognize that people are different, then you spend more time trying to draw out what was left in, rather than trying to stuff in what was left out. Let me give you an example of how we fail as bosses when we don't adhere to this concept. Suppose one of your members is having problems with a certain skill or job requirement. As the good boss, you set him or her down and say, "Look, I'm not happy with your performance in this particular area. You and I have to work on getting you from point A to point D." That's good, but here is where it falls apart if you don't understand the concept of talent and difference. The next part of the conversation goes something like this. "To get from point A to point D, I want you to do this, and this, and this." When we do that, we are telling the employee how we would get from point A to point D, whereas the employee, based on unique talents, mental filter, and understanding, may get to point D differently than we do. The question then is, Is it more important that the employee get to point D the same way we would or that they get to point D at all? The right way to address the employee may be, "Now that you understand where this weakness lies, how do we get to point D?" At this stage, the employee can provide feedback on how he or she can best get to point D, and together you can craft a plan that enhances the employee's performance by using what was left in, rather than trying to stuff in what was left out. It's complicated to evaluate and pick out the characteristics and capabilities of people, along with their talents. It takes time, effort, observation, and a pretty committed boss to make this work.

A tremendous amount of ink has been devoted to morale. We always hear whether morale is good or morale is bad. Give us company T-shirts—morale is good; don't give us T-shirts—morale is bad. Allow mustaches—morale good; demand that we shave—morale is bad. Give me a say in the way the work is done—morale is good; don't give me a say—morale is bad. Some people (managers, not leaders) don't believe that morale has anything whatsoever to do with productivity, commitment, drive, and alignment. I, on the contrary, believe that it is a fundamental building block of organizational relationships and the trust that comes with it.

The reason why some people don't consider morale is that they don't have a clue about what it is and when you cannot define or put your arms around something, it's difficult to admit it even exists. When I teach, I ask officers why they work for their particular fire departments. Aside from the facts that the money is pretty good, the benefits are good, and we give you a pillow and a blanket, there are other reasons. When people begin to describe why they come to work, reasons like "I enjoy helping people," "I really love the job," "I like the people I work with," and "I like to fight fire and run rescues" are given. All of these are human, heart-and-soul issues that don't have a thing to do with data.

Morale is a human feeling, and because it's about how a human being feels, morale is difficult to define. In essence, morale is how I, the individual, feel that the organization I work for values me and my talents and opinions. This then leads to the rhetorical question: What can I, the organization, do to show my personnel that I value their talents, input, and service? Interestingly, if you try to look too deep for an answer, you miss it altogether. For all its complexity, developing good morale is a simple surface issue. Giving the folks small things like modified grooming standards, a say in the uniforms (shorts, hats, patches, color of the shirt), improved livability (new lockers), and a say in the workplace (how we do business), as well as getting out and talking with your personnel (leadership by wandering around)—and dozens of other such little things—and they will work their hands to the bone for you. Consequently, morale will go up, and most of your members (you will never get them all on board) will be willing to go the extra mile for you. Sometimes making people feel good is enough of a reason. Most of the things that contribute to good morale are small in nature, but the momentum they develop, the attitude they nurture, and the enjoyment they provide to people have the potential to turn those people into just what is needed to change the course of your organization over the long run. It's like digging a hole: every shovelful is a very small amount, but each small shovelful eventually results in a very large and deep hole. Some people would say that the secret is, "When you find yourself in a hole, stop digging?" Maybe it is a bad analogy, but you got the point.

Even more important, though, is if you can give people little things that "boost morale," they will get on the bus with you for the long term, strategic stuff and do their part to make it successful. Stephen Covey calls it making deposits in the emotional bank account. Since some details of the strategic business plan take quite a long time to implement, the message we send our personnel when we give them little things is that change is possible. When people believe that change is possible, they will stay on board over the long haul. It was once said, "Organizational change is like a marathon race. The front of the pack is off and running, and the back of the pack has not even moved."

People are amazingly complex machines, on which our success—and very often in our business, our lives—depends. People always come before plans, since the best-developed plan will never be implemented unless the people tasked with accomplishing it believe in it and are willing to do what is necessary to accomplish it. People joke all the time about firefighters running in, when everyone else is running out. Stop and think about that. Do you think that you can get just anyone to do that? Do you think that just anyone has the level of belief that the job they are doing is so important that they are willing to risk everything for a stranger? Creating a vision in people is a function of morale, communications, belief, and basic human initiative, blended in a way that allows people to absorb it as if it were their own. When that happens, people will do whatever it takes to be successful. If it were easy, everyone would do it.

I could write volumes on people and what's organizationally important about understanding, working with, and helping them to succeed in order to help the organization succeed. It's one of those aspects of our job that is so important but so misunderstood. Like Jimmy Buffett says, "It was so simple like the jitterbug it plumb evaded me." As long as there is a fire service, there will be people, and it will always be a people business. Win people—win the game; lose people—lose the game. It's a game of relationships and trust—nothing more, nothing less.

Don't take your relationships with people for granted. It's only when we lose someone that we truly realize what we had. Value every smile, every handshake, every hug, and every talk, for when it's gone, you cannot regain what has been lost. I have lost two great mentors and dear friends in my life, Jim Page and Ray Downey. Both men were taken suddenly, without warning. Now I wish I had spent more time listening, learning, and asking when I had the opportunity. I just assumed they would be around forever. How wrong I was.

Leading Your B-Team Players

In a recent article, Nick Brunacini, from Phoenix, brought up the excellent point that not every member of the department can be a franchise player. Not everyone is a star, and to base all of your expectations on being a star sets you and the organization up for failure. As Nick indicated, if everyone were expected to play basketball at least as well as Michael Jordan, then there would be no teams.

The days of simply being a firefighter and being judged entirely on your ability to be a sled dog (pull hose, take heat, and blow black snot) are over. Certainly, these characteristics are as important as a surgeon's ability to cut cleanly and make the right repairs, but with all the aspects of our job now, being simply a blue-collar firefighter is no longer enough. By contrast, there are some minimum requirements that everyone has to meet to perform at the expected level; failing these, you have no business on the job.

Herein lies the conundrum for you as a leader. First, how do you avoid judging everyone based on a star player? Second, realizing that not everyone is a star (perhaps including yourself), how do you blend what you have to get the best out of your team?

In the previous paragraphs, I suggested that understanding people and learning to be a talent scout are critical. To maintain sanity, you must accept that your team will be composed of all different types of players, with different skills sets and capabilities. Never expect less than the minimum, and always strive to improve to the maximum. The vast majority of us are B-team players, or B-players. We are good at what we do; we show up at the right time, in the right uniform, at the right station; and we do our jobs to the very best of our abilities. When we find something that we like or are good at, we dig into it; when we find something that does not suit us, we try to avoid it—or if we are smart, we recognize it as a weakness and try to get as good at it as we can, nevertheless realizing that we will probably not become an expert.

"B-player" is perhaps a poor term, since a B in school, at least on my report cards, was reason to celebrate. I suppose the point is that your shift or organization will always be made up of a wide range of capabilities and styles. Here is a rundown of my perspective:

- *A-players.* Stars and go-to members. A-players have exceptional talent and are usually good at just about everything you give them. Our tendency is to overuse these members, since we can expect them to get any assigned task done. They can be trendsetters, and often work outside the boundaries.

- *B-players*. The vast majority of our members. B-players learn exceptionally well and do great things within their areas of expertise and ability. They are great for mentoring our younger members and for getting the business of the business done each day, including being dependable in an emergency.
- *C-players*. Not bad employees but won't give anything extra. C-players may consider the fire department their second job. Although they show up at right time, in the right place, and wearing the right uniform, they usually need some direction and sometimes a push to get started. They will not self-motivate unless you find something (a talent) that really excites them. Fast-food or big-box stores would love to have these folks!
- *D-players*. A very small percentage of our members (much less than 1%) who are usually below the minimum requirements. D-players are high maintenance because they are each a work in progress, but only for a limited time (a year or so). They need retraining and direction, documentation, and a mentor. If they don't get better, they need to find other employment.
- *F-player*. Not sure how they even get on the job; these are usually inherited members that no one ever took the time to evaluate properly. Usually a single or successive character flaws (violence, ethics, legal issues) makes them an F player. The F stands for "Fire them!"

The issue for us as officers is that, come transfer time, no one wants to give up an A-player, and no one wants a D- or F-player; consequently, many of the B- and C-players end up being moved around. Often we try to empire-build, with the expectation that we will be in our current supervisor role for ever; however, even though we might build an empire, we may not be around to see it function!

Your shift will be made up of different players. For the most part, you will always have members who can do a good job for you; sometimes you get dealt a work in progress, and sometimes you get a fix-it-or-fire-it case. Remember that people require lots of effort from a relationship perspective to function effectively.

If all this is true, then how does a supervisor pull together an unlikely group of collaborators? The following principles will help you accomplish just that:

- *Change it up.* Keep the work environment and the projects exciting and varied. You cannot come in every day and do the same thing day after day and expect people to be excited about what is going on. This is a real challenge, especially when you find yourself in a very slow work environment. It's a challenge to develop methods that excite, inspire, and challenge your team.
- *Present them with an irresistible challenge.* What the heck does this mean? Well think about taking on a high-profile challenge that the organization needs accomplished. Take on a process-improvement crusade, to fix or improve something that is not the best method of accomplishing business, or find a process or project that can make your team the winning underdogs.
- *Create a feeling of being more than themselves.* Capture a project that has a looming deadline, has a tight time frame, or provides unique and diverse challenges, to challenge people to expand their comfort zone. This allows you to use all of your team's unique tendencies and expertise, and you will find that switching from individual work to teamwork is not intellectual but emotional.

To succeed in these objectives, incorporate the following methods:

- *Share as much information as you can.* Let members know why the effort the team is putting into the project is so important. Let all of your team members be insiders, since most members want to be in the know. One of your key jobs as a leader is to identify reality, so lay the brutal facts on the table, good news or bad; rather than spin it, just give them the inside information.
- *Provide the right amount of guidance.* Use the power of ideas and different views, tendencies, weaknesses, and strengths to seek solutions. By doing this, you can discover what innovations each team member can offer to solve the issue at hand. This process and your guidance lay the foundation for collaboration among team members by applying, changing, and adjusting ideas and processes toward a common end. (This is discussed further in chap. 15 ["Team Decision Training"].) Let members draw from their experience (an excellent method), and don't micromanage the effort; that stifles effort and kills creativity and collaboration.

- *Stretch your members beyond their current skills.* Draw people in by allowing them to use skills they don't ordinarily use. I am not a tradesman, but I can tell you that every time I get a chance to work with a carpenter or plumber and use and learn these skills, I am far from the most efficient; nevertheless, it is exciting to learn and be watched by the expert. This allows team members to think fresh, which enhances their ability to collaborate with their peers.
- *Have some fun.* Find a method or a process direction that allows people to have fun. Make the results of the fun visible through rewards, changes, team lunches, dinners, and so forth.
- *Help people see and feel the challenge.* General Motors' Saturn division wanted to generate new ideas from their teams regarding a way to "surprise and delight" their customers. The company designed a core values training program in which each retail team built a bicycle to learn how best to work together. Next, the team had to design a delivery experience that would surprise the owners of their new bicycles. After the teams developed their strategy, the facilitators brought children from the local community into the room and presented them with the new bicycles. The teams not only surprised and delighted the kids but were also able to feel it themselves.[10]

Cain and Abel in the Workplace

Let's face it: not everyone in your organization is out to help you or anyone else. Yes, there are manipulative, evil people that work in every job, including the fire service. In the story of how Cain slew Abel, from the Bible, Cain provides an archetype of a kind of destructive behavior that is prevalent in certain people in the workplace. Gerry Lange and Todd Domke, in *Cain and Abel at Work*, have divulged how to recognize such Cain-like workers.

For our purposes, a few characteristics of these people, why you should watch out for them, and what make them so dangerous will all be described. VBF&R has a few of them, and because of poor decision making at certain levels, Cain-like behavior is able to thrive at higher levels as people move up!

Cain-like people in the workplace use politics and behavior to further their own agenda at the expense of others. This kind of person is a liar and backstabber, self-involved, manipulative, and ruthless. Such a person will use any method to control a situation—and to control people—for their own advancement and advantage. They use a wide range of tactics, and just about everyone at the senior staff level has run afoul of someone like this. Often Cain is recognized too late; in many instances, their manipulation and maneuvering are so skillful and subtle that they are never noticed.

A related problem is that the people who are responsible for judging such individuals' behavior may be poor judges of character or ability. Some senior staff members are selected for their management ability, not their leadership ability, and can end up directing and evaluating people who are performing jobs that the managers themselves are incompetent of doing.

In the following sections, the characteristics of these types of people are outlined, so that you can spot them early. However, don't expect that alone to solve your problem. If these people work for senior members of the staff, it is quite possible that they will be protected, as the manager they work for is unable or unwilling to see what is really going on.

Knowledge is power

People who display Cain-type behaviors will do all they can to amass knowledge and power and have no qualms about misusing either. Many times they will anchor themselves in a position in an organization and make every defense in the world why someone else could not do their job. When they get promoted, they will attempt to hold onto the jobs and power that they previously had or manipulate the job and the individual coming into the job to control both the position and the individual. They will often do this under the guise of sharing the knowledge or building a wider base.

For example, suppose Cain is the budget chief and gets promoted. First, Cain will do whatever he can to stay in the administrative positions that control the budget. Then Cain says, "Chief, it's best if I keep control of the budget for continuity, but let's bring in a battalion chief to do the other budget items." You add, "Let's limit the staff time of the budget battalion to two years. That way we can train more people in the budget."

The chief is ecstatic, but he's clearly been manipulated. First, if Cain gives up the budget, someone else will have the knowledge and power and might be able to point out that what has occurred in the past is not the best way for the organization to do business. Second, if the organization brings in someone for more than two years, he or she will develop a good understanding about what is going on now and what has occurred in the past and may well say "This is wrong" or "Why are we doing it that way?" Each of these actions shines a light on past practices that Cain does not want the chief to see. In the organization of today, data and information are blood. The biblical characters in the organization want nothing more than to gain access to information for themselves and deny access to everyone else, except their bosses at the right time.

Liars

Cain-type people will lie to maintain their position in the workplace. Along with the lies, they may claim others' ideas as their own. They may forget that they were provided the information that is delaying a process, by creating a bottleneck. Furthermore, whenever they can, they will fabricate information about others, to make themselves look good. Especially to external members, behind closed doors, and away from others, they will make statements such as "Chase is doing this wrong, and that's why we have these problems," or "Bob is manipulating these figures, but I caught it, and you can trust me to correct it." They have no data and no support, and they use their knowledge of the internal system to lie and baffle external members who have influence over events in the fire department.

Familiarity

Cain-type people will typically take jobs where assignments are conventional and easily accomplished and where success can be measured by simple objective standards. They are not innovators, because pursuing a different method of doing business that is better for the company puts them at risk of looking like a pest in the eyes of management. Fire service administrators would love for us to think that budgets are complicated beasts. The reality is that they are cumbersome in process because of all the bureaucratic rules placed on the organization by management services or whoever oversees the city budget process. In reality, your department budget is about 90% done already each year, since salaries, benefits, and constant cost of running the business plus inflation stay the same. These numbers roll over, so the actual margin of discretionary funds is only about 10%, if that. How difficult is that to figure out?

Manage it—don't prevent it

Cain-type people love to take credit for managing rather than preventing trouble. In fact, they will allow things to get screwed up expressly so they can step in and say, "Look, I saved the day again." When a crisis is allowed to occur, people feel threatened, and as a result, there is a high level of emotional energy resulting in anger, fear, jealousy, and other primitive emotions. Thus, anyone who can step in and eliminate this threat is considered a hero. Cain will take advantage of a crisis—either real or perceived—to blame others for its occurrence and will make others feel that he is their ally against the threat. In this manner, Cain-type people ensure that they are visible as part of the solution, even if they don't come up with a single useful new or specific solution. They will often avoid taking a specific position until it is obvious which way the wind is blowing.

Master of deniability

These people are experts at avoiding risk. Most live by the code that "Ability is less important than deniability."[11] They are also masters in the art of linguistic tango, or not taking a clear position until the outcome is clear. Cain-type people like it to appear that they are taking a stand, even while they avoid saying anything definitive that would associate them with either side. The following are examples of such careful statements:

- *For the most part, I agree with what you say* (which means that if things go wrong, it's the lesser part that will prove more important).
- *I see what you mean* (but I am not taking sides).
- *Okay, go ahead, but be careful.* (By definition, if you fail, you were not careful.)

It is important to realize that not everyone in the workplace has your best interest in mind, nor that of the organization. Understand that there usually are manipulative people, con artists, and liars on the job. Some of these individuals in their quest for power are just downright evil when it comes to damaging others' reputations, manipulating people and processes for their own good, and sowing the seeds of discontent among other rising stars in the organization to maintain their power base. They will stop at nothing to undermine and discredit anyone who gets in the way of their goals.

There you have it. It's about people in a people game. As long as there are people in the fire service, this will always be the case.

Notes

1. Buckingham, Marcus, and Curt Coffman. *First, Break All the Rules: What the World's Greatest Managers Do Differently.* New York: Simon & Schuster, 1999.
2. Ibid.
3. Ibid.
4. DiSC Personal Profile System, Dimensions of Behavior Instrument, Inscape Publishing, 1972.
5. Ibid.
6. Ibid.
7. Ibid.

8. Collins, Jim. *Good to Great: Why Some Companies Make the Leap…and Others Don't.* New York: HarperBusiness, 2001, p. 42.
9. Buckingham, Marcus, and Curt Coffman. *First, Break All the Rules: What the World's Greatest Managers Do Differently.* New York: Simon & Schuster, 1999.
10. Johnson, Lauren Keller. "Should You Play Favorites?" *Harvard Management Update*, September 2005, pp. 1–4.
11. Lange, Gerry, and Todd Domke. *Cain & Abel at Work: How to Overcome Office Politics and the People Who Stand between You and Success.* New York: Broadway Books, 2001.

Chapter 3
Universal Rules for Survival

> *You get yourself into a fix; you get yourself out of it. It's that simple. There is no way to run the sausage machine backward and get pigs out the other end.*
>
> —*Norman Augustine*

When it comes to leadership in the fire service, truer words have never been spoken. For decades and beyond, the fire service has ignored, pigeonholed, and been noncommittal to training, funding, and teaching our newest leaders. Consequently, we got what we deserved—a shamefully fragmented and mostly self-taught approach to leadership. It's a wonder that our organizations don't let more people die, get sued more, and experience more internal problems. Again by analogy to having children, it's as though we waited until our kids were 21 years old before we began raising them. It's time for a philosophical and functional change in the way the fire and rescue services treat leadership.

The reality of leadership is that it is a process of evolution—or as some call it, a Karma wheel. This concept is based on the inner-city projects. Because they know that sometime next week you may be helping them to deal with a major crisis, people in the projects give firefighter-paramedics very little grief.

Leadership has some of the same parameters. As a leader, you will be required to work for today but prepare for tomorrow. Providing a foundation for others to observe and therefore learn patterns of leadership, management, and values. This is much like raising children properly. In the long run, you see qualities that you may either like or dislike in your children; similarly, impressionable young firefighters and future leaders will be a product of a culture.

This chapter is about the foundational approach to everything we do. It's about values and the way we filter information. It's about being a soft leader at times and a hard leader at other times, as necessary. In a subtle way, it's also about teaching—not directly, but in an abstract way based on how people perceive you and your motives. Right now, as a new or aspiring leader, you are not even thinking about the generation to follow you. I know that as a young lieutenant, I was more concerned about surviving than worrying about who was going to lead next.

The final chapter of this book is entitled "When Leadership Fails." Short-circuiting the failure process begins right now, with your commitment not only to strive to lead in the most appropriate manner possible but also to espouse ideals and actions that set the ground for the next generation of leaders. Look around: You may well be looking at the future chief of a department. Take heed that your actions may well affect how he or she leads and may thus have an impact on entire organizations for decades to come.

Vision: The Essence of Leadership

What is vision? Certainly, we all need good vision to see each and every day. Some of us wear glasses, others wear contact lenses, and others still go straight to Lasik surgery! If only creating organizational vision were so easy.

Vision is the essence of leadership. People do not follow individuals; instead, they follow ideals that are communicated in such a manner that people feel they have some stake in the outcome. In many instances, vision boils down to having a stake in the outcome. A leader's ability to interest individuals in the outcome is directly related to the creation of an environment in which the follower believes the vision is both possible and probable. In turn, the follower should understand what needs to be done and what he or she personally needs to contribute in order that the personal stake pays off as the vision is turned into reality. Although they represent individual efforts originally, personal stakes, when played out time and again by multiple individuals within a given work unit, can become part of a global effort to reach a common goal.

Leaders who can convert vision into group commitment find themselves in a win-win situation, since the effective communication and implementation of vision are directly tied to morale. What does morale have to do with developing and simplifying a vision? Morale is how individual members of an organization feel they are valued—a purely humanistic explanation of how I feel you are treating me as a human being and a member of the organization. We all hear that morale is good or that morale is bad, followed by a variety of reasons in either case. When you list the reasons behind good or bad morale, three items are encountered most often:

1. Most morale issues are issues of the soul; they are how people feel based on how they believe they have been treated.
2. Improvement of morale is usually a matter of providing little items that make people feel good, connecting them to the group,

showing them that you are listening to them, improving their work environment, and/or showing them that they have value to you and to the organization.
3. Things that improve morale are therefore typically those that, through action, say to members of the organization, "I am listening, I am changing, and I recognize what has value to you."

In the end, communicating and implementing a vision means that most members of the organization are coming to work each day to do good things for you and for the organization. Conversely, you should make every attempt to do good things for them, and you have the magical power of trust!

The question now is, "What is the difference between a vision and a hallucination?" You could say that others see a vision, while a hallucination is seen only by you! Vision is difficult to define, but it could be considered a sense of direction, of knowing what has to be done and doing it.

People do not follow individuals. They follow the ideas—the vision that they create. This is the ultimate secret to leadership. Also, people do not follow those who have not proven themselves in the little day-to-day things during a crisis. Trust and belief in a vision are built over time.

The challenge for the company officer and even for the chief officer is to convert sometimes-nebulous statements into a vision with real drive. Let's examine this nebulous concept down the organizational chain of command. In Virginia Beach, the politicians assigned to the Safe City working group came up with the focus area, or as we might call it, the key point of customer service. They decided that the filter that would be used to measure the success of all public safety, from the courts to the fire department, would be to create a situation where citizens have the feeling of being safe anywhere, anytime. The question remains, What does that mean? The fire chief converts words and concepts into tangible action, thereby passing the vision relentlessly down the chain until it reaches the company officers.

Two movies come to mind: *Braveheart*, with Mel Gibson, and the second is *Gladiator*, with Russell Crowe. These films contain excellent examples of vision, leadership, and the desire of people to follow a vision created by individuals. In both movies, large groups of individuals follow someone they either do not know or barely know because of a vision, a common goal that they were able to communicate.

Why We Don't Listen and Hear

The most basic of all human needs is the need to understand and be understood. The best way to understand people is to listen to them.

—*Ralph Nichols*

***Inverted pyramid.** If we really want to hear and see, we need to change our way of looking at how information flows in the organization. Frontline personnel have a better handle on what is going on each day—what's working and what's not working—than those of us in the office.*

I would wager that some leadership or officer program you've attended espoused the value of being a good listener. It might have been listed on the board as a critical and desired characteristic of a leader; indeed, someone may have even instructed you on great listening skills. My question is, How many people do you know who listen but don't hear anything?

I've been accused of having selective hearing, especially when it comes to names, which I forget within two minutes of being introduced. It's embarrassing, so I have developed a system to pay closer attention: I equate the name with something I can remember and then repeat it to myself. The lesson we can extract from my failings is that by nature, we aren't very good listeners. Our eardrums are extremely sensitive to sound, rupturing at about one pound per square inch (psi); after years of being exposed to sound, they lose some of their effectiveness in the high and low ranges.

Another piece of human anatomy that affects our ability to listen and hear is the mouth. We spend a lot of time creating sounds that we push out of our mouth—some intelligent, others not so intelligent—but in the hierarchy of brain function, if our mouth is running, our ears tend to shut down and filter only muffled sounds.

We can also "hear" with our eyes. Think about how much more clearly and completely you hear when you are also watching someone's body language, expressing passion or anger or happiness. Our eyes are actually powerful hearing tools. To effectively hear, speak, and listen, we must unite ears, eyes, and mouth through one common nexus: the brain. The following sections should not be read as yet another speech about listening more and talking less; instead, the suggestions are intended to explain where you can find great information and how you can most effectively speak so as to convey the available information and ideas.

On the frontlines

Let's begin by considering how our communications systems are structured and how our current supervisors were trained. Many senior personnel were trained to use a very strict, military-style communications system where information, directives, and knowledge were passed down the chain of command. So we are victims of past wrongs, because hardly any officers, especially senior officers, ever had any formal training on being an officer. This sets up barriers to listening and hearing, thereby affecting our ability to help personnel, our organization, and our customers.

Suppose I'm the president of a widget company and today we're having the annual meeting of all the widget salespeople. During this meeting, a conversation takes place about how our cost measures aren't working, our sales territories are unfair, our policy is limiting the impact on our customer base, and so forth. In a nutshell, my operations personnel, who are responsible for making the company money, are telling me about the things that will have an impact on our business. The question that I would pose to you is, Who knows more about what has on impact on the customer? The personnel in the field who sell the widgets or the president of the company, sitting in his office? The answer is clear once you get past your ego. As Colin Powell once said, "Front-line personnel are always right until proven otherwise." If we fail to listen to and hear what our personnel are saying, we risk running ourselves right out of business. The lesson is that even though frontline personnel occupy the lowest block on the organizational chart, the feedback they provide is significantly more important.

To provide the best service possible, true leaders must understand key aspects of customer service. First, the status quo will always lose. Just because we've always done things one way doesn't mean it will be successful in the future. Second, great change has historically been driven from the bottom up, not from the top down. Thus, we must be especially attuned, listening to and hearing our frontline personnel.

Being so attuned is exceptionally difficult for a couple of reasons. First, distance separates us from the workforce on an everyday basis. Those of us in staff often find ourselves isolated, by geography and work schedule, from the very people we're supposed to lead. Company officers find themselves on the front line, acting as the intermediary for ideas that they relay up the chain of command in a timely and meaningful manner. Second, we're unable to deal effectively with our egos and understand that great ideas and messages have nothing to do with rank and everything to do with honest, hardworking people trying to provide solutions to everyday problems.

Middle management

To solve this communication problem, recall the time-management model. The frontline company officer's job is to provide feedback and communications about what works and what doesn't, as well as what needs improvement. The battalion officer's job is to listen to, filter, and evaluate that information, fixing what they can at the lowest level of the organization. Also, the battalion officer must look for trends, or a common message, emanating from multiple companies and officers. These trends often indicate an organizational issue that, if explored, crosses all barriers in the organization. Finally, the battalion officer should carry the information to the next level of command and provide sound counsel for making change, including the reason behind it and the necessity of it.

On paper, this process looks really good, but if the organization hasn't institutionalized it through training, expectations, and changing the senior staff's mentality about listening and hearing, then it will continue to meet resistance at the organization's highest level. The lesson is that listening and hearing must be accepted organizationally as sound and effective ways of doing business. They must be treated as a respected and anticipated cultural value.

It's not what you say, but how you say it

Occasionally, a great message is clouded by an inept delivery. What's important is to get to the grassroots issue in the message. Sometimes, if you ask *why* five times, you can whittle away the external issues and get to the real issue. Don't discount messages that upset or befuddle you as disgruntled or misaligned until you can look past the messenger and find the message.

"Sometimes people are simply angry, and their ravings have no true message." If you believe that, you're actually missing a very important message. Perhaps the information you're seeking is not in the message at all, but in the individual sending it. You have to ask from a humanistic standpoint, Why is the individual angry, or as one of my battalion chiefs, Larry Jarvis, says, "spewing vinegar"? Perhaps to listen to and hear this person, you must pay attention to the face, posture, and soul. Take the time to ask why some people are always bitter and at the same time seem to do good work for you. If you hear with both your eyes and your ears and take the time to discover the message hidden in the poor delivery, you can change the vinegar into something usable.

Your undivided attention

People expect attention when they're talking or when entertaining visitors. How many times have you stopped by the boss's office and asked something as he or she worked on the keyboard? Seemingly, your boss is able to talk and still keep typing. When someone comes to talk and you intend to listen, push away from the keyboard, turn, get up from behind the desk, and engage in a conversation.

During a recent visit to one of the fire stations that I oversee, I talked to nearly everyone, to see how they were doing. As a division chief, I'm tied to a cell phone, but using it has become so automatic, I don't even realize it anymore. Well, others do realize it, and it clearly takes away from the ability to communicate. Somewhere during our conversation, one of my captains looked at me and said, "Chief, can you not answer the phone while we're talking?" I had not even realized that on at least two occasions while engaged in a conversation, I had excused myself to answer the phone. Now, I simply don't answer it.

Two-way communication

When people talk, they need feedback, which may be a simple answer or a promise to look into an issue and provide some information. In many cases, feedback means taking ownership of an identified issue/problem, seeking a method to solve it, and assigning the resources to accomplish it. Nevertheless, we fail at this because we don't hear what's being said; we only listen.

For example, we recently conducted a battalion chiefs' assessment that included a tactical problem, a written problem, and an interview process in which candidates answered specific questions provided by the board. What we heard with our ears and our eyes was disappointing in many instances. It was clear that we, the organization, had not provided the training and direction necessary for our personnel to be successful at the expected level. (Of course, candidates are expected to prepare for the assessment as well.) My theory is that we failed to communicate effectively our desires and direction to the candidates prior to the assessment (speaking/expectations), and now we invite the consequences of that.

In at least one instance, we successfully heard what was happening. It was clear that our new members were not as successful as we would have liked in making the transfer from captain to battalion chief during tactical operations; performance during the tactical problem was not as we had expected. Once we heard this with our eyes and ears, we provided feedback by establishing a captain- and battalion-level tactical training program to provide our members with the tools they need in order to be successful.

If we're really listening and hearing, we must be aware of what was communicated at the previous staff meeting. The fire chief's feedback was, "The personnel that just completed the process were not looking global enough." So if we hear that and we're truly listening, then we, the organization, need to define "global" and develop a method to provide the tools, information, and training to our personnel to make them more successful in the future. This goes to the root of what I've been preaching against for years: We do not invest time and training in our officers across the organization.

In closing

Listening, hearing, and speaking with ears, mouth, and eyes are individual as well as organizational leadership skills. Change is always driven from the bottom up, and to be effective in managing change, creating a better organization, and meeting the challenges we face in all aspects of our jobs, we must be tuned in continually. Much like a NATO or National Security Agency (NSA) listening post, our officers and our organization must constantly be on alert for messages and information coming in all different forms and from all different sources. Your most valuable information doesn't always come from one-on-one communication. Sometimes an organizational process or procedure speaks to us, and we must attune ourselves to this phenomenon.

If we learn, as individuals and as organizations, to communicate openly, honestly, and frequently without putting a spin on the message, we will be much better off in the long run.

There is no such thing as a worthless conversation, provided you know what to listen for. And questions are the breath of life for a conversation.

—James Nathan Miller

Values-Based Decision Making

It's difficult for most fire and rescue service leaders to get away from the concept of having a rule or policy for everything. In part, we are victims or our environment and are deeply scarred by the 1%ers we have dealt with during our careers, who challenge everything and need a written policy for everything or continually push the envelope. Rules are a convenient and safe method for making sure we are all on the same page. Don't get me wrong, we need rules—especially when it comes to the health and welfare of our personnel and to provide standard methods of accomplishing emergency incidents, standard methods to ensure fairness in dealing with people problems, and a process to get people paid.

The reality is that fire departments have lots of goofy rules on the books. These exist because someone didn't follow an adult course of action; in response a rule had to be written. Most old-school officers have real problems leading with values other than their own values. For the sake of argument, let's review some definitions, which are examined in more depth in future chapters.

- *Rules*. These are courses of action that departments must follow owing to health and safety issues, legal rulings, legislative requirements, ordinances, and so forth.
- *Policies*. These are usually made by the department for department members and have been created because the organization decided that they are required.
- *Procedures*. These provide guidance to everyone on how to complete recurrent tasks. They may be staffing procedures or report procedures.
- *Guidelines*. These are best-practice methods. They are drawn from experience and trial and error over the years and are driven by changing science and technological breakthroughs. If the guidelines are followed, a standard outcome is expected.

In reality, though, our jobs are so complicated that different rules could be written for every particular situation. If we were to spend our time trying to anticipate every possible scenario, we would spend all of our time writing rules. Another pitfall—or more exactly, an assumption on our part—is that all of the personnel must be tightly controlled or they will run amuck. These are the same people we have entrusted with hundreds of thousands of dollars worth of equipment, to make life-and-death decisions for our citizens and our members on a daily basis, and we are worried about a few rules?

The real problem when you are rules driven, instead of values driven, is that you limit choices in favor of consistency. The real risk is that rules-driven people, especially senior members, place so much importance in following the rules that they lose sight of the goal of taking care of the customer.

***Values.** Values are those aspects of life that you hold dear; in the world, this is known as "experimental knowledge." As you make your choices—sometimes compromising, sometimes holding firm—you come to realize that certain aspects of your life are more important than others. These critical aspects become your values, guiding the choices you make in the future. Some of your values remain constant throughout your life. Others will change with time and reflection.*

If rules are not the ultimate answer, what can we use instead? How about values-based leadership? Values-based systems are ideal in that they allow both flexibility and rigidity in one great package.

In a values-based system, the philosophy is spun out of a set of core values that are a fundamental part of the decision-making process for the organization. The primary mission in a values-based organization is achieving a successful outcome and making decisions that are consistent with the values of the department.

We all have values that we use in our personal life for decision making. Not only do these values guide our decisions about what we believe in, who we will hang out with, and what we will or will not do, but they also function as teaching tools. We use them to raise our children and grandchildren—as something they can grasp when we explain life, why certain natural phenomena occur, and why we choose to live as we do.

Like everyone else, bad or evil people have values as well. Hitler, Stalin, Pol Pot, Charles Manson, and Jeffery Dahmer all had values; they were just skewed. Much like rules, values can be distorted, broken, and manipulated (as can any other concept).

Thus, values allow us to make choices. When we send our members out to accomplish a task, take care of Mr. or Mrs. Smith, deal with external agencies, and even deal with other members of our organization, we are asking them to make a choice about how they will act, respond, communicate, and accomplish and follow up a task.

__Core values.__ These are the things on which the members of the organization refuse to compromise.

Once established, core values provide the organization and its members with a decision-making tool to address all aspects of the job. It allows members to ask, "If I do it this way, does it meet our values system?" If the answer is yes, then the approach to the problem is certainly correct; if the answer is no, then another course of action, attitude, or decision needs to be undertaken. There are many methods to accomplish a given task, but as long as they are aligned with the values system, they are all correct.

When you trust your members and educate them up front, to obtain the best possible outcomes without compromising the organization's values, you let members choose among many different paths to excellence. Values allow for much more creativity and innovation from members and the organization.

Based on what you have read so far, you may be thinking, "Yeah, sure, values provide such a wide range of decisions that anyone can get away with anything." However, nothing could be further from the truth. Once the values have been established, people making decisions outside the values system are corrected immediately, in whatever manner is appropriate.

While leadership styles of rules- versus values-based organizations may well be at opposite poles, no organization is truly only one or the other. Even values-based organizations must have some rules, policies, and procedures to be effective. The difference is that the rules in a values-driven organization usually deal with time-honored behavior and health and safety issues; while the rules are as few and simple as possible, they apply to everyone and are strictly enforced.

The Moment of Truth

Think about the last time you went to a restaurant and your food was not good, or the service was terrible. What experience did you have? Did the staff fix it by comping your meal or removing an item from the tab? Did it happen at the front line of service, or did the waiter have to run back and forth to the manager?

What about purchasing a new car? Do you believe for one moment that the salesperson does not know exactly what he or she can sell you that car for? Do they have to run back and forth to the manager to get the deal? Of course not—and when they do that to me, they come back to an empty office, because quite frankly, I just don't tolerate it!

So what is the concept of the *moment of truth*? It's all about influence. It's about taking people you come in contact with during your job and influencing them to say, "Wow, what great people, what great service, what caring individuals."

Each time you make on-the-job contact with anyone, internal or external, take the opportunity to make an impression and thereby to influence what he or she thinks about you and the organization. Not all the customers you see or the members of other agencies that you work with will feel the same way. Yes, there can be jerks at any level of the organizational structure! Think about how you can do the very best you can each and every time you go out to work with someone or provide service. This is where the values-driven organization can really shine, since members can choose a path to have a positive impact.

The next interesting aspect of this concept is that although all members of our organization have daily moments of truth, the most junior members (firefighters and captains) on frontline suppression and EMS units have the most contact with—and thus the most influence on—our customers. Even the city manager's family, the city council's family, and very influential people get sick, do stupid things, get hurt, or need our help for some other reason from time to time. And guess what, the fire chief or senior officer cannot always be there when that happens!

That said, what do we care? First of all, if the frontline troops are really that in contact with our customers they will be the first ones to recognize what works and what does not work—what *wows* and what *pows*! So, if you want your finger on the pulse of customer service, then listen, listen, and listen to the frontline members.

Another critical point, especially for fire chiefs and other senior officers, is that influence means creating a personal relationship with the individual you are in contact with, even if it's just for a few moments. In recurring contact (as with city government figures), if you cannot build a personal relationship and understanding—not to be confused with like or trust—then you will never be effective.

Every day each member of our organization has multiple moments of truth, whether putting in a car seat, taking blood pressure, rescuing someone from a vehicle crash or house fire, or, if you're a member of senior staff, meeting on the budget or seeing the civic league. Every encounter is an opportunity to create and gain influence, and you never get a second chance to make a first impression.

Of Officers and Riverboat Captains

Organizations and officers could learn a lot from riverboat captains. They show up to work and are responsible for a multimillion-dollar vessel, dozens of crewmembers' lives, and hundreds, perhaps thousands, of travelers' lives. Riverboat captains are successful because they have a solid picture of their overall objective (global) yet apply a simplified approach (task level).

How many times during your career as an officer have you been told, "You just don't see the big picture"? I've always wanted to reply, "Well, your problem is you can't see the little picture," but somehow, everyone who ever mentioned the big picture to me outranked me by many horns.

Herein lies the difficulty of being a company officer in a fire-rescue organization: You lead at the task level, but you're expected to have a global understanding of the internal and external forces that affect your work environment, your members, your ability to deliver service, and yes, morale. The real question is, Where is the balance, and just how globally should we expect our frontline officers to think?

What does it all mean?

The solution does not involve rocket science, and it can be addressed with a series of approaches that provide education and understanding to our company officers. First, the organization must define "global." In other words, how much of the big picture should our frontline officers understand? Second, once we've defined this term, how do we provide the appropriate education and training for our members and show them how it can be functionally applied to their everyday activity? (This is the critical part of the formula, because simply educating them is useless if we don't show them how to apply what they've learned to their members, to the organization, and to the customer.) Finally, we must simplify these very complex, performance-based plans, objectives, and goals so that people can get their arms around them. It's not enough to simply send out a strategic plan and say, "Read this," and then become dumbfounded when you talk to your members about it and they look at you like you have a third eye. That is predictable behavior, and as Gordon Graham says, "If it's predictable, it's preventable."

Super models

How do we provide company officers—and thus our members, who represent 90% or more of our entire organization—with an opportunity to engage in global issues and see the big picture?

We all must understand that we work for something much larger than our department alone. A fire-rescue department is often part of a huge bureaucracy—one cog in a giant wheel and one sector of a city's or a county's business. As a result, many external influences affect our work. Moreover, politicians and, most assuredly, the budget people want to know what they are getting for their money. This first lesson is called "Why Things Happen the Way They Do," and it takes a broader look at the flow of government. We looked at this government model in the first chapter.

The model concludes with the service delivery section, which includes frontline personnel and officers. This group might consist of a few instructors from the training division; an engine, truck, medic, or squad company taking care of business; or fire educators caring for the youngest members of our community. Whoever it may be, they are responsible for delivering service at the task level. The members at the lowest section of the organizational food chain are responsible for implementing a plan they may never have seen or may not understand. Now, I don't know about you, but if I'm running this riverboat, I'm not willing to take the chance that the officers steering the ship, running the engines, taking care of the passengers, or responding to emergencies don't understand the plan. Thus, it becomes incumbent on me to educate my crew about the plan long before we set sail; otherwise, I must accept the consequences of their ignorance.

The bottom line is that the more educated you are as a frontline officer about the overall mission and strategy of the organization, the better decision maker you will become and the less toilet flushes you will experience.

Growing a good crop

There is a saying that has much truth to it: "As the officer does, so shall the men do." As a frontline officer or any officer within the organization, you must be able to accomplish what the organization needs done on a daily basis. You must also be good enough and smart enough to combine that with the talents, desires, and needs of your members and focus that energy and skill toward the common goal.

This answers, at least in part, the question of how an organization must define "global." If you don't decide what you want your frontline officers to know, how can you blame them for not knowing the right information? The organization, much like a farmer planting crops, is responsible for sowing the field, making sure it has the right water and nutrients, and taking care of it so that the crop will grow. If you want a crop of great riverboat captains or fire officers, you must raise them, help them, and let them grow.

On a mission

Now, we move on to the question of task-level work: How do you, as a frontline officer, convert these complex models into achievable, explainable tasks for firefighters to perform? Why is that even important to you? Furthermore, I would ask, Without this, how do you know what you're supposed to do? You might say, "Chase, you're so stupid. We do what we've always done." But is that what we want to do or should be doing? And how does all this figure into seeing the global picture but leading small? To answer these questions, let's use your mission statement (that dusty thing hanging on the wall) as an example. Your mission or vision statement is not an end goal; it's a way of doing business. Your mission statement should represent what you believe in and what you'll do to take care of your members and your customers. In other words, it's the filter that you, as an officer, should use to determine whether you're doing the right thing. If you use your mission statement in this way, it becomes the basis for making individual as well as operational decisions.

While I won't print our entire mission statement here, part of it says, "Make people feel safe anywhere, anytime." So how does that relate to what we do in the field? Like a riverboat captain, a frontline officer must break down complex plans into simple actions. Suppose the engine company is on their way back from a response at 0200 hours, and lo and behold, at the side of the road, they see Mrs. Smith, whose car has broken down. The captain must make a decision, but with no one there to tell him or her what to do, what drives that decision? Our mission statement tells us to make people feel safe anywhere, anytime—and you can bet Mrs. Smith is not feeling very safe. So the captain should stop the rig, fix Mrs. Smith's car or call a wrecker for Mrs. Smith and get her and her car off the road; failing that, the captain had better put Mrs. Smith in the shiny red fire truck and return her safely to her house. Not following such a course of action would be equivalent to not abiding by the mission statement.

Mission statements, mantras, strategic plans, and so forth are all useless without people who understand them and can convert them into action. No plan or saying ever took care of other human beings, their emotions, or their property; only other human beings can do that. If you look at your mission statement and can derive hundreds of examples of how to realize it at the task level, then you've taken a very complex issue and boiled it down to manageable tasks that have an actual impact on the people you serve and on the members of your organization.

The task at hand

When the riverboat captain shows up for work and has to travel from New Orleans to the Gulf of Mexico, he or she has a very complex task ahead; however, I would bet dollars to doughnuts that when the captain walks onto the bridge of the riverboat, he or she isn't thinking about the Gulf of Mexico. Because the riverboat captain understands the global aspects of the job, the captain has navigational charts, maps, and tools such as the Global Positioning System (GPS) at his or her disposal. Additionally, the officer and the crew have the training and experience that comes from making the journey before, as well as other journeys. Thus, when the riverboat is ready to leave the dock, the captain and his crewmembers are not focused on the Gulf of Mexico. Rather, they are focused on the task of getting the ship from the mooring, turning into the harbor, and safely moving into the great Mississippi River. Next, they are concerned about getting the ship into a channel in the river, avoiding the sunken ship to the right, and pointing to the first bend in the river. Once that portion of the journey is complete and the river opens up, they must be careful to avoid the snags to the left and the shallows to the right, keeping the boat in midstream. Never thinking about the Gulf of Mexico—but always understanding the final destination—they will undertake hundreds of tasks and decisions, simplifying a very complex plan until they finally reach the Gulf of Mexico.

My message to you

The message I have for the chiefs reading this article is that, if you want your frontline officers to accomplish their tasks and lead for you, then educate them and define "global" for them. The message I have for frontline officers is to be a riverboat captain: look large, lead small. Frontline officers are the most influential members of our department and accomplish so much for our organizations that we must invest in them, our members, and our future. Simply writing a plan or a mission statement is not enough. Without people committed to bringing the statement to life, it's no different than leaving one of Beethoven's symphonies unplayed.

I am currently at a cruising altitude of 36,000 feet, sitting on an airplane in a cramped, coach-class seat and heading to Ontario, California, to teach a Buddy to Boss class. Cross-country trips used to be fun; now they're characterized by small seats, pretzels, and extended time spent surrounded by strangers. So I spend a lot of time listening to people talk about business, family, and other topics. In many instances, when I overhear a complaint or a tale of failure, I can't help but think that someone did not think and lead like a riverboat captain. While I write this, I hope that my pilots are thinking like riverboat captains as we fly over the Colorado Rockies.

Understanding Your Duty as a Leader

"Duty" is a word that we have let slide from our organizational vocabulary. This book is really about duty: about doing what you are supposed to do; balancing your way, through being a friend; being a supervisor; and being aggressive at times and being defensive at other times, as on the fire ground.

Duty is defined by Webster's as follows: "1. An act or a course of action required by custom or law; 2. Moral obligation or the compulsion to meet such an obligation."

Duty is usually invoked in military and paramilitary organizations, but duty applies to any individual working for an organization. Young firefighters need to understand their duty—not just the duty to act aggressively and professionally under extreme circumstances, but to show up on time, work hard and diligently, and put in a good day's work for a good day's pay. This duty needs to be expressed as part of the organization's values on the very first day that a new employee comes to work. It should be a daily expectation of the organization and its leadership.

The organization also has a duty to the employee. This duty is to ensure a safe and comfortable place to work, adequate recognition of and advancement in exchange for quality work, and a workplace where they can thrive. It is a quid pro quo relationship—or something for something. Those who do their duty should expect to be treated fairly and work in an environment that offers challenges and job satisfaction; those who do not do their duty should expect that they will not be employed very long.

Officers have a duty to act effectively and professionally at all times. They have a duty to lead from the front at all times and create an environment where employees can thrive. Officers have a duty to make sure employees are safe, and when in harm's way, the officer should be the first in and the last out. Officers also have a duty to do the part of the job that is not glamorous or fun, including discipline and ensuring that expectations are met.

Being a leader is tough work, because every day you have to come in and by your example, your actions alone, you have to do the right thing. Once you say yes to the promotion, you cannot look back. If you don't want to do what we are talking about here, save yourself, the organization, and its members a lot of heartache and just say no before you promote.

Duty is hard, especially when working for poor leadership or when lacking leadership at senior levels of the organization. It's also difficult when you have to discipline or correct one of your members or when you have to deal with a fringe employee. (Much more is said on fringe employees later, since I have devoted an entire depressing chapter to dealing with halfhearted employees.)

What does duty mean, and how do you choose people who can do their duty? Here is an excellent set of guidelines, from Colin Powell:

> **Look for intelligence and judgment, and most critically, a capacity to anticipate, to see around corners. Also look for loyalty, integrity, a high energy drive, a balanced ego and the drive to get things done right.**

"Intelligence" is a tough word to sell. Many people think a formal education is the key to intelligence, but I know many an educated idiot. Intelligence in a leader means not only some form of formal education but also the abilities to see trends, to adapt to changing environments, to spot and apply talent, to fight when it is right, and to know when to stop fighting when the risks outweigh the benefits. Leaders are politically intelligent as well, knowing who and what drives the agenda and how to influence it or stay out of its way.

Judgment is another characteristic of people who can do their duty. Judgment takes on many forms, in both emergency and nonemergency settings. Using good judgment in making adult decisions is difficult to teach. Furthermore, knowing when to use judgment to not support a losing position or to discipline or counsel one of your members when they make bad choices, rather than to try to defend a losing position or bad choice, is an admirable trait. Judgment on the fire and rescue ground is critical to survival and success, and poor judgment can often make the difference between life and death.

Anticipation, the ability to see around corners, is part of your duty. The fire and rescue service is rapidly changing, as are the expectations we place on our officers and our members. Duty-bound leaders discern trends as they arise and anticipate what it will take to address a given trend. To expect something to solve itself or to simply say "It's someone else's problem" amounts to sticking one's head in the sand and is emphatically not doing one's duty. Anticipation on the fire ground comes with experience; this includes the abilities to project what is going on in the building, to picture in your mind the fire's effect on the trusses and the stairwells, to realize what your actions will do to eliminate hazards, and to estimate how much time you have to be successful. A final component of this trait is the ability to detect in advance the needs of one's boss.

Loyalty is often thought of as a blind quality, but nothing could be further from the truth. There have been admirable examples of blind loyalty in the past, but closer examination reveals that many have been blindly loyal to an unethical or dangerous cause. "Loyalty" means speaking up when necessary and in the right environment, as well as standing by your position when you know it to be right and standing by your boss, your peers, and your members when you know they are right. "Loyalty and duty" mean arguing your position until the decision has been made and then getting behind the decision, regardless whether your case has won or lost, executing the decision as if it were your own. When you find that you are no longer able to be loyal to an organization, it's time to find something else to do.

Integrity is an integral part of doing your duty. People don't much care for what you say, but they care a lot about what you do—your actions scream for you! The other day one of the acting deputy chiefs of our organization tried to really screw a battalion chief, informing him, "You need to pay more attention to what we are telling you." The battalion chief replied, "No, I don't. We have an employee/employer relationship, so what you say is not very important; it's what you do to me that counts." Integrity means doing the same thing, making the same decision even when no one will know exactly who made it. Integrity is about what you stand for, about being committed to something more than yourself—to have a purpose and a set of values and ideals, so that your actions honestly reflect your convictions.

> *Leadership is a potent blend of strategy and character. But if you must be without one, be without the strategy.*
>
> —*H. Norman Swarzkopf*

Doing your duty means that you are driven. Drive is a characteristic that expresses itself in making events move faster and getting things done as quickly as possible. Doing your duty means not waiting passively for marching orders, but seeing what needs to be done and doing it. It means not whining about how extreme the pressure is, or how heavy the workload is, or how hard it is to catch one's breath around here. Duty requires making a commitment to getting the job done quicker, easier, and more efficiently and then seeking the next challenge. Set objectives beyond your formally assigned goals. Be impatient with the status quo and turn over some stones to improve performance.

We all have egos, some bigger than others. The secret is to learn to balance the ego. Don't apologize for being good at what you do or for being competent; instead, recognize how long it took you to reach that pinnacle, and don't expect everyone else to reach the same level immediately. Be self-assured but don't be overly impressed with your own status or importance; ego balance implies a level of self-awareness. You need to know when to blast ahead and when to pause, to regroup and enlist new allies with complementary skills. (Recall the earlier discussion of tendencies.)

The Irish have a saying:

> *If you want to know how much you will be missed put your hand in a bucket of water and pull it out. The hole that's left is just how much you will be missed!*

The same applies to being missed from an organization.

Finally—and most important—is character! This includes what you stand for and how others, especially people who choose to follow your lead, take cues from you. Character is about what you do, rather than what you say. It's about standing for something or falling for everything. When character is absent, leaders seem to stand up for whatever is politically or financially expedient at the time, even if their decisions seem fuzzy, contradictory, or self-serving.

In leading young people, either as a parent or a teacher – you can't lecture them as to what they are supposed to do. The way they really learn what the right things to do in life is by watching. They're not always listening; they're not always paying attention to what you're saying. In fact, they take every opportunity not to pay attention to what you're saying, but they're always watching.

—Colin Powell

Sometimes on the fire ground and in the administrative arena, doing your duty means more than simply meeting the requirements we have talked about in this section. There is a famous story about duty called "A Message to Garcia," written by Elbert Hubbard in 1899, which stands as a testimony to duty, integrity, character, courage, and all the aspects we have discussed. I would like to share this story with you, to wrap up our discussion of the characteristics that make a leader.

A Message to Garcia

In all this Cuban business there is one man that stands out on the horizon of my memory like Mars at perihelion. When war broke out between Spain and the United States, it was very necessary to communicate quickly with the leader of the Insurgents. Garcia was somewhere in the mountain vastness of Cuba—no one knew where. No mail nor telegraph message could reach him. The President must secure his cooperation, and quickly.

What to do!

Someone said to the President, "There's a fellow by the name of Rowan who will find Garcia for you, if anybody can."

Rowan was sent for and given a letter to be delivered to Garcia. How "the fellow by the name of Rowan" took the letter, sealed it up in an oil-skin pouch, strapped it over his heart, in four days landed by night off the coast of Cuba from an open boat, disappeared into the jungle, and in three weeks came out on the other side of the Island, having traversed a hostile country on foot, and delivered his letter to Garcia, are things I have no special desire now to tell in detail.

The point I wish to make is this: McKinley gave Rowan a letter to be delivered to Garcia; Rowan took the letter and did not ask, "Where is he at?" By the Eternal! there is a man whose form should be cast in deathless bronze and the statue placed in every college of the land. It is not book-learning young men need, nor instruction about this and that, but a stiffening of the vertebrae which will cause them to be loyal to a trust, to act promptly, concentrate their energies: do the thing—"Carry a message to Garcia!"

General Garcia is dead now, but there are other Garcias.

No man, who has endeavored to carry out an enterprise where many hands were needed, but has been well nigh appalled at times by the imbecility of the average man—the inability or unwillingness to concentrate on a thing and do it. Slip-shod assistance, foolish inattention, dowdy indifference, and half-hearted work seem the rule; and no man succeeds, unless by hook or crook, or threat, he forces or bribes other men to assist him; or mayhap, God in His goodness performs a miracle, and sends him an Angel of Light for an assistant. You, reader, put this matter to a test: You are sitting now in your office—six clerks are within call.

Summon any one and make this request: "Please look in the encyclopedia and make a brief memorandum for me concerning the life of Correggio."

Will the clerk quietly say, "Yes, sir," and go do the task?

On your life, he will not. He will look at you out of a fishy eye and ask one or more of the following questions:

Who was he?

Which encyclopedia?

Where is the encyclopedia?

Was I hired for that?

Don't you mean Bismarck?

What's the matter with Charlie doing it?

Is he dead?

Is there any hurry?

Shan't I bring you the book and let you look it up yourself?

What do you want to know for?

And I will lay you 10 to 1 that after you have answered the questions, and explained how to find the information, and why you want it, the clerk will go off and get one of the other clerks to help him try to find Correggio—and then come back and tell you there is no such man. Of course I may lose my bet, but according to the Law of Average, I will not.

Now if you are wise you will not bother to explain to your "assistant" that Correggio is indexed under the Cs, not in the Ks, but you will smile sweetly and say, "Never mind," and go look it up yourself.

And this incapacity for independent action, this moral stupidity, this infirmity of the will, this unwillingness to cheerfully catch hold and lift, are the things that put pure Socialism so far into the future. If men will not act for themselves, what will they do when the benefit of their effort is for all? A first-mate with knotted club seems necessary; and the dread of getting "the bounce" Saturday night, holds many a worker to his place.

Advertise for a stenographer, and 9 out of 10 who apply can neither spell nor punctuate—and do not think it necessary to.
Can such a one write a letter to Garcia?

"You see that bookkeeper," said the foreman to me in a large factory.

"Yes, what about him?"

"Well he's a fine accountant, but if I'd send him uptown on an errand, he might accomplish the errand all right, and on the other hand, might stop at four saloons on the way, and when he got to Main Street, would forget what he had been sent for."

Can such a man be entrusted to carry a message to Garcia?

We have recently been hearing much maudlin sympathy expressed for the "downtrodden denizen of the sweat-shop" and the "homeless wanderer searching for honest employment," and with it all often go many hard words for the men in power.

Nothing is said about the employer who grows old before his time in a vain attempt to get frowsy ne'er-do-wells to do intelligent work; and his long patient striving with "help" that does nothing but loaf when his back is turned. In every store and factory there is a constant weeding-out process going on. The employer is constantly sending away "help" that have shown their incapacity to further the interests of the business, and others are being taken on. No matter how good times are, this sorting continues, only if times are hard and work is scarce, the sorting is done finer—but out and forever out, the incompetent and unworthy go.

It is the survival of the fittest. Self-interest prompts every employer to keep the best—those who can carry a message to Garcia.

I know one man of really brilliant parts who has not the ability to manage a business of his own, and yet who is absolutely worthless to any one else, because he carries with him constantly the insane suspicion that his employer is oppressing, or intending to oppress him. He cannot give orders; and he will not receive them. Should a message be given him to take to Garcia, his answer would probably be, "Take it yourself."

Tonight this man walks the streets looking for work, the wind whistling through his threadbare coat. No one who knows him dare employ him, for he is a regular fire-brand of discontent. He is impervious to reason, and the only thing that can impress him is the toe of a thick-soled No. 9 boot.

Of course I know that one so morally deformed is no less to be pitied than a physical cripple; but in our pitying, let us drop a tear, too, for the men who are striving to carry on a great enterprise, whose working hours are not limited by the whistle, and whose hair is fast turning white through the struggle to hold in line dowdy indifference, slip-shod imbecility, and the heartless ingratitude, which, but for their enterprise, would be both hungry and homeless.

Have I put the matter too strongly? Possibly I have; but when all the world has gone a-slumming I wish to speak a word of sympathy for the man who succeeds—the man who, against great odds has directed the efforts of others, and having succeeded, finds there's nothing in it: nothing but bare board and clothes.

I have carried a dinner pail and worked for day's wages, and I have also been an employer of labor, and I know there is something to be said on both sides. There is no excellence, per se, in poverty; rags are no recommendation; and all employers are not rapacious and high-handed, any more than all poor men are virtuous.

My heart goes out to the man who does his work when the "boss" is away, as well as when he is at home. And the man who, when given a letter for Garcia, quietly takes the missive, without asking any idiotic questions, and with no lurking intention of chucking it into the nearest sewer, or of doing aught else but deliver it, never gets "laid off," nor has to go on a strike for higher wages. Civilization is one long anxious search for just such individuals. Anything such a man asks shall be granted; his kind is so rare that no employer can afford to let him go. He is wanted in every city, town and village—in every office, shop, store and factory. The world cries out for such: he is needed, and needed badly—the man who can carry a message to Garcia.

The Universal Three

There are three critical aspects of leadership that apply no matter what the situation. Together, these rules are called the Universal Three.

1. *When in charge, take charge.* People want a leader to lead; they want someone to take charge, direct, seek counsel, create a plan, and implement it. No one comes to work to sit around aimlessly while the boat drifts on the tide. People want to be masters of their own destinies; they do not want someone deciding for them. They want, crave, and need leadership.
2. *People want you to take charge, but when you do so, they will resist it.* People want to be led, but it is a quirk of human nature to resist the will of the leader. People will challenge your position, so be ready to defend it with good sound logic and reason. People will challenge your decisions; listen to what they say and use it in your decision-making process, and then make the decision.
3. *It's a natural thing; get over it.* Don't get disappointed or angry when people challenge you. That's exactly what you want: a noisy, bubbly organization full of ideas and people who are not afraid to speak up.

The ABCs of Leadership

The ABCs of leadership comprise a timeless principle to keep with you throughout your career. Fundamentally, if you fail in implementing these principles, you stand a pretty good change of ending up as a manager and not a leader.

A. Trust your subordinates

If you ask someone to name the characteristic traits of a leader, trust is one item that always makes the top of the list. Even though we talk a lot about trust, we infrequency discuss what it means, and we hardly ever define it.

How can you expect your subordinates to go all out for you when they don't believe you trust them? Trusting is easier than you might imagine, but it brings with it significant risk, and for risk-averse managers (notice I didn't say leaders, because true leaders are not risk adverse), it becomes very difficult. Fundamentally, trust means, "I believe you are doing good things for me and for the organization." When I trust you, I believe that you have my best interest in mind and that you are leading us down the organizational road simultaneously.

The old theory that you have to earn trust is a little weak when it comes to a supervisor/subordinate relationship. There is a distinct difference between, on the one hand, trusting that a human being wants to do the right thing and will do the right thing and, on the other hand, trusting that they have the KSAs to do the job. You can trust new rookies or new officers, but you still have to provide mentorship, keeping them close while you teach them, until they gain the KSAs to function safely. Consider suspicion as a permanent condition: if you don't give trust up front, will a day ever come when you wake up and say, "I didn't trust Bob yesterday, but today I trust him"?

There are several points that you need to bear in mind about giving trust as a leader. First, when you give trust to human beings, someone is eventually going to let you down and disappoint you. You have to expect this, but that is no reason to not give trust up front. When someone disappoints you, sit them down, outline the expectations, and move on. Most people don't disappoint you on purpose or set out to dissolve your trust; it's usually a function of making a mistake or doing the wrong thing.

Micromanagement is the first sign that there is a distinct lack of trust between parties. There is a difference between mentoring or teaching and digging down deep into someone else's job to dissect every little perceivable weakness. If you have adequately prepared your members with training and practical experience, given them clear parameters that guide them in accomplishing the task, and given them the proper tools to accomplish it, there is then no reason to micromanage them.

As I mentioned earlier, just because you give trust as a leader, don't expect the same courtesy in return. The reality is that you are going to have to earn your subordinates' trust by your actions on the fire ground, through your handling of personnel matters, and in everyday station and organizational operations. Welcome to leadership.

B. Develop a vision

An entire text could be written on this, so here we will break this down into the bare necessities. First and foremost, you have to focus on the challenge at hand. Depending on your position in the organization, that challenge changes. Where do you want to take your company or organization, and what part does the team play in that? How do I carry out the plan that gets me from here to there, and then from there to the next destination. The higher one climbs in the organization, the more and more complicated this becomes, and what we call vision requires different approaches at different levels of the organizational structure.

For the company officer, this means make sure you understand your role and your team's role, where you are supposed to be spending your time. It also means that you have to take time to plan effectively to accomplish everything you and your team are expected to accomplish on a daily, weekly, monthly, and yearly basis. As a leader, your focus or challenge might be a crisis, a chronic problem, or a sense of drift from your members. These challenges have to be met with a sense of purpose; in other words, the solution must come from your head.

Charlie Grimm was the manager of the Chicago Cubs in the 1930s and '40s. One of his scouts, excited about a young new pitcher, came to him and said, "Charlie, this is the greatest young pitcher I have ever seen. He struck out 27 in a row—on one even got a foul off of him until two were out in the ninth. I have him here with me now. What should I do?" Charlie said, "Sign up the guy who got the foul. We're looking for hitters." Now that's focusing on first things first.

C. Keep your cool

When it comes to the emergency scene, the station, personnel matters, or a crisis, keep you cool. No one wants someone who is supposed to be leading screaming into the microphone, running around with no sense of direction, scolding people in public, and jumping to conclusions. What they want, as I have said countless times, is for you to stand there, sun in your face, with Viking helmet on, and with a calm voice, make a decision, providing direction and adjusting as necessary

When dealing with human beings and the issues associated with them, don't rush to conclusions or be stampeded by first reports. It is not unusual for initial information about the fire ground or about someone's actions to be 80% wrong. Take your time in sifting through the information, and keep your cool while collecting all the facts.

On the fire ground, there is no event that I have been involved with—and I have been involved with some pretty big ones—that was ever cured by screaming into the radio! On the contrary, screaming into the microphone disrupts the operational Karma of the event, disturbs your members, and destabilizes emotions. Take several deep breaths, look at it, and remember, "This is what they pay me to do!"

D. Encourage risk

The safety folks who are reading this are probably about to jump out of their socks! But what I'm talking about is allowing some freedom in the workplace. If you want accountability, you have to let people to take risks when it comes to work and the aspects of accomplishing organizational missions. Risks might involve nontraditional methods when the traditional method is unusable. Had Chief Harry Diezel not allowed us to take a risk in developing and funding a technical rescue team by using money outside the normal budget process, we would have never had a Federal Emergency Management Agency (FEMA) task force or a world-class technical rescue team and training facility in Virginia Beach.

Taking a risk also means removing some of the constraints and rules that organizations seem so fond of—and without rules, we will have anarchy! What if I were to suggest that the organization do away with the quota system for hydrants, inspections, pre-incident plans, and so forth. What if I were to give my officers the number that they have in their area and say, "You know what needs to be done, and you know what your workload is; now set your own quota for getting these done"? I would expect that the manager in you is on the verge of a stroke on hearing these words.

The point is that if you are going to trust your personnel and if you are going to allow them to use their talents, then you have to take a risk. People are not going to go all out for you if they think that the least little failure might jeopardize their career. In some ways, taking risks mean driving fear out of the organizational structure.

Finally, be prepared to take risks on the fire and rescue ground. The bottom line is to not threaten firefighters' lives in a needless and reckless manner; nevertheless, don't be afraid to implement something like vent, enter search, or search above the fire without a line when appropriate—or any of the other high-risk tasks we do when implementing a given tactic might save a life. You are paid to take risks; they are what the job demands and are necessary sometimes—just be smart.

E. Be an expert

When you go to the doctor, you expect an expert; when go to a lawyer, you expect an expert. Your expectation is for these individuals to be proficient at their given professions and that whatever they do not know, they have the ability to find out to address your health needs or to provide you with good counsel. So it is with the people that you lead: from the boardroom to the mail room, you had better be an expert. I can't tell you how many times I have heard from firefighters and officers who say, "Chief So-and-So forgot where they came from!" My reply has usually been, "You're wrong; they didn't forget where they came from because they were never there!"

Members want a no-nonsense approach to doing business. If you earned your reputation as a sled dog and a good firefighter, then go ahead and use it, but if you spent little to no time in a company prior to your promotion, then don't try to con people into thinking you know what you are talking about. The bottom line is to admit what you don't know and stand by what you do know. Remember that none of us is as smart as all of us, so surround yourself with great people who have inclusive talent. Machiavelli said, "The first method for estimating the intelligence of a ruler is to look at the men he has around him."

As you move through the organization, your expertise changes. What was once your expertise—because you did it all the time, trained at it all the time, and taught it all the time—changes. That doesn't mean you don't understand it anymore and that you can't do it anymore; it simply means that as you move through the organization, your focus changes as your job changes, and thus, your true expertise changes.

To lead, however, you need to have made your name as a sled dog, spending time in the companies, making the long hallways, doing rescues, cutting roofs, performing technical rescues, and managing and commanding the fire and rescue ground. If you don't have this foundation, you end up as just another bureaucratic manager who is better suited to the world of corporate business than the fire service.

F. Invite dissent

If people are not speaking up, especially when important things are on the table, silence becomes a lie.

You need a bubbling cauldron of discussion if you are really going to reach the best solutions. Organizations that stifle discussion and stifle creativity for the sake of the status quo will eventually fail.

Collectively, a group is always smarter than an individual. Expect—demand that people speak up. And be prepared to be angry, upset, and often amazed when you let people's ideas flow.

G. Simplify

As a leader, you have to be able to take very complex issues and ideas and break them down into digestible portions. You have to understand the big picture yourself before you can explain the interactions with and impacts on your work group to your members. We talked about putting first things first, but if you are going to communicate a vision and set a course for your team members, you have to be able to simplify.

My experience has been that company-level personnel couldn't give two hoots about the strategic plan. What they want is to be listened to, to get the tools they need to do the job, to be respected, to be communicated with frequently, honestly and openly; and to be given a say in their work environment and rules. That does not make the strategic plan any less important, but it means that you have to take those very complex and interlocked ideas and simplify them, breaking them down into task-level actions every day you come to work.

There you have the ABCs of leadership. They are simple and easy to read, but not so easy to undertake. Fail these and fail as a leader; master these and you are at the least setting a solid foundation on which to build your leadership.

Methods to Improve Your Leadership

Here are some simple, practical ways to improve your leadership every day you come to work:

- *Strive to do small things well.* The fire service is full of dues-paying jobs—mundane maintenance and organizational stuff that just has to be done. These are sort of like the chores you do around the house. They are not glamorous and sometimes can be downright boring. Find a way to make them fun and to change it up. Above all, excellence in small things breeds excellence in big things.
- *Be a doer and a self-starter.* You know what has to be done, both in your little corner of the world and in your response area. You know what your members need, you know what your community needs, and you know what the organization needs. Go out and do it. Don't wait for marching orders. Be aggressive and take the initiative. Also, take time to plan the functions around your house or organization so that people know where they are going from day to day and week to week. Set the course, and be the captain.
- *Strive for self-improvement through constant self-evaluation.* Take the time to be reverse-evaluated by your members. I always say, if you want to know how a leader is doing, ask the followers, not the boss! Don't be afraid to say so when you make a mistake; learn from it and continually evaluate what you are doing and how you are doing it.
- *Never be satisfied; always ask what we could have done better.* Take those few precious moments at the end of even the most routine calls to ask yourself and your crew both What did we do well? and What can we do better next time? Even little tweaks here and there can produce amazing results when added together.
- *Don't overinspect or oversupervise.* Teach your members, and then let them go and do it. Allow them to make mistakes from time to time; it's a learning tool. Errors made in

training and reinforced lead to better operations. Remember, micromanagement is the first sign of distrust. If you don't trust someone to get the job done or to carry out the plan, you had better deal with it long before this point.

- *Keep personnel informed.* Don't let the dreaded rumor mill become the primary method of communications. Give your members the news now: Good or bad, don't hold back. Don't try to spin something that is negative into something that is positive; life is full of both, as are organizations. Keeping your members up to date about what, where, why, and how is a critical aspect of earning trust.

- *The harder the training is, the more they will brag.* Get your rear ends off the couch and stop looking at videos. Go out and train, train, train. Get wet, muddy, dirty, hot, and tired; get back to whatever you don't do very often. Simple training on basic items can save your life when you need to go on autopilot. When was the last time you got a master stream stretched 300 feet and in service? When was the last time you did a ladder rescue or a covered mask drill? Train to live, and live to train.

- *Be enthusiastic and fair and have physical and moral courage.* Come to work jazzed and push it through the ranks.

Optimism and pessimism are infections and they spread more rapidly from the head downward than in any direction.

—Dwight D. Eisenhower

Ask yourself if you will have the physical and moral courage to do what is necessary when called on: When you have to risk your life, and perhaps give it, are you ready? Can you do what was done by hundreds in the Twin Towers on 9/11, as well as by countless individuals throughout history? This is a question you had better answer right now!

- *Showmanship.* Know when it's time to be a showperson, to lead using strong impressions and to stand up and be recognized.

Know when to use humor to lighten the mood or to poke fun at yourself. Know when to be reverent and pay respects to whatever deity you worship.
- *Speak and write well.* You must be able to speak effectively and precisely, and you must be able to write well. Without these two skills, you will find yourself in a shark pit, swimming in chum.
- *Have consideration for others.* Not everyone is like you or responds in the same manner you do. Consider that all members have value and respect and be considerate of others' positions and emotions.
- *Yelling detracts from your dignity.* If it's an emergency to you, then who on earth do I call! You are paid to work in the worst conditions possible, making decisions in much too little time based on much too little information. Don't be the football player who does a dance every time they get in the end zone; instead, act like you have been there before. Don't chastise and counsel people in public: praise in public; scold in private.
- *Understand and use judgment.* Once the decision has been made, get behind it and support the decision, regardless of your original position. Don't talk bad about or undermine the decision. Know what to fight for and know when it's a loser. Don't teach your members bad habits by defending everything they do under the guise of "taking care of my people" when you know they are wrong. Educate and correct when necessary.
- *Stay ahead of your boss.* A good boss will make clear his or her expectations at the first time you meet. Stay in front and make him or her look good; do more than is required, and do it effectively.

The Best Leaders I Ever Had: Characteristics to Embrace

Captain Earl Stanton retired today. I know that doesn't mean much to you, but Earl was a fixture around the Virginia Beach Fire and Rescue organization; he was one of our original hires, officers, and investigators. As a firefighter, he always did his duty, showing up on time and in the right uniform. He respected his bosses, and his peers respected him. Earl was also an excellent officer. He had a well-rounded personality, and although he was never known as a ball of fire or a spark plug, he was respected, trusted, loved, and revered by many in the department.

How much we loved and respected Earl was clear when more than 150 members showed up to toast and roast him at an impromptu retirement party. What made the night so special was that the people who attended did so because they wanted to and because they cared about Earl and what he meant to them and our organization.

Sitting around sipping a cool beer, I sidled up to retired Captain Keith White, and we started reminiscing. Some of the best learning experiences and the best times I ever had as a young firefighter took place while I worked for Keith. He was a tough boss and a world-class tradesman. Eight-hour days meant nothing to him; he worked at the fire department like he did in his construction business: going 100 miles per hour and expecting 100% perfection.

The entire event made me think about the best leaders I ever worked for, and I realized I was blessed: From Department Chief Harry Diezel, to Battalion Chief Paul Mauch, to Captain Keith White, I've had my share of great supervisors. Around my fifth beer (I took a taxi home), I thought about the characteristics these folks shared; common approaches, albeit with different styles, made working for them fun and challenging. Ultimately, they created a learning environment that allowed me to further my professional goals. But what was it specifically that endeared these teachers to me: even while they disciplined me and worked me hard, how had I managed to keep my respect for them all these years?

Caring for the job

While I could talk about the touchy-feely side of this subject, that's not what I mean to do. To put it simply, each of these officers cared about their job. In fact, it wasn't just a job to them; it was a calling, a need, a desire, an adventure. Many of them were so talented, they could have made plenty of money doing something else, but they were professionals and loved what they did, no matter what. They worked hard, always stayed out in front, directed effectively, delegated competently, and trusted their employees, while always looking out for their safety and survival.

My best leaders cared about much more than themselves, and it rubbed off both on the firefighters they worked with and taught and on the organizations they worked for.

Technical competence

Chief Harry Diezel, the first fire chief I ever worked for, although a commensurate politician, didn't desert his people in the process; he remained a firefighter's fire chief. He was technically competent enough to know which aspects of the game he wasn't part of and left them to those who were. He never relinquished authority, and he used his talents wisely.

Like Chief Diezel, my best leaders were competent in every task. If they were on the fire ground, they excelled at either the task level or the command level. If they performed routine administrative work, they were good at planning, getting it done on time, and keeping it in order. If they performed routine maintenance, inspections, or any other dues-paying task, they strove to be good at it, because they understood that every part of the job is important.

Their competency was infectious and had the momentum of a giant snowball. This encouraged and enabled us to take on any challenge and created a learning environment that allowed subordinates to grow, learn, and practice. It raised the bar for the entire shift. As Captain Abshoft says, "A rising tide lifts all boats."

Will Rogers once said, "It isn't what you know that will get you into trouble, it's what you know that isn't true." The best leaders I ever had were smart enough to know what they didn't know! They realized that a person can be good at only so many things; providing a learning environment for people with talents in other areas was critical to their own success.

Understanding people

My best leaders never forgot the human element within business. They recognized my talents and weaknesses and worked on both of them. They knew when I was up, when I was down, and when I needed a good kick in seat of the pants. When I did well, they told me so, and when I messed up, they weren't afraid to teach me what I should've done, never letting it get personal. They cared about your success—and they sure let you know about your failures. They cared enough to realize that not everyone has the same talents, but they were smart enough to recognize individual talent and help to focus it in a direction that would benefit the team. They would mentor you in skills you didn't have while busting your chops with comments like, "Chase, can't you even drive a nail?" or "If you had as much water coming out of the nozzle as you did the relief valve, you would have one heck of a pump job working!"

The men and women I learned best from recognized that to succeed, you must surround yourself with talented people. Often that means creating an environment where people can thrive and learn to be great, even if they weren't so great when they first walked through the door.

They made sure we had fun, while also making sure that we were pure business when it was critical to success. They constantly reminded us that we existed solely to take care of people and that no matter what their social stratum, everyone deserves to be treated with dignity and respect.

Expecting the best

My bosses made their expectations clear and did not compromise: They expected excellence. They understood that, as Jim Page said, "In municipal government, minimum qualifications become maximum expectations," but they didn't live their organizational lives that way.

They never accepted "good" as good enough; if something wasn't done right, we did it again, whether we were performing a training evolution or cleaning the fire station. My former captain, Keith White, made us take apart wooden shelves twice because they didn't meet his standards. They held us to high standards that many times exceeded written expectations or those outlined in human resources documents. Those who didn't like excellence left for somewhere that would accept mediocrity. Those who stayed became better firefighters, EMS providers, and supervisors and learned a variety of skills.

These bosses set the same high standards for themselves. They never asked their crew to do something they wouldn't do. If it was 100°F outside and we had to paint hydrants, they were shoulder to shoulder with us. If they expected us to train in the rain, the battalion chief was right there taking a look and spending some time with us.

Taking a risk

I know what you're thinking, and the safety folks are probably having palpitations. Yes, we took calculated risks designed to get results on the fire ground. But my best supervisors were never foolish in their approach to action or carelessly exposed the team to harm.

My leaders were progressive risk takers who weren't afraid to speak their minds and break from the old ways of doing business. They allowed for and listened to new ideas and weren't afraid to try new techniques and operations. By breaking with the norm, they exposed themselves to possible criticism and to the risk of failure and the associated stigma. Keith once allowed us to create a "hose bed ladder carrier," which I wrote up and sent to *Firehouse* magazine; it was the first thing I ever got published, and Keith was really proud that our shift was in a national magazine. My leaders never stopped taking risks that might improve customer service, erring on the side of progress. When we failed, we learned; but we succeeded many more times than we failed, and we became good at achieving success. We regarded failure as a stone in our shoe that we had to stop and dump out before walking onward.

They were also willing to pass down their skills, allowing us to do some of the same things they did, running the risk that we might do something better than they did, which in turn challenged them to stay ahead of us. It took many years before Keith allowed anyone other than himself to use the sewing machine provided by the city, but he finally relinquished some control and found out that Wayne was pretty good, which gave Keith time to perform other officer-related things.

A final word

My best leaders had all these characteristics and many, many more. Not all of them possessed the same number of talents or such varied methods of conducting business, but when I look back now, it seems as clear as day: Each of my leaders taught me without knowing it. Their actions, rather than their words, had the greatest impact on my learning experience. Even if I never told them, I will be forever grateful to the Earl Stantons, the Keith Whites, the Paul Mauchs, the Harry Diezels, and the Steve Covers of the world. Thanks, guys, for giving me such a strong foundation to build on and pass down as a model to someone else.

Chapter 4
Keeping Humanity in Your Leadership

> *Organization doesn't really accomplish anything. Plans don't accomplish anything either. Theories or management don't accomplish anything either. Endeavors succeed for fail because of the people involved.*
>
> —*Hyman Rickover*

In an organization that depends almost entirely on people every day—to do the right thing for the customer, to make the right decision under enormous pressure, and to direct the lives of their crew, as well as those of civilians—I am surprised at how little attention is paid to the human side of doing business. In most fire service organizations, because of either labor or management (or a combination of both), the human side of the equation inevitably takes a back seat to tangible details that they really think make people happy and productive.

People are strange creatures and, as I have said previously, the most complex machines ever created. It is nearly impossible to understand the human mind to the extent that we would like, but there are numerous behaviors and characteristics about people that we can figure out without being Freud.

First, people like to have fun. They would rather be having fun than having their teeth extracted or undergoing a rectal examination! Next, people want to feel as if they are contributing something to the process; call it work, customer service, or whatever you like, people want to feel as thought they make a difference. It has also been my experience that people like to be told when they are doing good for their leaders and for the organization, and in that context, they don't mind being pulled aside and told how to improve.

In addition, people want to be kept informed, because that allows them to have some say over their destinies, or at least to better fight to stay alive. They want to be treated fairly, and they don't mind the rules of the game changing; however, they expect to be told when the rules change, to be given the information, and then to be allowed to make the switch. This is sort of like instant replay: if you going to add it into the game, that's fine; just let people know how it's going to be used, and then give the teams time to adjust to it. Your members want to be respected for who they are and what they bring to the table.

So, how do we keep the humanity in our leadership? Of the many ways possible, I am going to recommend a few in particular in this chapter. To begin with, here are some fundamental rules that apply to human relationships and human nature.

The Paradoxical Commandments

1. People are illogical, unreasonable, and self-centered. Love them anyway.

2. If you do good, people will accuse you of selfish ulterior motives. Do good anyway.

3. If you are successful, you will win false friends and true enemies. Succeed anyway.

4. The good you do today will be forgotten tomorrow. Do good anyway.

5. Honesty and frankness make you vulnerable. Be honest and frank anyway.

6. The biggest men and women with the biggest ideas can be shot down by the smallest men and women with the smallest minds. Think big anyway.

7. People favor underdogs but follow only top dogs. Fight for a few underdogs anyway.

8. What you spend years building may be destroyed overnight. Build anyway.

9. People really need help but may attack you if you do help them. Help people anyway.

10. Give the world the best you have and you'll get kicked in the teeth. Give the world the best you have anyway.

Copyright© Kent M. Keith 1968, renewed 2001

People-Based Organizations

It shocks me when I hear an executive fire officer remark, "I am just not a people person." That's unfortunate in a business that is all about people. Our members (people) complete the vast majority of the projects, see to the customer service, and develop and implement the ideas that make us successful. Aside from that, we are in the business of fixing other people's problems. When their children are in crisis, they call us to fix it. When their house is experiencing some chemical reaction that is quickly consuming it, they call us to fix it. If the new car seat they just bought is too complicated for them to install (imagine that), they come to us to fix it. We are the social, human-being fix-it crew, with the big red trucks (or, if yours are lime green, then leave them on the ramp; when the sun hits them, they will ripen and turn red!), clean T-shirts, and complicated toys, which we have for one reason and one reason alone, to fix people's problems.

Taking care of people, understanding people, and being able to interact with people are leadership skills that you must master. Your ability to get things done, no matter what your rank, is based entirely on your relationships with other people and your ability to manage that relationship in such a way that people want to follow you. You can manage money, you can manage pencils, and you can even manage apparatus. By contrast, you cannot manage people; you have to lead people. People are like a rope: you cannot push it; you have to pull it. Thus, if you have no desire to interact with people and are not a people person, then you are in the wrong business.

I remember a conversation with my current chief that took place at a senior fire leadership meeting. We were discussing the issue of getting things done, and I commented that he had sent a battalion chief to the field (operations) who had no personal interaction skills. The firefighters would say, "Chief, come on over and have dinner," and without acknowledging them, he would walk in the office and close the door. My question was, "How can you expect someone to lead when they won't even interact with people on a personal basis?" The chief's response was, "That's your fault, Chief Sargent. You have to teach him to be a people person." No matter that this battalion chief has been on the job for 15 plus years and has been this way since day one. Don't get me wrong: he is highly intelligent, perhaps one of the best research people I have ever known, an excellent planner, and follows the rules to the nth degree; still, he cannot deal with people. The moral of the story is "Never assume malicious intent when ignorance is sufficient to explain it"; one nonperson individual was directing me to train another nonperson individual in the fine art of personal relationships. When you don't understand people, relationships, and trust, among other concepts, you just don't get it! In this example, we do not have the right person on the bus, the wrong person(s) off the bus, and the right people in the right seats on the bus; with that as the case, no amount of pseudo education is ever going to solve this problem.

David Cottrell has written a magnificent book called *Monday Morning Leadership*.[1] Consider the following exerpt from his book as we talk about people-based organizations:

- *Your scorecard as a leader is the result of your team.* You are needed; you are important. But you get paid for what your subordinates do, not necessarily what you do.
- *You need your team more than your team needs you.* You need each other, but cumulatively, the 17 people on your team accomplish much more than you do!

As you begin to think about your role as a leader in a people-based organization, try this on for size. As a leader, what percentage of the work that needs to be done is getting done while you're in a meeting? If your answer is that about 95% is being accomplished even when you're not there, realize that some may say 105%. One day, the firefighters I lead came in the office and stood at my door. "Chief," they said, "you have led enough for one day; if you go home right now, we can save the organization!"

Now what happens if I put the 17 team members who make up your shift or staff in a meeting and leave you in the office to do work—what percentage of the work would you accomplish? If you answered not much, you are right. The point to be made is if you accomplish only 5% or 10% and the team accomplishes the other 95%, then who needs who more? It's clear that you need each other, but the point is to never forget that your job is to help each other. The team members have entrusted a portion of their life and their success to you. It's your job to help them grow—personally, professionally, and financially. Remember that when you begin to think that you are the center of activity as a leader.

People before Plans

Quite a bit of this book has already been devoted to strategic plans—how they work and how you can look big and lead small. Planning effectively is so important that a later chapter will be dedicated to examining what it takes to become an effective planner.

No plan ever accomplished anything without people who were willing to engage the plan—to turn it into a living, breathing action plan—and who understand why they are doing what they are doing! When you begin to assign crews at the fire ground, look at the plan from the crew's perspective. When you assign an interior hose line, or a search-and-rescue team, or a roof ventilation team, it certainly has something to do with the overall strategy and tactics of the event. However, if the crews you are assigning did not have any pre-incident education—on why they do what they do, how to do it, why doing it is important, and the ultimate outcome of their contribution—then you set yourself up for failure right out of the gate. Relate this to any strategic plan or to any endeavor you undertake with your people and you will see why developing a plan and simply sending it out for people to read and understand is a surefire way to fail. Moreover, as with our professional skills, our conception of the plan and our actions and expectations need to be updated in the classroom periodically, if we are to succeed.

So you want your members to do the right thing for you. Getting them engaged is about understanding what drives people and using their talents. It's also about everyone knowing what's important—what the "main thing" is. There are a couple of basics that you, as a leader, need to know.

First, most people quit people, before they quit an organization. Want to find out why people leave your fire department, your shift, your battalion, or your work group? If so, the first question to ask in an exit interview is *why*. What you will discover is that most people leave because of people, not because of the organization. I am a perfect example. After 26 1/2 years with a marvelous department, I found that I could no longer tolerate my fire chief, who lacked integrity and character in just about everything he did. I left him, not the department. Alternatively, perhaps I simply did not develop the kind of relationship that I needed with my boss to understand what his main thing was. Nevertheless, the point remains that I left a person, not an organization.

Next, people need to know the main things—the critical aspects of our work that the team must achieve. People often have different perspectives on what these main things are. Additionally, because of various circumstances, the main things could change. The main things must be communicated and supported through actions, resources, and communications. Your job is to help people at all levels of the organization in getting the training, tools, resources, and assistance they need to accomplish the main things.

If you're a day hiker, camper, or traveler, you know that your backpack, camper, or luggage must be packed right to meet the expected challenges. Day walkers venturing in the out-country can end up in real trouble should they fail to bring the basic necessities for the environment (it's only a day hike, what could go wrong?). Suppose that everyone who works with you wears a backpack that can be filled with motivation. That backpack can be filled part of the way, or it can be overflowing, or it can be empty, in need of filling. Over a period of time, the backpack may begin to show its age, acquiring tears, leaks, and holes.

Everyone in the organization has the ability to add or remove items. Some people in the organization love to put stuff into other people's backpacks on a continual basis—in the form of negativism, cynicism, confusion, stress, fear, or any other form of confrontation or direction that constrains motivation. As a leader, it is your job to keep everyone's backpack filled with the right stuff for the journey. The question is, How do you keep people's backpacks full, without making them too full?

- *A full backpack, designed for the journey must be filled with important items, the main things to do the job.* To be successful, people must know the main thing(s). A good leader fills backpacks with focus and direction, while a leader who creates confusion and inconsistency is constantly taking things out of people's backpacks.
- *To keep backpacks full, you need to let people know how they are doing.* Performance appraisals do not provide the kind of feedback and motivation needed to keep people's backpacks full. You have to be sincere in your appreciation of people's actions, and you have to be specific in your feedback. The feedback you give people must be timely, and it has to be aligned to the receiver's values. Don't try to fill a backpack with something that is important to you but not to them.
- *To keep backpacks full, let people know you care about them and the job they do.* Paychecks are a temporary method to fill someone's backpack, but once given and spent, they are forgotten.

There is one cardinal principle which must always be remembered: one must never make a show of false emotions to one's men. The ordinary soldier has a surprisingly good nose for what is true and what is false.

—Erwin Rommel

Each individual has certain things that really fill up their backpacks (remember the concepts discussed in chap. 3).
- *Tell them how the backpacking expedition is going.* Keep your team informed about how they are doing as a team.

Keeping in Touch

This may sound funny, but you have to figure out how to stay in touch with your members. I can vouch from 30 years' experience that the further you move up the chain of command, the harder it becomes to keep in touch.

I can almost guarantee that when you do something wrong—or perceived to be wrong—you will hear about it. By contrast, how often do we hear about it when we do something good? My gut feeling is that collectively we do more things right then we do wrong. I would bet that if you looked at what the organization does every day, our members do 99.9% of the main things (and other things) right! This poses the question, Why don't we celebrate what we do right every day? As a leader at any level, you need to create methods, processes, and ideas that enable you and your team to celebrate success. It may be as simple as buying sandwiches for lunch, to say thanks, or as complicated as developing an awards program, to celebrate our heroism, our good works, and our everyday lives.

The following edicts will keep you in touch with team members:

- *Help people fix small problems.* What a simple concept: The reality is that most times people come to you with small problems and ask for your help. If you can help people to fix the small problems, they are more likely to come to you with the big problems. Stephen Covey refers to this as making deposits in the emotional bank account. Consequently, when—not if, but when—you inevitably have to tell them "No, I cannot fix that" or "I will not fix that," they will be more likely to understand.
- *Be approachable; look for results, not salutes.* If you think that you can lead and that your team will follow because you have gold, you have just made a big mistake. People follow you because you give them a reason to, because you have earned their trust and respect, because they believe you are doing the right thing. There is a time and a place for formality in a paramilitary organization, but the bottom line is to get results from everyone regardless of rank. When you point to your collar and tell people that you are in charge, you are more than likely not!

- *Treat newbies well (raise your kids early).* I recollect well how we used to treat our new recruits. When you came on the job, did you feel the slightest trepidation when you walked into the fire station on your first tour? Did we send you all around town looking for a water hammer, or something worse? I remember putting a rookie on the flagpole, wrapping him with two-and-a-half-inch-gauge hose, and charging it. It very quickly became obvious that I might have as well put him in a cage with a giant anaconda that had not eaten for a week. As he turned blue and his eyes puffed, I screamed to cut off the water, rushed over, pulled the hose from his smiling but concerned face, and said in my best leadership tone, "Don't tell anyone." Although new members will still be given their fair share of toilet and dish cleaning and will be told to "wait for a year until you can have an opinion," the reality is that most of the folks we hire today are smarter, faster, and better educated then their predecessors. Keep the newbies close, mentor them, treat them like you would want your son or daughter, niece or nephew, or any other relative treated if it were his or her first day and night in the fire station.
- *Be the rising tide that lifts all boats.* Some people in the organization will shun you for trying to lead, be fair, and treat people with respect. Their assumption is that it has always been done a certain way and that if you deviate from that, you must be out of the mainstream. Don't worry about what others say; instead, do what you know to be the right thing, and eventually, when it's clear what the right thing is, everyone will have to follow suit or be left behind. Morale, esprit de corps, and integrity are all valid concepts that must be espoused by officers, because success is a collective endeavor.
- *Manage for talent.* Not everyone can be everything to everybody. Many fire service bosses mistakenly assume that everyone is capable of doing every job. As nice as it might be to expect everyone to have the same talents and the same capabilities, it's just not that way. Some members are great administrators but horrible incident commanders; other members are great at training but horrible at administration. While there is a minimum set of qualifications, both technical and administrative, that everyone needs to have, people thrive in different environments. Putting the wrong person in the wrong spot creates a spiral of failure for the individual, as well as for organization and its members.

Here are a few additional steps to stay in touch:

- Involve team members in major decisions. Take the time to listen to them, because they often have the best ideas.
- Know the team members, their families, and their current status. Care about them.
- Send cards and notes of condolences or congratulations when appropriate. This communicates the message that "we know that you have something going on"—good or bad—"and we want to make sure you know we are here for you."
- Let your superstars mentor members of the shift who need it. Let them become teachers and leaders.
- Post pictures of events that demonstrate the team in action, showing both the successes and the humorous side. Keep it light but informed.
- Treat people as they wish to be treated.
- Spend time with the team. Take time for "leadership by walking around"; just being seen can indicate that you care.

Want a great way to stay in touch with your boss? Bosses want self-reliant leaders who accomplish what needs to be done. Stay ahead of your boss and predict what needs to be done, based on the main thing. If you do this, you will build up your boss and in turn create trust.

The Second Chance

How many of us have pulled a boneheaded stunt, fallen below the line, or let someone down during our careers or in our personal life? The answer is just about all of us. Now the question is, When that happened, did you get a second chance from the person(s) affected? When you ask people to take a risk for you, it sometimes means failure. When people are left to make decisions, sometimes they make the wrong decision. Actions based on adult decisions lead to adult consequences. Giving a second chance does not mean that there will not be consequences (discipline, expectations, actions); rather, it means that you when it's over, you put them back in the saddle and get on with business.

People who make the same mistakes over and over again are compulsive, addictive, stupid, or ignorant. There is a limit to the number of chances that you can give someone when they have been continually reminded of the expectations.

When you have given team member second chances, use these as moments of learning. Find out what was going on mentally the first time; let them express their perceptions. Make sure you let them know what you think and share with them the organization's values and perception of their action or inaction. Point out how it could be done better or how it should have been done; use the opportunity to correct the course and get the values compass realigned with the final destination.

Remember that it's a big Karma wheel: what comes around goes around. Look past your anger, frustration, and disappointment and figure out how to salvage the relationship and the individual.

Some Final Words about People

We spend lots of time screening, mentoring, and training our members. The vast majority of people we hire are competent. You can count on the native intelligence and ingenuity of your members to get things done and move the organization forward if they believe in you. Sometimes people will let you down, but don't expect or tolerate that as part of normal practice.

Don't take yourself too seriously. Every task in the organization is important if it has been aligned with the main thing(s). Everyone in your organization has a job to do, and everyone is linked together by success or failure. If you convey to your members that they are high performance, they are more likely to be high performance. Headquarters is there to support the troops, and the people closest to the customer are usually right.

There will always be malcontents, and there will always be the Cains of the organization, who are out for themselves no matter what you do. Even though they are the anomaly, they should still be weeded out, so that they don't weigh down everyone else's backpack with their negativity.

Remember that the vast majority of the people who make it through the hiring process are competent and committed. Our members show up every day to do good things for their leaders and for the organization. As human beings, we sometimes fall off course and need some redirection. As a true leader, your job is to keep the ship and its crew on course.

Notes

1. Cottrell, David. *Monday Morning Leadership: 8 Mentoring Sessions You Can't Afford to Miss.* Dallas: CornerStone Leadership Institute, 2002.

Chapter 5
Being Tested as a New Leader

> *Treat a man as he is, he will remain so. Treat a man the way he can be and ought to be, and he will become as he can be and should be.*
>
> —Goethe

No matter where you go in an organization—whether you've been an officer for 20 years, just been promoted, or just inherited a new work crew or shift—people will test you. It's human nature to find out where other people draw the line, to determine their level of integrity, how they're going to function as a boss, and how to push their buttons. Firefighters are especially adept at pushing and prodding you to see just where you draw the line before they step back, giving you a hangdog look that says, "What? I didn't do anything." (I just wish someone had told me this before I was promoted!)

How Personnel Test a New Leader

First day on the job

People will initially test your resolve and expectations in at least four ways:

1. Station procedures and informal policy
2. Formal policy, as well as your interpretation and enforcement
3. Response guidelines and expectations
4. Past supervisor actions

I remember my very first shift as a lieutenant: my shirt was pressed, my gold was shining, and my ego was at an all-time high. I arrived at Fire Station 1 ready to change the world. Unfortunately, the officer training I received (none) didn't match the shine of my badge or the width of my ego that day. I had just moved from one of the busiest companies in the city, where I had been a firefighter-paramedic, to one of the slowest and oldest stations. I had no way of knowing that I was about to inherit much more than I had bargained for.

Every story requires some background information so that you can really enjoy the pain faced by the participants. If you look hard enough at any situation, you'll always find clues that let you know when you're standing in the middle of an interstate at rush hour. Anyway, after my initial meeting with the firefighters of my new station, I missed the very first clue. I arrived and met my shift, two master firefighters and a firefighter (none of whom I knew very well) in a single-engine house. Being well versed and highly educated, I began to line out my expectations: "I know I'm new, so if we have problems, let's work them out in house. If I push you too hard, let me know, and let's work together." How inspiring! What charm, what leadership—what a fool I was!

Sitting at my desk afterward, I found and opened an envelope from the previous lieutenant. Inside was a matchbook with a single match and a very short note that read, "Rookie, use this match to set the world on fire. When it goes out, quit!" I thought it was strange, but stranger still was that I didn't realize the significance of the note or the relationship it had to the former supervisor's actions.

Well guess what?

At the time of my promotion, we were changing shifts at night (1800 hours), and policy stated that members had to be out of bed at 0700 hours and working by 0800 hours. After that first day, I bounded out of bed at 0600 hours to shower and then start familiarizing myself with the filing system and my office. Around 0830 hours, I noticed only Gary Reich, one of the master firefighters, up and about. "Gary, where are the others?" He made a sleeping motion, so off I went to the bunks, where I found two sleeping beauties snug in their beds. I picked up the ends of both beds and dropped them, shaking the occupants awake. Although I cannot repeat the conversation that followed here, I will say that my professional and personal life were immediately called into question.

"Get up," I replied. "It's way past 0700."

"Well," they responded, "Dickey never made us get up this early."

"Well guess what? Dickey don't work here no more!" I shouted. Grumbling, they got up, cleaned themselves, and started working. Later, while I was sitting at my desk, one of my late-sleeping geniuses came out of the bathroom and slammed a mop down. I jumped up from my desk, "What's wrong, Dave?" I asked.

"Dickey never made us do morning cleanup!" he said. It was all I could do to keep myself from leaping on him and choking him. "Well guess what? Dickey don't work here no more!" I shouted again.

The prior work ethic and supervision were dealing me a losing hand. Clearly, I had failed to establish my expectations. I called the shift together and laid out my work plan, which included cleaning the ceilings (old stations allowed you to smoke, but no one had ever cleaned the ceiling), painting the bay floors, cleaning the closets, and stripping the floors, as well as a rigorous training plan. The looks on their faces said it all: this was a foreign way of doing business to them (except for Gary, who was the station's only saving grace). As I dug through the next few shifts, I discovered hose testing to be two years behind, vehicle maintenance records to be incomplete, and a variety of other operational and administrative issues to be less than satisfactory.

The truth was that I had stepped into a company where mediocrity was considered a value. The previous supervisor had been lazy, failed to do his duty, and surrounded himself with people who relished that lifestyle. It saturated everything we did in the months that followed. Their fire ground skills were weak from lack of training, supported by a lack of responses. Their EMS skills, administrative skills, policy adherence—and just about everything else the company did—were weak, weak, weak.

The initial reaction from certain members of the shift was that I was not like Dickey. They judged me against the previous supervisor's work ethic, and they didn't like the change. As it happens, this is a two-sided coin. If you inherit a work group, a shift, or a division that has had previous good leadership, supervision, and relationships, then your job becomes much easier; you can even learn and improve leadership skills by evaluating what the previous leader left in place and by adopting those positions, programs, and initiatives that have real value. However, if you inherit a house where it was previously okay to violate policies, where a strong work ethic wasn't valued, and where "Wipe down, sweep down, sit down" was the house motto, you have a real problem as an officer; you must establish clear expectations of how personnel should act and how you will respond.

I knew it was time to leave when, after 8 1/2 years in staff as the special operations chief, the little voice inside my head said, "Time to go back to the field." Although my job afforded me lots of operational time and plenty of fire responses, it still wasn't shift work, and it wasn't an operational battalion slot.

Bright and early one morning, I walked into the office of my boss, Deputy Chief James Carter, and told him how I felt. He understood fully and said, "Name your assignment, and I'll see what I can do."

The ballers

Having never been a battalion chief in the first battalion (the high-rise, oceanfront, tourism district), I chose that assignment, and before long, I found myself on my first shift. The station housed two engines, one ladder, two captains, and 10 firefighters, most of whom were veterans.

VBF&R has a long-standing reputation of providing physical fitness training. We used to play "felony" volleyball, basketball, and football, and as one might expect, we had more than our share of members who were on crutches and had sustained black eyes, broken fingers, and other bodily injuries. As a result, ball sports are no longer authorized activities for our department.

On the evening of the first shift, after a good workout in the gym, I decided to go for my evening run. As I rounded the corner, I noticed a group of highly motivated members of our department in working uniforms, shooting baskets with a portable net. Mind you, this was no fierce competition, certainly not felony ball, just a leisurely group around a basket displaying the manners and respect one would expect at an English tea party as they retrieved and handed the basketball to each other. Now these are intelligent, educated veterans of the department. Was I to believe they had no idea that there has been a ban on ball sports for more than 20 years? Or were they trying to see what the new battalion chief's reaction was going to be: Will he continue to run by and pay no attention to us, or will he come up and see what is going on? As I walked up the ramp, they smiled, and I asked, "Hey guys, what's going on?" Their reply was classic (always appear stupid in the face of an obvious question): "We're shooting baskets, Chief."

"I know what you are doing," I said, "but why?"

"Well boss," they began, "you know there is no specific SOP [standard operating procedure] prohibiting—"

I cut them off in midsentence. "Guys, you know you cannot play ball sports," I said. "Take the net back over to EMS and put up the ball."

"Yes, sir," came the reply. With no fuss, they went off, and I continued my run.

After this cardiovascular workout, I decided to visit the two captains, who had been conspicuously absent from the basketball game. Captains Stanton and Thompson were in the office talking when I knocked on the door. "I was just out running and saw the most curious thing," I said. "The shift was on the back ramp playing basketball." I'll never forget their look or their reply. With straight faces, astonished that their personnel were involved in such a grievous endeavor, they looked at me and said, "No!" The conversation that ensued went something like this:

"This morning I told you I had no intention of running your stations, and if I have to, you are going to be unhappy. So let's make sure we have our expectations straight. Do we need to talk about this?"

"No, sir," came the reply, and we were back to normal just that fast.

The bottom line is that they were testing the new chief. They clearly wanted to see how I would react to their behavior. Would I simply ignore it, or would I hold my ground and expect them to meet my organizational expectations? My response set the precedent for our relationship from that point onward. Granted, it was a little thing, and I don't believe for one moment that any of these members meant disrespect or were attempting to violate policy. It was a test—nothing more, nothing less. However, in sum, little things can set the stage for bigger things, and if you're not consistent in your expectations and choose between the policies you enforce and those you ignore, then you set yourself up for real failure as a leader.

The routine call

Most of us run the "smells-and-bells" calls quite often. Since these alarm calls are frequently to the same location, it's fairly common to get to know the occupants by their first names—and to know the names of their pets and relatives too. On one day, we were the smells-and-bells kings! I pulled up to the location and watched the units go to work. A truck company from another battalion got off the truck, with the acting officer in front. Their coats were unbuttoned and they had no SCBA, hoods, or gloves as they walked toward the front of the building. I motioned for the acting officer to come over and asked him, "What's up?"

"Fire alarm, Chief," he replied.

"Well, what a coincidence—that's why I'm here!" I said. "I mean, what's up with the gear?"

"Chief, it's a fire alarm. We come here several times a week."

I contained myself and thought, "When you're angry, don't do anything that makes you feel good," over and over. "Let's get this straight," I said. "When you show up, I expect you to be ready for battle. That means geared and focused for a fire. As the company officer, when you get off the truck looking like this, you send a message to your crew and rookies that this is acceptable, and it's not. If you ever show up like this again, there is going to be a problem. Do we understand each other?"

From the look in his eyes, I could tell he was thinking, "Who does this guy think he is?" But he thought better of it and simply said, "Yes, sir," and moved away. They never showed up without gear again.

The source of the problem was that someone had allowed this kind of behavior to go on over a period of time. What occurred between the acting officer and me was a clear communication of my expectations, which created a mutual understanding of how he and his crew should perform their duties. He didn't have to like it; he just had to do it! Just ask the members of the Memphis Fire Department about fire alarms and the costs of getting complacent.

Actions speak louder…

Suppose your organization prohibits members from growing Fu Manchu mustaches or mustaches that extend below the bottom of the mouth. Mustaches are pretty visible things, so when they're too long, it's obvious and it's an easy thing to say, "Hey, cut that mustache." Now, also suppose that you issue physical-training (PT) gear to personnel and that you have a policy that says PT gear can be worn as casual station wear after 2000 hours at night. Suppose we start PT around 1630 hours and work out until 1730 hours. Since I can't wear my PT gear until around 2000 hours, the correct thing to do is change clothing. But in the meantime there's dinner to cook and the kitchen to clean afterward; and if there are a couple of calls in the evening, it's 2000 hours before you know it. So your crew may be in their PT gear from around 1630 hours until 2000 hours. Will Mrs. Smith die because of this egregious breach of policy? Is the world ending? In both cases, the answer is no. However, I'm not talking about the logic behind the policy; rather, I'm using it to make a point.

Suppose one day you approach one of your employees to request he cut his mustache. He respectfully replies, "Yes, sir," and as you turn, content in your omnipotence, the employee says, "Excuse me, Captain, does this mean we're going to start enforcing policy?"

Stunned, you turn and reply, "What are you talking about?"

"Well, Captain, you know you're making me cut my mustache, but you let the guys run around in PT gear all afternoon in violation of policy. Does this mean that from this point on we are going to enforce policy?"

If you can think of a good comeback, please share it, because this member has cornered you in an indefensible position.

The lesson here is that if you want to establish clear expectations, you had better practice fair and universal enforcement of policy. You can't pick and choose which polices you enforce and which you don't enforce. Your actions are much more powerful than your words.

This leads to the next lesson: Support your expectations with your personal actions. I made a point to wear my turnout gear and usually my SCBA when I was in command, even when I was working the board at the back of the battalion. Other battalion chiefs would get irritated because of the comments I would get from the firefighters about being the only chief who wore gear on scene. These included, "It's not practical for me to wear my gear when I am in command," and "I'm running command, not fighting fire!" To this day, I encounter resistance, even from some of my officers, who would rather stand at the board in their street clothes. They can do what they want, while I do what I know to be right, and when I have to call someone out for not wearing personal protective gear, it can't be said by anyone that I didn't wear mine. What you do, not what you say, is the ultimate test of your credibility, reinforcing your expectations. Finally, I confess that there was a hidden motive for being in my gear: when an operations officer or fire attack officer was needed forward at the fire (away from the command board and up where the action is!), by already being dressed, I was naturally the perfect candidate.

Keeping the Kittens in the Box

If you're going to keep the kittens in the box, you'd better have a strategy for doing so. I have boiled it down to six critical steps to establish a strong leadership role:

1. *State your expectations of members early.* Make sure your members know what you expect and what you stand for. Nature abhors a vacuum; thus, if you don't fill up the expectations bucket, something or someone else will. You can't expect people to follow you if they don't understand what you stand for; moreover, if you plan to hold people accountable, they must be aware of the standard.

2. *Identify your members' expectations of you early.* If during initial discussions, you find that your and members' expectations are divergent, resolve that immediately. Doing so will also give you a good feeling for how (legitimately or falsely) your members are going to judge you.

3. *Enforce policies universally and fairly.* As an officer, you don't have the luxury of choosing which policies you want to enforce. If a policy is foolish, not enforceable, or doesn't meet your needs, change it through the proper process. When you pick and choose what to enforce, you risk creating favoritism and gaining a reputation as someone who sides with the most favorable position as it benefits you.

4. *Maintain open and constant communication.* Take the time to reinforce your expectations, to live up to the expectations of your members, and to make changes when necessary. Use failures or policy breaches as moments of learning, to teach officers what should have occurred and reinforce expectations. Listen to your crew, and if changes need to be made, especially when they pertain to you personally, make them. If we take the time to listen more, talk less, and communicate frequently, we can avoid

misunderstanding, stop the snowball effect that silence has on issues and problems, and effect meaningful change.

5. *Plan effectively.* We do a pitiful job of teaching our officers how to plan, so when our expectations are not met, we often blame them for not keeping on top of matters. If you're going to accomplish all your tasks, you had better be able to plan, track, and account for what you do and who does it. That means you must be able to track multiple projects and project timetables, establish beginning and ending dates, and manage your day-to-day activities in support of the shift, the customer, and the organization.

6. *Mentor rookies and personnel with a need.* You've already read about the importance of being a talent scout as an officer. If you want people to meet your expectations, you may need to provide training, mentoring, reinforcement, and assistance. If you provide these consistently to improve people's talents and their commitment, you'll help the team to meet your expectations. Likewise, if you continually do this and someone just doesn't get it, there's no harm in helping that individual find another job somewhere else.

These are the fundamental steps to take in keeping the kittens in the box. If you're smart, you will do this up front, to manage effectively throughout your leadership career; it's much easier than trying to do it after the kittens have gotten out of the box and are crawling all over the room.

Giving Orders

In my entire 26-1/2-year career with the fire department, I don't think I ever said "That's an order" more than three times. Although we all give orders every day, most of them are phrased such that we get them done without pointing to our collars. When you say "Truck 9, open the roof," that's an order. When, during you morning meeting with the shift, you say "Today we need to get the vehicles done, we have windows for the month, and I want to get about five inspections in this afternoon," that's an order as well. Most of the time, we call them directives or direction, and very seldom do we yell, "That's an order."

As a leader, giving orders and supervising are included in your job, so when you give an order or directive, you are keeping the kittens in the box, so to speak. To better understand orders, we need to break down what an order accomplishes and how to use them.

If you are directing a work project, for example, it might be wise first to give an example of what is not expected. If you are painting the station, you might indicate that the current walls look horrible. This does not manage how it should be done, but instead how it should not be after it is redone.

Consider the following aspects of orders and their content:

Intent

This is what needs to be done. On the fire ground, it might be an indication of a strategic, tactical, or task-level direction supported by people or apparatus; around the station, it might be a general overview of what needs to be accomplished for the day and in what the order.

End state

This is an indication of what the end product, the outcome of the action, should be. On the fire ground, it might be "search the second floor, or open the roof," but by your direction, you indicate what you want the end product to be. If you send a truck company to the roof, you might, on the one hand, want them to ventilate it; on the other hand, you might want them to simply open up a second means of egress. Make sure everyone knows what the end state of an order is supposed to be.

Main effort

The main effort is the most important item that needs to be accomplished at a given time. The main effort is important because everyone on the team needs to know when and how it changes. Anyone with military experience, especially in combat arms, knows that if you are engaged with the enemy and they attempt to flank you, the main effort becomes to stop the flanking maneuver. Around the fire station, time and process may dictate what the main effort is on a given day or week. Suppose you have five tasks to accomplish over the day, but at 1300 hours the fire chief and the city council are coming to cut the ribbon and dedicate the new toilet (that took you five years to get). If the entire shift realizes that 1300 is the witching hour, then at about 1230 they will all (hopefully) converge on the main effort to get it done despite everything else that is going on.

Suppose you are engaged in a multifamily fire with multiple lines in service and searches going on and suddenly you get the message "5-B to command. Mayday, Mayday, Mayday," followed by air, situation, location, intent, and personal accountability report (ASLIP): "We are trapped on the second floor by a collapse, we are at 50% air, we are in the A/B corner, and we are going to attempt to make the rear window for escape!" Has the main effort of the fire ground just changed? Of course it has, and everyone on the team had better know that, as well as what role they now play.

Rules of engagement

A rule of engagement is not what to do but what to avoid doing in the process of doing whatever you are assigned to do. Although that may sound somewhat convoluted, rest assured that rules of engagement can save lots of property and aggravation if they are well understood and executed. Suppose there is a fire alarm from a four-story building of assisted-living apartments. The engine company reports "light smoke on the second floor, investigating," and you order Truck 3 to the third floor to check the floor above the fire. With your knowledge of what the engine company relayed, you might want to tell Truck 3 not to force any doors; this is because after about two minutes, the engine company will discover the burned-up toast in Mrs. Smith's toaster oven (about $1.50 worth of damage), but you will have caused five thousand dollars' worth of damage on the third floor by forcing doors and K-tooling locks.

As a young lieutenant I learned this the hard way. We responded to a structure fire, and according to the map, there was a hydrant right in front of the house. On approaching the block and noticing the heavy smoke (that's a clue), I failed to see the hydrant where I had expected it to be. Not wanting to be without water, I directed the operator to hook up to the hydrant at the corner, and we laid 500 feet of five-inch-gauge hose. After extinguishing the fire, I started searching for the elusive hydrant and found it hidden deep inside the owner's oleander bush. As I was on annual leave as of the next shift, I instructed my shift to "clear out the bush next shift, so we can see the hydrant" Not that I expected the house to burn twice, but we still needed the hydrant clear.

What I failed to indicate (breaking the rules of engagement) was what not to do. I didn't tell them "Don't hook the brush truck up with chain and pull the bush out!" In the end, the hydrant was as visible as any I have seen; however, Mrs. Smith was steamed, the toilet flushed, and Lieutenant Sargent spent $24.95 on a new bush and learned how to plant. Boundary lines and constraints are important to recognize.

The Bank Account

Stephen Covey, in *The 7 Habits of Highly Effective People*,[1] suggests that everyone has an emotional bank account. As a leader, you need to make as many deposits in these accounts as you can, because eventually you will have to make withdrawals. The question is, How do you make deposits? Once again—like everything that has to do with leadership—it's about actions. You make deposits in people's bank accounts through visible actions that have an impact on situations, people, problems, or organizational processes. This is one method to keep the kittens in the box.

Go first

One way to make a deposit is by going first. That may seems vague, but there are numerous examples that of very simple actions that just require you to think and then go do. Suppose today you are short staffed at the station, and it's station cleanup time. Then get out of the office and go empty trash cans, sweep the bays, and assist the team in whatever tasks need to be accomplished. Admittedly, there are days when you have to stay in the office to do planning or complete paperwork, but when necessary, get out and lend a hand.

I remember when I was a battalion chief doing hydrant maintenance with one of my engine companies. It was August in Virginia Beach, and it was hot and humid. All the firefighters were out painting, greasing, and clearing away the debris, and I noticed the company officer sitting in the front seat of the fire engine. At first, I thought he was marking the hydrant on the map before getting out, but I pulled over anyway, to take a closer look. I then watched in astonishment as he simply sat in the cab and let the others do the work. Such actions will not endear you to your members.

On another occasion, I was deployed to the Atlanta Olympics as a task force leader for Virginia Task Force 2 (VATF-2). We were locked down in the Cobb County Fairgrounds on a 10-minute recall, so that we could respond by ground or air to events at any Olympic venues. (The night after we demobilized, the bomb went off in Olympic Park. How's that for timing?) Alongside three other task forces, we trained, worked on operations and ate food that had been catered in for the Forest Service. After a week of this, VATF-2 needed some motivation and a morale boost, so we went to the Incident Support Team (IST) command tent to ask permission to have a pig pickin'.

Now a good old Virginia pig pickin' is done one of two ways: entire pigs are slow cooked over a charcoal fire and basted in a barbeque sauce made of vinegar, rock salt, and crushed red pepper; or you can use Boston butts, an easier method. Our intrepid crew asked permission for the pig pickin', but we were told no for several reasons, the first and foremost being that we did not have any certified food handlers. To this, we had a reply already: "Oh yes we do. In fact we have four." (Rule one is to always anticipate the questions. Rule 2 is to have people on your task force who do this part time!) The answer was still that we could not have a pig pickin', but now the reason given was that the Metro-Dade Task Force (FLTF-1) had had a pig pickin' and buried three pig carcasses on the fairground properties, the disposal of which had become a problem they didn't want to have repeated. We replied that since we would be using Boston Butts, there would be nothing to worry about. However, the answer remained no, because of the pig carcasses buried on the fairground by FLTF-1.

Now, I am not an exceptionally smart man, but I could see that I was getting nowhere through arguing, so I decided to negotiate instead: "If I take care of the pig carcasses—if I dig them up and get rid of them—then can I have the pig pickin'?" There was a stunned look, but no answer, so before they could digest the proposal, I took that as a yes, collected my staff, and scurried back to our base of operations (BOO). Once there, I said, "Get the component managers together and have Mackey start building the cooking pit!"

Now, first of all, picture the scene. We were in Georgia, where it had been 105°F every day since the pig carcasses had been buried more than two weeks ago. Our job was to dig them up, put them in very large plastic bags, and dispose of them. With this unwelcome task at hand, I gathered the task force and told them, "I have good news, and I have bad news: the good news is, we can have the pig pickin'; the bad news is, before we can do that, we have three rotten pigs to dig up!" So, who had to be the first one with a shovel in the ground? You bet: me. I took one shovelful, threw it, and ran away, screaming back to the rest, "The Dumpster is in the back!" The point is that you have to go first—first in the building, last out—both to help and to show your team you will do what they do.

Stay in touch

The next way to make deposits in emotional bank accounts is by learning how to stay in touch. Depending on what level of the organization you are at, you must use different techniques. I can tell you this, though: the most powerful way to keep in touch is by your physical presence. I say so because if you promote to staff or to a high-ranking officer's position, you are going to find it very difficult to get out and see people in operations. You must, however, ensure that you find the time to do this, at all costs. I remember when I was promoted to division chief, I made a pledge to myself that I would get out and see people. I put nights and some days aside for just that. The most telling thing was that the first five stations I visited, the first words from the firefighters' and officers' mouths were "Who's in trouble?" In law enforcement, they would consider that a clue—about what when on in the past and about what your members expect.

As a company officer, every morning when you come to work, you should dedicate at least 15 minutes to getting the shift together and giving a daily briefing of what you want to have accomplished during the shift, to make assignments, and to go over any directives or e-mails that have arrived since the last shift. You also need to take the time to look at your folks—for example, making sure that Steve Lesinski didn't come to work sick, because it's in his nature to show up no matter what. The briefing provides an opportunity to keep in touch and make sure the ship is headed where you want to go, and it also gives you a chance to keep your personnel informed as to what, when, where, and who. This is a surefire way to keep in touch.

There are dozens of other methods to connect with people and let them know you are there. When an individual team member is going through a personal triumph or a crisis, send a congratulations or sympathy cards. When there has been a breakdown in communication, provide a rumor-control document outlining what has been going on and put together the pieces of the puzzle that each shift, person, or group has; fill the vacuum with truth before rumor and innuendos fill it.

Make meaning

Find a way to make meaning daily. This is really the essence of leading and, in some ways, is the core message of this book. People want to be part of something good, something worthwhile; more to the point, they want to contribute to that end. You must create an environment that allows people the freedom to use their talents, to learn new and challenging things, and to thrive, enjoy, and have fun with their work. This is especially true in the fire service, where we can be faced with horrific inhumanity in a mere instant.

Teach through stories

Become a storyteller, for your own well-being and for the safety and survival of others. Teaching with stories has been done for centuries; in ancient times, people traveled the known world to tell tales and spread news around open fires and large stone fireplaces. Those open fires and hearthstone places have been supplanted today in the fire station by the kitchen table. Use stories to impress upon our new members the necessity of doing certain things—to relate how doing something one way may save your life while doing another may cost you your life. Use your experience by spending the time to share it through stories.

A Story I Will Never Forget

I lost many dear friends in the World Trade Center terrorist attack: Ray Downey, my mentor for over 15 years; Chris Blackwell, from the rescue companies and a good friend; Tommy Langone, from the Emergency Services Unit (ESU), not just a friend but one of my instructors for Spec Rescue; Jack Fanning, a chief with the Fire Department of New York (FDNY) and a member of my FEMA IST; and the list goes on far too long to print. These were sit-alongside-my-family, eat-with-me friends, who on any other day, in any other location, would have been right beside me doing the same thing I was doing on those dark days.

I spent lots of time in the first weeks telling surviving members of the New York City Police Department (NYPD) and FDNY—some of whom I knew and others I didn't know—how sorry I was. But after a few weeks, it just didn't have any meaning anymore, so I stopped.

One early morning, around 0230, we shut down the pile for heavy-equipment operations. I was physically and emotionally beat up, so I decided to take a break and headed for the 10 engine/10 truck fire station. I headed up to the roof to have a cigar. On the roof was a FDNY firefighter who was probably 55 years old, and it was readily apparent he had been on the job for a long while. He was sitting in a chair looking out at the pile, with the same idea I had, taking a break and decompressing. Not wanting to say I was sorry, I offered him a cigar, which he took. I cut it, lit it, and sat back puffing my own Macanudo when he began to speak.

"You know," he said, "we lost 343 people here. I lost uncles, cousins, family, friends I went to school with, served with in companies, taught with, and went to the academy with, and it's almost too much to grasp, an incredible tragedy. But you know what else we lost though? Fifteen hundred years of experience is buried in that pile. You see," he said, "in New York when you fight fire in the Bronx, or Manhattan, or Harlem, or Hell's Kitchen, or Brooklyn, it's all different. We can only teach you so much in the academy, so we depend on the experienced guys to take the young guys and teach them, tell them stories of what they have seen and how the buildings in their area respond to fire and collapse."

He went on: "If the people buried there did not take the time to teach, to tell stories to give the information to the young members, it's lost forever, and we will never get it back. This loss is almost too much to bear, but I wonder what our fire fatality rate for firefighters is going to be in the next 20 years if the members who were killed here never gave up the information to our younger kids?"

I was stunned to the point of tears. What a prophetic observation that was—and what a poignant reason to make storytelling an everyday event. This point did not hit me until many weeks later, but it is a story from a firefighter whose name I don't even know that I will never, ever forget!

Reflect on moments of learning

Finally, teach through moments of learning. Every time you return from a response, take the time—no matter how small or trivial the call might have been—to ask yourself and your crew, What could we have done better? The answer might be just a little thing, like a little less kink in the soft sleeve or a quicker way to get up the stairs or to the incident itself, but use good and bad events as learning endeavors. History has a way of repeating itself, and if you learn from history, you are will be a survivor. For readers who have sat through Chief Billy Goldfeder's course or looked at FirefightersCloseCalls.com, what they are giving you is history, using events as moments of learning. There is no better way to learn and to stay alive and be successful at the same time.

Big Stuff, Little Stuff—You Would Be Surprised

This is a story about how failing to keep the kittens in the box cost a firefighter his assignment. There are lots of lessons to learn here, so pay attention. I don't have every detail of the event, since it was a personnel issue and happened just as I was retiring. However, I pieced together enough from many of the parties involved to recognize that officers made mistakes that cost someone else. The names have been changed to protect the innocent and guilty both.

It all starts with emergency medical technician (EMT) recertification. As the department had done for almost 18 years, it sent notices along the chain of command and directly to individual firefighters about the upcoming two-day EMT refresher course that they were required to attend. As fate would have it, about half a dozen of our intrepid members did not show up. Since they were not at their assigned work location when they were supposed to be, they were, for all intents and purposes, absent without leave (AWOL).

The first question you should ask (if you have been paying attention) is, "Is this a problem with the system, or did these folks simply not show up because they didn't plan it, mark it, and follow through?" You don't want to hammer people for an organizational misstep. In this case, however, the notices were sent to both the chain of command and the individuals (just as they had been in the past), and the firefighters simply failed to show up—AWOL.

The next question is, "Is there a policy that deals with AWOL?" As luck would have it, there is a clear and precise policy that outlines the ramifications of not showing up where you are supposed to be, on time and at the right location. The first offense results in a written reprimand. For consistency, everyone should be subject to the same consequences, except in cases of illness or other true emergency precluding attendance; this is something that you should also check out, as in fact it was the case in this particular example.

The unfortunate thing is that the frontline officers of the members involved did not see it that way and instead of simply handling it at their level and addressing it, they thought of every reason why there shouldn't be consequences for their actions. In and of itself, this is an issue, but it also speaks to the culture of, "Take care of our members at any cost, even when they are wrong"; this mentality has permeated the fire service for hundreds of years. No officer likes to discipline members. I certainly didn't like to, but if you take an officer's job, you can wager that if you are doing your duty and your job as a supervisor, eventually someone is going to make the wrong choice and will have to pay for adult choices with adult consequences. This is the first chance to keep the kittens in the box. Failure to address the problem properly at this point lets the kittens out into the living and dining rooms, from which they can move freely about the house.

The subsequent problem is that the issue gets kicked upstairs to the battalion and division level, which is never good. A global look at the issue addressed two points: (1) Does the policy apply, and were these people AWOL? This is a basic supervisory question that is now being asked two levels above where the question should have been asked. (2) If the answer is yes and there is a policy violation, then for consistency, everyone for whom this is a first offense should get the exact same discipline—namely, a written reprimand.

The division chiefs (myself included) reviewed this, discussed it, and came to a consensus on the basis of all the information we have that our members were AWOL and that consistent discipline needed to be administered. Why we had to come to this conclusion is another story; it should have been reached long before it got to us.

Are you with me so far? Because the story gets even more interesting very soon. So far, we have members who made choices and frontline officers who did not want to do what was necessary to correct these choices in accordance with policy; we have a problem that has been kicked upstairs to the division and battalion level to be evaluated and handled; and we have the decision that, yes, these folks were AWOL, and discipline is appropriate. So the chain of command directed the captains to issue written reprimands to the members who failed to show up.

As you can imagine, this caused quite a stir, since the officers did not want to do anything in the first place. They disagreed with the decision, and some emotional conversations followed between captains, battalions, and divisions. After all was said and done, the direction was given, and once the decision has been made, you don't have to like it, you just have to do it. If employees feel unjustly disciplined, a process by which they can remedy this; this is called a grievance procedure, and it is in place to protect employees from potential abuses of power.

Enter Smertz, from the hazardous materials (Hazmat) team. An acting captain, on the promotion list, due for promotion, and an excellent employee. Fine firefighter, excellent instructor, no past history of problems, a quality guy. It just so happens that he is the acting supervisor of one of the members who did not show up and has been ordered to write the reprimand, with which he still does not agree.

Now enter his battalion chief, Snodgrass. Captain Smertz writes the reprimand, but puts on the reprimand—in writing, mind you—that he does not agree with this action. The rule as an officer is that you can disagree, you can discuss, you can even argue behind closed doors with the people, but you do not ever as an officer put on an official reprimand or other discipline that you, the officer, do not agree.

First, this sends the wrong message: it says that you are not willing to do you job. Behind closed doors, you can tell your employee that you don't agree, if you are so inclined. You can advise him or her of the right to grieve the reprimand. You can even help him fill out the paperwork, but you cannot countermand a disciplinary action in writing at the same time you are giving it.

Because of their affinity for their members, frontline officers sometimes, especially when they have not been trained effectively, make this mistake. I have made it, others have made it, and it always has the same outcome. That is why we have battalion chiefs—to be a little wiser, a little smarter, to protect us from ourselves when we so something that will cost us, regardless of right or wrong.

In this case, however, Battalion Chief Snodgrass did not do that. Instead of seeing that statement and saying to Captain Smertz, "I know what you are saying, and I cannot say I disagree with you, but the decision has been made. You cannot send this paperwork up with that comment on it. Take it off," instead he said "I agree. Leave it on there" (at least, his actions indicate that). Here is the second chance to keep the kittens in the box, as well as a potential moment of learning. Now the kittens are out of the box, out of the room, and heading out the front door into the yard!

Battalion Chief Snodgrass failed to protect Captain Smertz by not teaching him what he should and should not do and what he can and cannot get away with. As a result, however, it will not be the battalion chief who pays for this error in the long run.

To make matters worse still, Battalion Chief Snodgrass fails to follow the process and copy this to the chain of command. Consequently, it goes right by the division chief (a third lost opportunity to potentially herd the cats back into the box before they run into the road) and lands square on the fire chief's desk. After this gives him a minor transcendental ischemic attack, he asks the deputy chief "What is this, and what is an officer doing writing this?" And as we saw in chapter 1, the toilet flushes.

So, what's the end result? The division chief has a mild infarct, transfers the acting captain to another station, and the station, the shift, and many department members are left to wonder what happened? The kittens made it into the street, and were all run down by a tractor-trailer.

What are the lessons here about keeping the kittens in the box?

- *Officer training is critical.* The first step to keeping the kittens in the box is to define the box for them. You must teach your officers—acting or not—what is expected of them and what their duty is. They must be willing to evaluate and handle policy violations at the lowest level. Failure to do this, to make the adult choice, takes it out of their hands.
- *Battalion chiefs are supposed to teach and be smarter than that.* The battalion chief should have had the captain remove the statement. Agreeing with the acting officer in a show of solidarity and allowing him to send the paperwork up in its current form is tantamount to allowing him to execute himself. The battalion chief could have saved the acting officer from being penalized for making a boneheaded error in judgment. Instead the battalion chief exercised his own poor judgment, and the acting officer ended up being the sacrificial lamb.
- *Give a second chance.* I am not sure under what parameters the division chief was working. His orders to transfer the acting officer may have been given to him from well above, or he may have made the decision himself, as a show of power or based on past performance. The question here, with regard to deposits in the emotional bank account, is, Would it have sufficed to give both the acting officer and the battalion chief a good chewing out, behind closed doors, and then returning them to their jobs with a firm set of future expectations and a direction of "Now go back to work, and do good things"? In other words, did we really need to sacrifice the acting officer?
- *Officers are supposed to be filters.* Officers must evaluate choices made by employees and the consequences of those choices. As an officer, you cannot support and defend actions that are, for a lack

of better term, wrong. Your job is to filter and teach what is right and wrong (as a parent does) and to deal with it at the lowest level. If you are a company officer or a battalion chief—or for that matter any officer—and you continually defend actions that are nondefendable, you will lose your credibility. Then, when your members really need you to stand by them on a righteous issue, your support is dismissed as unconsidered and meaningless.

Sometimes not keeping the kittens in the box has consequences for members other than yourself. You job is to set expectations, to enforce policy fairly and honestly, to make daily deposits in the emotional bank account, to stay in touch, and to use every day as a moment of learning for you and your members. Herding kittens is tough work, requiring patience, experience, the willingness to make the tough calls, and a desire to teach even when some pain comes with it. The real problem is that when we don't keep the kittens in the box, it's the kittens that end up getting hurt.

Notes

1. Covey, Stephen R. *The 7 Habits of Highly Effective People*. New York: Free Press, 2004.

Chapter 6
Maintaining Technical Competence

It's not what you know that will get you into trouble; it's what you know that isn't true.

—*Will Rogers*

Education—What's It Worth?

For years the fire service has debated the value of education, both formal and informal, and there is no doubt that both have benefits and detractions. Education can better be broken down into two categories: training and experience. For success, there must be a unique and balanced blend of both. Classroom lecture without any didactic component—touching, seeing, and doing—is a poor method of learning. Likewise, training in practical skills without providing some foundational knowledge, including the theory behind the practice, creates robots with no thinking skills.

It is fair to say that all the degrees in the world alone will not make you a good fire officer or firefighter. Nevertherless, the discipline to undergo some formal education is required in order to get ahead in the modern fire service.

Two essential characteristics of a great fire and rescue service are training and discipline. In successful organizations, these characteristics are apparent in every achievement in both the emergency and the nonemergency setting.

If training is so important, why is it so often neglected? Why, in many organizations, are other many other day-to-day operational aspects of life in the fire service accorded higher priority than training? Even when these organizations train, they focus on advancing hoses, ventilation, and extrication, among other operational necessities. Seldom do they address the training in personnel issues, administrative procedures, maintenance, and safety.

Training and education are not insatiable enemies that consume an individual shift, officer, or organization's time. Training should be embraced as a means by which the organization, its members, and its officers can explore the diverse problems and missions that confront them. To be a successful officer, you have to understand and be able to function on the fire and rescue ground strategically and tactically. That capability comes from formal training and experience in the operational setting. Together with this, you must also have the management skills needed to survive amid the bureaucracy and to complement your ability as an officer.

Credibility—Earning It and Keeping It

If you expect members of your organization to follow you into extremely dangerous situations as a frontline operations officer or to stay with you for the long haul as a member of senior staff, you had better have their trust and respect. In other words, they must be aware that you know what you're doing. This is a fundamental principle from the ABCs of leadership, which were discussed in a previously chapter. If you are the chief of an organization, your members had better believe that you know what you are doing administratively, politically, financially, and operationally and that you have their well-being and best interests in mind. If you are an operational officer, your members had better know, in their hearts, that you are a sound strategist and tactician and that you know what you are doing, even down to the task level, on the fire and rescue ground. If they do not believe this, you will lose or fail to gain credibility. There is a simple formula for acquiring credibility on the basis of technical competence, derived from training and experience.

Trust, Respect, and Other Concepts

"I believe you are doing good things for me and for the organization." This statement provides a concise definition of trust. In other words, I believe that you are doing good things for me and the organization, no matter what—whether you are leading down a long hot hallway or whether you are fighting for me at the budget table. I am willing to follow you right until the end. I trust you because I believe that you have the technical competence to do the job and that the decisions and choices you are making will do all they can to protect me, make the organization better, and fulfill my needs, as well as those of the organization.

"I believe you know what to do and are doing it." From a fire and rescue ground perspective, respect means I believe you have identified the hazards and the chief problem and that you are applying the strategy and tactics required to resolve the situation. I will gladly follow you and implement these tasks to meet that end, even at great risk to myself.

Your credibility exists solely because people believe you do the right things on the basis of technical competence, derived from training and experience. Thus, to anticipate what might happen next and plan, train, and prepare accordingly is a sacred trust. The primary task of a frontline officer is to ensure that the crew is ready to respond to and safely mitigate any event. In the fire service, training is not just about the company level: from a leadership perspective, it must be embraced by the organization and its leaders, and sufficient resources, both fiscal and human, need to be allotted to ensure that it's the number-one organizational priority. People at senior levels of the organization are tempted to cut training (and the associated costs) when times are hard, and that is a critical mistake, driven by a lack of competence and a lack of training and experience on their part.

Job Maturity

Some people equate time on the job with competence, which is a big mistake. Rather, competence is having the KSAs to do all of the job, not just part of it. When someone says he or she has 20 years on the job, this poses the question, "Do you have 20 years 1 time, or 1 year 20 times?" Some people just haven't kept up, have retired on active duty (ROAD), and have flown under the radar, and unfortunately some of these people are officers. Certainly time on the job should (and most often does) equal training and experience; nevertheless, I know some folks that have 20 years' experience and I wouldn't let them touch a booster line, let alone anything else—and we took booster lines off our rigs 15 years ago!

Credibility as an officer is largely based on expertise, and that expertise is displayed in two ways. First, it's displayed in your actions that are required of your job—not just the emergency aspects, but all the administrative, human resources, planning, and other duties that are required of you. Second, it's displayed through how you train your personnel, the methods you use to pass on the KSAs your members need in order to grow, thrive, and survive; ultimately, it's about training first and then experience. If you find yourself assigned to a specialty company, you will end up doing double duty, keeping all of your specialty skills honed while still practicing all of the basic skills that you might not use all the time.

If time and staffing constraints were to prioritize service delivery, a list could be made of the three critical aspects of the job a fire department does:

- *Response.* Go out and take care of the customer.
- *Department training.* Prepare every day for what you do know and what you can expect to encounter, as well as for what might happen.

- *Physical training.* Keep the service delivery unit (the human) in shape; after all, the fire service and EMS is a cardiovascular, muscular-endurance, and mentally challenging team sport.

One complaint I hear from my volunteer brethren is that "We don't do enough training, and it's not realistic enough." If you're a volunteer, one method to keep your troops engaged, interested, and safe is to create drills that are realistic and meaningful. Engage the personnel in learning and keep them sharp on the basics, and the outcome around the station and on the fire/EMS ground will be much better.

Another feature of the volunteer fire service is the way in which officers are appointed, elected, or otherwise selected. I originally rose through the volunteer ranks, and after I retired, I went back to the volunteers. Since election is the most commonly used method of selecting volunteer officers, I urge all volunteer companies to set minimum qualifications. Not just anyone should be able to run for operational or administrative officer positions. If you meet the minimum qualifications, say firefighter 1 and 2 and officer 1, then you should be eligible. Anyone who has not shown up for training in 2 years, or has certificates that are 10 years old with no new training, or lacks certificates should not be eligible for the voting.

I just recently reviewed a new book on leadership by one of the brightest minds in our business. After one section is a story about a group of lost firefighters in a commercial structure who were following a hose line out, running out of air, and disoriented save for the connection with the hose line. The officer was absolutely convinced that he had led his troops further into the structure, into almost-certain death, until he came upon a coupling, found the female end, and realized that they were indeed heading the right way. Like a pilot following his instruments in deep fog and bad weather, when all his senses are saying no and his instruments are saying yes, this team was saved by the very basic concepts of training. Performing covered-mask drills in the station, following the hose, and reading couplings (not rocket science, nor even science) saved the entire crew. How profound these couple of hours in the station turned out to be. Yet it's a good bet that on the day of the drill, at least someone (perhaps several team members) was complaining about it.

No video and no lecture can provide this type of preparedness; this comes only by taking the attitude that you must get off the couch, put on your gear, get out in the hot (or cold), wet (or dry) environment in which you work, and put your hands on the business end. An officer's primary responsibility is for the health and safety of personnel.

One critical aspect of keeping your members as safe as possible is to "Train like you fight, and fight like you train." Survival requires muscle and brain memory. We do not always have time to think in situations where our lives are on the line; we must respond automatically or suffer the consequences of not recognizing what is happening and of failing to act quickly enough.

For example, suppose you are trapped and disoriented in a structure fire. If you have to think about your Mayday procedures—if you do not immediately and automatically recognize that you are in trouble, accepting it and immediately taking the necessary steps to mitigate the danger—you will certainly end up dead. The actions that should be automatic include calling the Mayday, activating your personal alert locator (PAL), declaring your ASLIP, and doing something to help yourself survive.

Readers with past military experience in combat arms or in law enforcement will be familiar with these as *transition drills*, from long gun to handgun. There is no time to think. Just do it right now; don't and you're dead.

Organizations must embrace training, just as they do leadership, for it to be successful. In the previous example, an organization would have to establish firm and specific instances in which personnel will call a Mayday, with no questions asked—without thinking maybe, maybe not. We cannot leave it up to our members to determine whether they will call a Mayday. As with a military pilot's ejection protocols, there can be no weighing the parameters, to determine what to do. If policy is not linked to critical training, too much is left to ego-driven machines who by their very nature will fight to the last moment, never admitting they are in trouble and never admitting they cannot deal with it themselves until it is too late.

Organizations fail to provide adequate training for officers, so it's no wonder that many of them do not have the skills to be trainers themselves. Smart organizations understand the value of well-trained and -schooled officers. This begins with an officer-candidate school for those who have made the list; this is followed, on promotion, by officer-development school, to provide management skills; and finally, the trend is continued by holding quarterly officer in-services, to teach new skills and establish expectations over the course of an entire career.

Getting into trouble

Organizations—and consequently members who are delegated either full-time permanent or acting supervision responsibility—are risk managers. They are responsible, like it or not, for ensuring the safety of their members and of the public and for ensuring that rules and regulations are enforced equitably all the time.

How does that relate to technical competence? As a supervisor, failing to understand the basic principle of technical competence places one in potential liability. Interestingly, you do not have to be present or directly involved to be liable. An employer is vicariously liable for negligent acts or omissions by employees over the course of employment, regardless whether such acts or omissions were specifically authorized. Vicarious liability (substituted or indirect) is the liability that any organization or its officers and supervisors have for the acts or omissions of those whom they supervise. This book is not intended to provide instruction on the law; however, as an officer, realize that you accept a great deal or responsibility, and because of the environment in which the fire service operates, your exposure to legal matters is significant.

Rules and regulations from a wide range of sources, including the National Fire Protection Association (NFPA) consensus standards, the Occupational Safety and Health Administration (OSHA) rules, state statues, city laws, and the National Institute of Occupational Health and Safety (NIOSH), exhibit some disturbing similarities. First, they are designed to provide a solid, accepted method of practice that will keep you safe, injury free, and clear of liability. Second, they outline how you can best accomplish that. Here are some common themes related to training from these documents that, if ignored, will get you into real trouble:

- *Negligent appointment.* This is a failure to check the background of a person, including his or her qualifications prior to employment or prior to an assignment that requires specific KSAs. It might also apply to allowing someone to do a job without providing the proper training.
- *Negligent retention.* This is keeping a person in a job or position that he or she has demonstrated an inability to do, once you have ascertained that he or she cannot do it. It may even be the case that you identified the problem and sent the person back through training, but he or she still could not effectively meet the required standard.
- *Negligent assignment.* This involves the assignment of a person to undertake a job that he or she cannot do, is not qualified to do, or has not been trained to do.
- *Negligent entrustment.* This is ordering or allowing someone to use or operate a piece of equipment or device that he or she is not adequately trained on or has not been trained to competently and safely use.
- *Failure to train.* This is when the organization or the supervisor does not provide training to personnel according to their duties, assignments, work tasks, and so on.
- *Failure to supervise.* This can be global in scale (organizational) or can be isolated to specific divisions or supervisors. In either

case, this results in members under your command who act in an unsupervised manner at any time, or at any place, or in any condition where supervision is required in order to maintain the safety of all parties involved.
- *Failure to direct.* This can result from a failure to provide rules, regulations, guidelines, SOPs, and/or instruction, as well as the failure to enforce the same, in relation to training and operations or other specified activities.

If organizations or officers allow any of these conditions to exist, they may find themselves liable and involved in criminal or civil action.

Liability—how we create it or avoid it

Here I'm going to borrow from Gordon Graham, who has provided the best description I have ever seen regarding risk management and why things go right and why they go wrong. All the tasks that we do can be divided among four categories:[1]
- High risk/low frequency
- High risk/high frequency
- Low risk/low frequency
- Low risk/high frequency

What does that have to do with me and with training? Consider that human beings achieve proficiency at things by doing them and by seeking instruction from people who have done them before. If you want to drive a golf ball a country mile, you had better be out there hitting balls all the time. If you want to be an ace shot with that Kimber 1911 you just purchased, you had better be putting in time at the range. In addition, you had better take some lessons—in golf or shooting and in anything else—from someone who has the credibility to teach.

There are two methods that we can use to do it. First, in a really busy company that responds to a half-dozen codes and another half-dozen structure fires each week, we will get pretty good. By contrast, at a less busy company or one whose call volume is focused on other aspects of the job, we had better spend a greater amount of time on training. Again, the first part of the formula for making things go right is training and experience, and what you lack in one you had better make up with the other.

Most of us lack experience or the levels of experience required in order to get really good at everything that is asked of us, so training offers the only stopgap to achieve a level of proficiency. How many of us can truthfully say that we see lots of working structure fires? For some people, this may be a high-risk/low-frequency event, whereas for others, it may be a high-risk/high-frequency event; I daresay that for the majority of the fire service, it is the former. Therefore, we must commit, as officers, to training that is *documented, verifiable, ongoing, and realistic*.[2] How do you think all of the military special operations teams operating around the world—and especially in Afghanistan and Iraq—are so successful? Had most of the operators ever been to war before? No, but they trained all the time using ongoing, documented, verifiable, and realistic methods.

So, the principle is that you can never get enough training. Unless you find yourself in a war zone where you are doing it all the time or the city is burning down—like the Bronx in the '60s, or Detroit in the '90s—you will never have enough experience except in very isolated areas of knowledge.. These aspects of training are critical not only for your safety but also from a liability standpoint, for both you and the organization.

One of my sidelines is expert witnessing, which means reviewing depositions, writing up case reports, reviewing actions of parties involved, testifying in courts, and, you guessed it, reviewing training records of personnel involved in these events. The very first things I ask the attorney to subpoena are the training records of the individuals involved and the training records of the individuals who trained them. You would not believe what I see; sometimes it so comical that it almost borders on criminal.

Documented training means that you have established an accepted standard and that you can cross-reference that training to some KSA level. It also entails having more than a simple roster with names attached; Documentation encompasses cross-referencing to a specific standard having the instructional outline attached to the roster for cross-reference.

VBF&R and many other departments realized just how important this issue of documentation was when the National Incident Management System (NIMS) requirements were disseminated. VBF&R has been using incident command capably for decades. The department continually trained and updated members and officers, developed sound policy, and worked hard at creating a culture of solid command. However, where we failed, not of our own accord, was in documenting the specific KSAs obtained through teaching and training sessions and therefore in equating those to specific standards. As a result, we had to spend money and time to put the entire department back through NIMS 100, 200, and (for a few selected members) 300—to train the trainer so that we could without hesitation show that we had met the requirements, had attached the KSAs, and were in compliance with the standard.

Verification involves matching your training to a verifiable standard, code, ordinance, requirement, or skill that is accepted nationally, regionally, or locally. Most EMS agencies have different protocols about what paramedics are allowed to do. Some agencies have much wider skill ranges than others, but the records of the individual and the organization always contain a verifiable track, tied to specific KSAs.

Ongoing means just that. You cannot be trained on a subject only once and expect that to suffice for your entire career. With technology being what it is, our world is changing rapidly, often quicker than we can keep up with it. The vehicle extrication course you took 15 years ago is relatively useless to you today with the different constructions, hybrids, fuel systems, and myriad other issues associated with modern vehicles. Structure fires offer many hazards not previously encountered. Laws and regulations governing the way we enforce codes and ordinances change. Doctors know this concept very well, for what they learned about diagnosis and treatment five years ago in medical school is old science already, and no one wants to go to a doctor who practices old science.

Realistic means you have to do training that is relevant, covers the KSAs needed to accomplish the job, and has relevance to the potential hazards or tasks that you will encounter. People don't show up to work and say "Today I am going to burn this house down," or "I am going to wreck the truck. I have spent all night thinking about it, and today this is my job." No, such things happen because we take well-meaning people, given limited training and experience, and put them in very complex situations and compress the time in which they have to make a decision; as a result, they make a mistake.

There is a lesson here for all officers, and it reflects both sides of the coin. First, when you are limited in the amount of time that you have to make a critical decision, it had better be made on the basis of training and experience. Second, when you are in a situation that is not life threatening and you have time to make the decision, take the time!

Here is where that failure to train effectively and realistically comes into play. When you arrive at an incident, your very first job, after ensuring that your members are going to be safe is to determine what the main problem is. That may sound simple, but how many times have you been dispatched to structure fires with people trapped, only to find a barbeque grill in the backyard? Determining the main problem is so critical because it drives all other decisions thereafter. Once you determine the problem, you choose a strategy, which drives tactics, which drive tasks. Moreover, when things go bad at the task level, bad things happen all because we failed, based on our training, to identify the true problem.

Suppose you show up at a structure fire and it's a two-story, wood frame, with heavy fire coming out of the second floor. The main problem appears to be that the house is on fire, so you direct your attack lines inside and upstairs to extinguish it. But wait, did we do what realistic and ongoing training tells us to do as an officer? Don't assume that the main problem is a house fire until you do your walk around and notice, Oh my, Mr. and Mrs. Smith are hanging out the rear window, with heavy smoke and heat pushing over top of them! Now, the main problem no longer appears to be a house fire; in fact, if you put a line in the room, you push the fire on top of Mr. and Mrs. Smith. Based on action and reinforced by training, you would determine that the real problem is a rescue. The lesson here is that ongoing training reinforces the aspects of our job that keep us from misidentifying the true problem and choosing the wrong course of action.

Finally, let's talk about how we train. Most career departments have a mandated level of training that must be accomplished each shift. However, I have also seen departments that have no required training initiatives, and these scare me. Volunteer companies deal with this quandary in a variety of ways—from required drill dates, to a specific number of hours required each year, to, you guessed it, requiring no continuing education at all. No matter what the process, you need to get rid of the training videos and take a hands-on approach. Holding a video drill once in a blue moon is okay, but as the mainstay of your training program, it is worthless. Computer-based training and virtual reality (VR) training do have their place and are extremely useful tools for training in incident command and large-scale events.

Continuing education should be two pronged. First and foremost, frontline officers and battalion chiefs play a critical role by developing and presenting real training and multiple-company training. Second, mass training is completed by the organization in the form of in-service. Consider the following concepts when developing a company-level and battalion-level training plan; above all else, realize that these should be conducted off the couch, with as much hands-on time as possible.

- *Train on the basics from time to time.* Covered-mask drills, following the hose line, ground ladders, water operations, drafting, ventilation, vehicle extrication, SOP review, and updates—any and all of the bread-and-butter items that will reduce loss of life or save your own life—make ideal drills. Base the commitment on how often you do these things in the real world.
- *High-risk/low-frequency issues.* Spend time on saving-our-own techniques, remote deck gun operations, RIT setup and activation, awareness-level reviews for special operations events, flashover simulations, and in some instances, live burns. Identify skills and responses that you have not made for a while and hone them.

- *Multiple-company drills.* If you're a battalion officer or even the company captain, spend time on drilling with multiple companies at the training center or at a location where you can run several scenarios.

Training is one of the most important aspects of any officer's job. Making it meaningful and realistic requires time and planning well in advance. On more than one occasion, the most innocuous training has saved firefighters' lives or the lives of those we serve. My blood boils when I hear "it's too cold to train," "it's too hot to train," "it's too dark to train," or "it's too wet to train." All of these are environments in which you will have to operate when the flag goes up, so why are personnel so reluctant to commit to training in those environments—because it's uncomfortable, or because of safety? Come on, you can manage safety and tolerate the exposure by being smart, but you can never duplicate the environment in which you will be expected to work from the inside of a fire station. Get up, plan, train, and make sure it documented, verifiable, ongoing, and realistic every time you train.

"People Are My Most Important Asset"

How many times have I heard that line? The reality is that it's not simply people who matter but trained, competent members who have documented, verifiable, ongoing, and realistic training under their belts. In the changing environment in which we work, people with limited skills and abilities need to apply elsewhere.

Many career and volunteer companies have significant problems with this. Ask most volunteer companies how many active members they have, and they will say, "We have 40 on the roster but only about 15 who are active and trained." Training is the key ingredient to saving lives and property and keeping our members safe. So the reality is that trained, competent people are your most valuable asset; if people were that important, then anyone could do the job, despite lacking any significant technical competence.

A Sad Tale of One Organization's Failure to Hold Itself and Its Supervisors Accountable

It was one of the slower stations, but it covered some valuable real estate. The officer on the shift was not known for his ability to train his people or for his tactical prowess. As fate would have it he has off this shift. The shift—composed of a master firefighter riding up front, supposedly trained; a rookie in the jump seat; and a questionable driver-operator—was in station when the alarm bells sounded for a house fire: Multiple calls came with reports of people trapped—the worst possible scenario for any firefighter. On arrival, the driver operator could not find a hydrant, even though it was right in front of him, because it was covered with sand! Of course, with the small number of hydrants in this response area, a hydrant book should have been made, but that's another story.

A girl was trapped inside the house, and her mother had gone back in to rescue her. The master firefighter jumped down from the front seat but provided no direction to the rookie, who heard the people in the building screaming. The driver-operator had only his tank water, and the acting officer gave no orders. The rookie, doing what he thought was right in the absence of direction, took a line to the second floor, on a wooden deck in front of a plate glass window. Thankfully, he had all his personal protective equipment (PPE) in place and his mask on, because when the fire flashed from below, it enveloped him on the porch and set him on fire. Coming down the steps on fire literally burned the soles of his boots off, and he was extinguished by a battalion chief who had arrived on the scene. The fire continued to burn, the people inside were not rescued, and a rookie was on the way to the burn unit.

The real tragedy is this. The organization fired no one for this grievous breech of capability in the most serious situation firefighters can find themselves in. They instead reprimanded the rookie for being in the wrong place, despite the fact that the acting officer was responsible for his well-being and for the development of the initial tactics. The driver-operator went back to driving and had a similar problem months later. The officer of the house returned, was never effectively dealt with, and continues to work there.

Why? you ask. Because even while this organization preached accountability and even "reach out and touch" people for policy violations, when it came to one of the most negligent acts that one can fathom, on the fire ground, they were afraid to admit anything because they recognized the liability of the organization, of the city, and of themselves. Put simply, it amounted to a cover-up.

So how do you ensure that people are trained, competent, and up to date in the real world? Admittedly, it's hard, because it takes time, money, and resources to ensure that the continual cycle of training is documented, ongoing, verifiable and realistic. Following are some suggestions about keeping your people trained.

At the company level

- *Ask.* Training is not just about fire and rescue skills; in this day and age, there a wide range of skills necessary to function at a professional level. Company officers need to sit down with their members and ask what training they need. Only by taking a one-on-one approach will you meet the needs of all individuals, through identification of their weaknesses and their strengths. You may discover some specific training needs that the individual would never confess to you in front of the remainder of the company. Keep a spreadsheet of each member's needs and work on getting them the training they need.

- *Career dig.* I would bet that no one ever asked you what you wanted to do in the department. That may sound funny, but there are many more opportunities than there are firefighters in most modern fire service organizations. If you can uncover the career aspirations of your members, you can more effectively get them headed in the right direction, adding to their portfolio of KSAs. Officer, Hazmat technician, inspector, investigator, training instructor, technical rescue team, paramedic, chief of department. If you can discover someone's goals, then as an officer, you can help him or her to get the training to accomplish those goals.

- *Teaching is learning twice.* Talent scout your members and assign them drills. Don't settle for poorly developed or initiated drills; instead, make it clear that you want training that inspires and challenges people. Use the skills, life experiences, and technical capabilities that people bring with them to work to enhance your training program. Give everyone on the job the ability to teach at the company level, and groom them to be better instructors, who can search for the necessary knowledge to present the program. In this age of Google and other Internet search engines, there is nothing that cannot be found, packaged, and presented.

- *Teach by assignment.* Rotate the administrative tasks between your members periodically, perhaps quarterly. This gives everyone the chance to learn the administrative skills, time frames, and processes that are necessary to the effective running of a station. By the time someone leaves your supervision, he or she should at least possess an understanding of every task that he or she might be required to undertake.

- *Dedicate.* Dedicate time for company training and physical training. These aspects are critical and should take a backseat to no other initiatives besides response. Plan effectively to ensure that priorities are met.

At the organizational level

Organizations are responsible for ensuring that we all know what we are supposed to do when we come to work. As silly as that may sound, you know what I am talking about when I say that we have the A-shift, B-shift, and C-shift fire departments. From a cultural perspective, that is somewhat unavoidable, but from a training and policy perspective, it certainly is not. What can organizations do to make training a priority?

- *Let them know what the main thing is.* Large organizations have so much to accomplish at so many levels that lots of details get lost in the shuffle. Despite strategic plans, if someone is not shepherding it and if we have not simplified our chief expectations, then we will fail to train effectively. The main thing encompasses those priorities that an organization believes to be the most important things to accomplish daily, weekly, monthly, and yearly in order to be successful. Training should always be one of the main things; it needs to be culturally repeated; and officers and senior staff must support it at all levels.
- *Adjust.* If training is important, then the organization needs to be willing to adjust as opportunities becomes available. Some training opportunities present themselves at inconvenient times, and often the organization or a person in a position of power is unwilling to change the schedule or plans to accommodate the opportunity. We expect to put as much planning and forethought as possible into organizing and permitting training opportunities, but sometimes events interfere: a building becomes available for training on short notice; training spots open up in a class or a brochure arrives for a program at the last moment; or perhaps something happens that dictates a need to focus on a specific area of training. Don't be afraid to adjust to get training accomplished.
- *Lay it out.* Years ago VBF&R had a firefighter in training named Mark Johnston. Mark left for greener pastures many

years ago, but when he was here, he developed the *Mark Johnston Training Manual*. It provided a set number of drills that every company, on every shift in the city, would do. There were always open shifts in the schedule, allowing the company officer to address specific needs, since every company is different. But the manual's most remarkable achievement was to ensure that we were doing basic things the same way, every time. Each month all of the ladder companies should be doing the same ladder drill or same rope drill, so that when we work together we are doing things the same. Each engine, squad, and medic unit should likewise have some standard drills that express how the organization wants specific skills done.

- *In-service*. There are skills or mandates that everyone in the organization must have. These may be legal mandates, or they may be high-priority skills that offer little room or margin for error. VBF&R was a pioneer in the development of in-service training programs. Every year you need to pull your companies in for live fire training, as well as for programs focusing on other mandatory skills.
- *Specialty training*. Specialty companies require specialty training. Ladders, squads, marine units, Hazmat units, technical rescue teams, and investigators, among other specialties, all require specific KSAs to function. You must support this in both planning and implementation.
- *Officer training*. This is the biggest gap that most organizations have. They do not train their officers before they are promoted, and after they have promoted, the organizations wonder why things don't go as well as planned. Have an officer-candidate school and follow up with continuing education for officers.
- *Rookie manual*. Every rookie coming to the field should have a standard set of KSAs in a book that the officer needs to ensure are accomplished before their rookie year is over. This ensures that everyone is doing the same thing and sets a standard that everyone will adhere to and be evaluated against.

Training requires a significant commitment of time, money, and resources and should be related to everything a unit or the organization may do. It's a priority second only to response.

Sharpening the Saw

Training is not just about keeping our KSAs and organizational direction. There is an added component to training and technical competence that enhances everything we do. Stephen Covey, in his book *The 7 Habits of Highly Effective People*, likens this to sharpening a saw.[3] In other words, although book smarts and technical skills are critical, if we are going to send human machines to see and smell everything while undertaking all the rough assignments that are expected of us in the fire service, we had better recognize other aspects that contribute to overall competence.

Physical competence

It's amazing to me that we lose more firefighters each year to heart attack and stroke, yet still people undercut the value of and need for a physical-fitness program. Firefighting and EMS make cardiovascular and muscular demands and require great endurance. Look around during conferences or even at your own department to see the sad shape that many of our colleagues are in.

To do the job, you have to keep in shape. That means you have to be able to pull your own weight for extended periods of time, recover only briefly, and go and do it again. I am not suggesting that you have to be Arnold Schwarzenegger or Hercules, but you do have to have a certain level of cardiopulmonary capability as well as some muscular ability to accomplish the physical aspects of the blue-collar job that is firefighting and EMS.

We eat terribly—and I am no exception: I love to eat, and I am sure that if I didn't undertake the kind of physical-fitness program that I do, I would look like Jabba the Hutt! I could stand to lose a few pounds, and I know that I have to make a commitment to do whatever is necessary to accomplish that. That is exactly what this is: a commitment to keep the machine in working condition. You would not run your car without oil, gasoline, or transmission fluid for every long, so what makes you think you can run the human body, in the environment that the fire service expects you to perform in, without keeping it well tuned.

Mental competence

Firefighting and EMS is a thinking person's game. You have to keep your mind sharp, to synthesize everything that gets thrown at you in times of crisis and make intelligent decisions quickly. Exercising your mind can be done in a variety of individually chosen ways that can be both mentally stimulating and pleasing: Reading, game playing, puzzles, writing, debating, and creative thinking are just a few options.

Spiritual competence

However you choose to achieve this, it is critical to your well-being. Whereas for some this is about religious connection, about praying and meditating, it does not have to be: it can be making a connection with the earth, or it can be experiencing inspiring works of art, literature, or people. Believe me when I tell you that good and evil do exist in this world, and you need to be prepared in your own way to recognize and deal with these forces, especially in the profession you have chosen.

Social and emotional competence

First, you have to maintain some sort of life outside the job. Never confuse your job with your life. One of the most blatant mistakes I have made in my own life has been to associate myself so closely with my job that there became very little distinction between Chase Sargent the fire chief and Chase Sargent the man. Find some social contact, connection and interest that takes you away from the job.

This is a highly emotional job. You are going to have to prepare your body and mind to see things that you cannot even comprehend. Suffering, injury, pain, emotional distress, and all the other devastation that humankind brings upon itself during our everyday work. If you are not emotionally prepared to deal with this, the life will be sucked out of you by the events you witness.

Likewise, you have to realize that the people you work with and around will need emotional support from time to time. Thus, you will again need to make deposits in the emotional bank accounts, as someone who understands what they have seen and experienced.

A Final Word on Training

One of the things I have been fortunate enough to do in my career is expert witnessing and casework. I can't even begin to describe every aspect of these cases, but they do demonstrate some very salient points. When I sit down and look at the depositions associated with cases, many times I cannot believe what I am reading: It often borders on the bizarre, mixed with the comical, blended with the ignorant, and covered in a slight sauce of tragedy. Although I am not an attorney, I would like to pass on what I have learned from being in the courtroom several dozen times, working on liability or personal injury cases.

First, whenever I review a case, I always subpoena the training records of the individuals involved in the event and of the individuals who trained the personnel involved in the event. As mentioned previously, I am ascertaining whether documented, verifiable, ongoing, and realistic training was provided the individuals involved, and I am trying to establish the credibility and technical expertise. I want to specifically see the following:

1. *What is the exact course content, and what are the KSAs?* I want to see exactly what was taught and what KSAs the student was supposed to leave with. The reason why I ask for this will soon be made clear.
2. *Are the instructors qualified to teach it?* I want to determine what qualifications the instructional staff have to teach this particular subject, including but not limited to professional qualifications (instructor certifications) and technical expertise if the course or subject is in the special operations field.
3. *What standard does this follow?* I want, whenever possible, to link the training guide back to an acceptable method of doing business.

I have discovered that it makes very little difference to juries—or to judges—that you are not an OSHA state or the NFPA standards are consensus standards. What they are interested in is whether the individuals involved followed some accepted practice that in their chosen area of work (fire and rescue) could be duplicated, accepted, and used in other locations in a safe manner. In other words, was the training or the standard that was used verifiable with regards to its national or regional cousins?

Finally, I want to establish how long it has been since the individuals practicing a given skill (or skill set) have been trained or retrained on skill. If the last training they had was 10 years ago and the accepted standards of practice today have since changed significantly (most assuredly, they have), then the training they have had is no longer valid, safe, and applicable to today's environment.

We talk at length about policy, procedure, rules, and guidelines in another chapter, but an important point bears mentioning here. When I am finished looking at the training records, I want to see the policy that outlined how the conclusions were reached that led to the implementation of the action that caused the incident. The policy had better be up to date, because it is not an acceptable defense to you say, "We haven't done it that way in a long time."

There is an old saying: "How can you tell there is going to be a tragedy in the Southern fire station? It's always precluded by these words: 'Hey y'all, watch this!'"

The more you sweat during peace,
the less you bleed during combat.
—Source unknown

Notes

1. Graham, Gordon. *7 Pillars of Liability and Success.*
2. Ibid.
3. Covey, Stephen R. *The 7 Habits of Highly Effective People.* New York: Free Press, 2004.

Chapter 7
Understanding and Enforcing Policy

> *Policy making is not as easy as slurping down cabbage soup. Caution should be combined with decisiveness. As the saying goes, before going into a room, make sure you can get out again.*
>
> —*Yegor Ligachev*

The Culture of Policy and Procedure Making

All organizations have policy and procedures. Throughout its history as a paramilitary organization, the fire service has been famous—and occasionally infamous—for policy and procedures. Times change and so do policies and procedures. This chapter is about learning, living, changing, and enforcing policy.

Because we live and work within an organization that has policy and procedures, we are personally affected by them. As a supervisor, one is expected to understand and enforce them. Moreover, in a very good organization, one may be asked to tamper with them from time to time.

Fire service personnel are curious by nature, and consequently, they are just a little bit sneaky. Inevitably, they—we—will try to figure a way around a policy: how to have influence and change a policy, how to interpret a policy, and how to fill the vacuum when no policy exists.

Policy is part of the organizational culture, and plays a big part in how it is created. Individual and sometimes group values may be injected into policy, which makes for an interesting mix. The question when this happens is, "Does the individual value align with the organizational value?" You cannot afford to send mixed messages.

Not all policy and procedure is fair, intelligent, or even wise. Sometimes it is driven by past experience or fear, and there is real opposition against changing how it has always been done. When individuals with lots of power get involved in the process of making policy, without the input of stakeholders or basing their decisions on personal values, there will be a stormy ride ahead. In terms of both the creation and the enforcement of policy, there is a saying that applies to such people: "Sometimes when you can't change people, you have to change people." In other words, sometimes people will not change, no matter how much you reason with them; sometimes people will just have to go away through retirement, transfer, movement, or demotion.

Perhaps the first major topic to delve into is how this creature we call policy came into existence. I guarantee that no matter what fire and rescue service you belong to, policy and procedure are crafted the same way. There are good methods to create policy, but again, there are as many bad methods to create policy.

First, understand that I am using "policy" as an all-encompassing word for everything we have to assist us in determining what we should be doing. To simplify this section, I have used policy to express an overarching creation that is usually in writing, that provides guidance, and that is expected to be applied universally, throughout the organization. In a later section of this chapter, true policy is differentiated from other types of directives.

Your name on this policy

It's no secret that many of the policies that are in place were created because someone screwed up. The "Won't make T-shirts" policy of Chase Sargent and the "Don't back the truck up without a backer" policy of so-and-so are prime examples. In general, such polices are created because the organization simply overlooked what should have been in place years ago; they represent a rapid and convenient method of imposing order, in response to a specific violation in a limited area. If you were to sit down with anyone in any fire station, in any city in any country, and go through the policy book, they would tell you, "That policy came about in 1995 when Bob Smertz did this or that."

This is a very poor method of creating policy. First, if a safety issue is at stake, it should fall under another category of policy making, as we explore later in this chapter. Second, the real solution to isolated issues, violations, and problems is to go directly to the source and correct the problem with the individual or the shift. This kind of policy making is inherent in the old mentality that supposes, "The beatings will continue until morale improves."

I'm the king or queen

When you are the boss—especially the big boss—you can pretty much do as you see fit. How you handle this speaks directly to your credibility and integrity. Nevertheless, you can do what you want, and you can impose your will on others.

I had a boss who came into the organization and immediately changed the grooming standards. His policy was, "No more Fu Manchu mustaches"— or any mustaches that extended below the base of the mouth. Certainly, that was his right as the boss. The problem in this case was how he approached the issue. Instead of simply saying, "I don't like that look," he repeatedly stated, in person and on the department video, that it was "an OSHA violation and a violation of the mask fit test requirements." Anyone with half a brain could tell that this was untrue, since fit test is qualitative or quantitative in nature. When called on this multiple times by members of his staff, he even when so far as to say, "I will write OSHA and get a ruling on this." Of course he never did, never changed his position, and well, it was apparent to everyone what was going on.

The moral is that when you are the king or queen, you have to be careful about using your power to fit your own personal agenda. Focus on the agenda and desires of the organization, as long as they are safe and prudent. Again, this is a horrible method to apply when choosing how to make policy.

Good sound policy

Most of these policies have to do with the health, safety, and welfare of our members and our customers. These polices make sense, they support the main thing, and are sound in their development and application. The following are a few examples of good sound polices or procedures:

- Incident management policy
- Mayday policy
- Operation of emergency vehicles
- PPE use and maintenance
- Vehicle maintenance
- EMS protocols
- Ride-along policy
- Tactical guidelines

Good sound policy is created with two principles in mind. First and foremost is the safety and survival of our members and our customers. Second, a service or product should be delivered in the most effective manner and in a consistent manner, to ensure that when we work together we are all singing from the same book. This is an excellent manner in which to make policy.

Statutes, laws, ordinances, and consensus standards

Many of our polices, especially on the administrative side, are driven by legal mandate or by liability concerns. The legal system has enacted laws to protect workers' health, work environment, and civil rights. The reason we have workplace violence procedures, equal employment guidelines, and sexual harassment polices has more to do with meeting legal requirements than with doing the right thing. Mask fit testing, apparatus design, and response guidelines ensure compliance with OSHA, NFPA, American National Standards Institute (ANSI), or NIOSH standards.

In many instances, legislative acts drive policy that is unfunded. As a result, the organization must comply while at the same time finding a funding source to support the initiative. For example, an NFPA consensus standard is applied to policy, construction, or operations because of the real chance that failing to attempt to comply will result in liability for the organization and its members.

This process can be an excellent method to create policy, since it mandates specific health and safety standards, as well as human rights and protection for our members. The pitfall, however, is the unanticipated funding required in order to abide by laws such as the Americans with Disabilities Act (ADA). We can complain all we like, but the organization must nevertheless comply with this method of policy development.

Crisis policy

Some people call them blood policies, or crisis/reaction decision making. This kind of policy results from a situation where a void in policy or procedure is recognized and is filled in after the fact. In some instances, it was unanticipated and could not have been predicted; in others, it was clearly identified but ignored.

I have a dear friend who is a union president in a major city. For years the union complained about the lack of response they received from their city garage in charge of repairing the trucks. The city ignored the pleas, until a ladder truck responding to a call lost its brakes, went careening down a hill, and crashed into a commercial structure, severely injuring several firefighters. Even after the accident, the city attempted to shift blame and responsibility. It was clear that a vehicle maintenance policy would have prevented the injuries and that this need had been brought to the city's attention multiple times. They now have a policy governing this process. It should go without saying that this is a horrible method to develop and implement policy.

Political policy direction

Sometimes policy is driven by the political system. This can be good or bad, depending on the type of policy direction and whose side you are on at the time of development and implementation.

For example, take a large city where the EMS system had been all volunteer for years. The fire service was running a large volume of EMS calls at the basic life support (BLS) level but had many firefighters who had received advanced life support (ALS) certification from the volunteer rescue squad or from other career or volunteer organizations. However, because of political considerations, the firefighter-paramedics were prohibited from acting as ALS providers when they came to work for the city, unless they volunteered.

Understanding and Enforcing Policy

As a consequence of this policy, many patients in need of ALS who could have been treated by firefighter-paramedics were not. This battle raged for years, and the city council and city manager hired a fire chief whose job was to keep the status quo and not rock the boat. It was obvious that the fire chief was not going to push the issue and acquiesced to anything that the volunteer system and the manager wanted. This action, an inability to work with the union, and other issues ultimately led to a political solution.

The union formed a coalition of police officers, firefighters, and teachers and elected 8 of the next 11 council members. Within two years, the department had their own ALS personnel, ladder company officers, and city third-service EMS providers on the street to support the volunteers. The tide of political support had changed, and neither the fire chief, city manager, EMS chief, or budget director would be able to stop the change.

It's unfortunate to have to make policy in this manner. If intelligent people with the best interest of the customer would sit down and discuss these kinds of things rationally, such a situation can be avoided. However, the nature of most political processes does not lend itself to street-level rational thinking.

I am sure you can think of even more ways that policy is created. Of all of these, only two are good methods to develop policy, yet all of them come into play. This will not change in the near future, so the question that begs to be answered is this: If this is how policy is made, then how can I influence policy direction?

Influence

How can I influence policy and have an impact on policy direction? The answer is that you have to get in a position to influence the decision. Based on location and position in the organization, we each have a certain level of influence, as well as a level of concern. While we would like to control and influence everything, we know that is not a rational thought. Again, I will fall back on Stephen Covey, who likens this to a circle: in the organization, we each have a circle of influence and a circle of concern (fig. 7–1).[1]

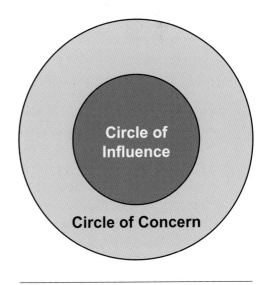

Fig. 7–1   Circle of influence

Your circle of influence can be bigger or smaller depending on rank, location, work environment, boss, and political connections. This is where you have control—or at least have a say in the things that go on and affect the area. Your circle of influence might be your engine or truck company, or it might be your work unit, battalion, or division.

Events that occur in your circle of concern affect your ability to get things done or directly influence your direction. However, for any number of reasons, you don't have control of influence over them. This can be frustrating, and it may constrain your ability to accomplish what you think needs to get done. The question then is, How do I expand my circle of influence?

Methods to Expand Your Influence

The key methods of expanding influence are as follows:

- *Promote*. Promotion ensures the expansion of your circle of influence. While it may not expand as widely as you might have hoped, the higher you go up the chain of command, the more say you have on events, people, policy, and direction.
- *Committee*. Benjamin Franklin once said, "If you live in a country run by committee, you had better be on the committee." If you want to influence the kind of breathing apparatus or PPE you purchase, then find a way to get on the committee that does the research and makes the final recommendation. There is an old saying: "Those that show up and speak up have a say, those that show up and don't speak up have no say, and those that don't show up have to live with it!"
- *Expertise*. Become the expert and smart people will seek your advice on critical aspects that influence events. When purchasing rescue rope and hardware, hazmat suits, or monitoring and detection equipment, you seek out the experts in the organization to make a recommendation. There is just one catch: Make sure that you become an expert in something that no one else in the organization already has a lock on.
- *Change geography*. Your physical location and work schedule have a lot to do with your influence. In the fire service, the everyday, nuts-and-bolts running of the business is done during the day. That's not to suggest that the response aspect, when the doors go up and the trucks go out, is unimportant; not at all, since it is, after all, the bulk of what we do. However, the

fiscal, directive, and administrative work making all of that possible is usually accomplished by 40-hour-a-week employees who work staff jobs or by shift personnel who come back on their days off. (If you're an exempt employee, you know exactly what I mean.) Let's face it: a captain working in the fire chief's office—who has direct access day after day and sits in meeting after meeting—has more influence on organizational direction than a captain working the engine company that works one on and two off. The bottom line is that if you can whisper in the ear of the king or queen on a daily basis, your influence expands. Geography and work schedule have a lot more to do with influence than you might believe.

- *Political.* The union president (or anyone holding a similar position) has a significant influence on events regardless of official rank in the department. In my EMS story from the previous section, the union president was a firefighter, but highly effective and skilled in the political arena, thus having a tremendous influence well above the level enjoyed by members with much higher official ranks. If you are actively engaged in the political process and are good at it, you can wield incredible influence.

As with policy making, I am sure you can identify further methods to increase your sphere of influence The gist of what I am saying, though, is that you will never expand your circle of influence, no matter how much power you accumulate, to the point where you no longer have a circle of concern. In fact, the higher you go, the more your circle of concerns grows, encompassing more global, expensive, and political issues that you wish you could control. Also, influence can fade over time or based on your position, rank, political standing, or personal relationships.

Understanding the difference

There are lots of words to describe methods used to create an organization that is moving in the same direction, ensuring accountability, and being consistent in the methods it uses to accomplish the business of the business. Any group of your peers will hold a plethora of diverse interpretations. Therefore, I have chosen to provide you with what I believe are some starting points for discussion, when it comes to clarifying policy, procedures, guidelines, and so forth.

Rules are typically defined as those things that we have to follow. This is usually because of a law or a statute that regulates what we can and cannot do.

Legislative requirements, statutes, ordinances, laws, and regulations drive our ability to do what we would like to do, balanced against what we have to do.

Policies are usually made by the department or organization. Personnel are expected to adhere to these polices because the organization has decided that they are required for the good order of business.

Procedures explain how to complete recurrent tasks that we may undertake, such as administrative paperwork or the approach to a car fire. To complete the task in the correct manner—and most often the most efficient and safe manner—you need to follow the procedure, or in the case of EMS, the protocol.

Guidelines provide the best-known process to accomplish a particular task or set of tasks, to reach a desired outcome. The guidelines people follow will normally be based on the outcome that has been identified as beneficial or desired.

The Story of the Monkeys

Suppose we were to take four monkeys and put them in a large cage. At the center of the cage, we place a large set of steps, and from the ceiling, we hang a large bunch of fresh bananas. Now, understanding the affinity that monkeys have for bananas, it would be a sure bet that before long, one or more of the monkeys is going to get to the steps and make a play for the bananas.

When this occurs, we instantly douse the entire cage and all of the monkeys in ice-cold water. After a little while, they think, "Okay, let's try again," and make a play up the steps for the bananas. We repeat the process, dousing all of the monkeys in the cage with ice-cold water. As long as the monkeys try to accomplish this, we will continue to douse them with ice-cold water.

Now, let's take one of the original monkeys out of the cage and replace him with a new monkey. Rest assured that the new monkey, not understanding the consequences of his actions, will make a play for the bananas. An amazing thing happens. We do not even have to douse the monkeys with ice-cold water, because all of the original monkeys, knowing what is about to happen, will jump on the newcomer the minute he makes a play for the stairs and deliver a merciless beating. Battered and bruised, the newcomer will sulk to the corner, trying to figure out what happened. Almost assuredly, if he gains the confidence to repeat his actions, he will get another beating.

We can replace another of the original monkeys and watch previous scenario repeat itself with the newcomer. However, this time a curious thing happens. The previous newcomer will joyfully—yes, gleefully—join in the beating of the newest monkey as he tries for the bananas, without even having a clue as to why he is beating him!

> *We can repeat this process until all of the original monkeys, who had been doused with ice-cold water, are gone and all that remains are monkeys who have never been doused with ice-cold water. Regardless of what monkey we now put in the cage, each time the newcomer attempts to go for the bananas, they will be beaten. None of the monkeys know why they are beating the newcomer, having never been personally doused with ice-cold water.*
>
> *But do you know why it happens? Because that's the way it has always been done around here!*

The Consequences of Nonenforcement

We are going to clarify the differences in verbiage—between policy, procedures, guidelines, and so on—in just a little bit. Before we get there, it is important to understand what your duty as an officer is when it comes to policy enforcement. First, you don't have to—nor will you ever—agree with every rule and regulation the organization puts in place. We have already spoken about how to expand your influence, but sometimes it is what it is.

Expectations of an officer are different than those of the men and women you supervise. Here are the golden rules regarding policy, procedures, and rule enforcement. If you cannot live by these—and they are not always fun or pleasant—then you need to consider whether you can effectively function as an officer.

Argue at the right time and place

If you think something is wrong or is an inappropriate rule, then discuss or even debate the what, why, and how of the issue. Never stand idly by and let something that is fundamentally against what you believe to be right sail slowly by. Argue your point right up until the decision has been made, and then get behind it and support the decision. Even after the decision is made, you can still work to change it through the proper channels. I worked, expressed, and pleaded to change the mustache policy right up until the day I retired. I thought it foolish, but despite that, I enforced—and expected my officers to enforce—the grooming policy.

Don't talk bad about policy in front of the troops

You might not like it—in fact you might think it's the stupidest thing on earth—but keep it to yourself. In *Saving Private Ryan*, once the assignment is given to go find Ryan, as the squad are walking across a field, the captain (Tom Hanks) is listening to one of the soldiers complain about the mission. After having his say, the soldier turns to the captain and asks, "What do you have to complain about?" He replies, "I don't complain to soldiers below me, only to people above me. You complain to me; I complain to them—that's the way it works." It's hard to do, but sometimes you just cannot have a comment.

You don't have to like it; you just have to do it

Sometimes you will be expected to enforce policy that you feel is stupid. This requires an understanding of how to inform people as to how to comply only until such time as you can get the policy changed.

Know the process and the policy

It's incumbent on you to know the policy, its intent, and any loopholes that lend you a degree of flexibility when it comes to interpretation and enforcement. The Dalai Lama says, "Know the rules so you know how to break them properly."

We had an employee who showed up for work with a skin disease. The station went nuts, screaming about mattress contamination; of course, crisis policy decision making and policy action was not far off. Within a week, there was a directive that all beds would have a mattress cover on them (the same as are used for childhood bed wetters).

Talk about an uproar. People did not like the mattress covers: they were hot in the summer, cold in the winter, and slippery during all seasons. A shift would put them on the beds; B shift would come in and pull them off the beds; and C shift—well, C shift was just trying to find the bed. It was a shift war, and officers were being forced to spend time enforcing a policy that was ridiculous.

I was at my wits' end when I began to research it, but I found that it was a directive, not a policy. As a result, I could change it without violating policy. I immediately distributed a memo informing my shift that anyone who wanted a mattress cover could have one individually issued and all others did not need to sleep on one. A week later, only 3 members of the 400-person department had requested a mattress cover; case closed.

Don't pick and choose the policies you want to enforce

Let's say that you have a grooming policy that says no mustaches may extend below the corner of the lip. Let's also say you have a policy on physical-fitness clothing. Don't get me wrong: I am not taking sides here, and I am not saying that either of these polices is wise or even right; I am using them simply as examples. The mustache issue is easy to enforce, because when someone has a long mustache, you can tell him to cut it. By contrast, the physical-training (PT) policy—along the lines that PT clothes are to be worn during PT and then removed, but that PT clothes can be used as casual wear around the station after 2000 hours at night—may be much more difficult to enforce.

Your credibility is a valuable currency that takes time to build up. As discussed earlier (see chap. 5), you do not want to be caught picking and choosing what you will and will not enforce. If you enforce only the easier of these rules, mustache trimming, it may give the impression that you are singling out an employee—and that will be communicated up the chain of command office so fast you won't even be able to say "transfer"! Start enforcing policy universally, so that your credibility is not challenged.

Don't allow or have a "look over the fence" mentality

Just because you do the right thing does not mean that everyone will. Sometimes you work next to a supervisor who seems to be the antithesis of everything you are trying to accomplish. There are certain ways to handle that, as we talk about shortly, but it creates the temptation for staff to take a "look over the fence" mentality: "A shift does it, so why can't we?" Or, "Captain So-and-So does not enforce that stupid policy." Influence from across the fence is like a bad neighbor: it's easy to let your kids get caught up in it. Don't allow your personnel or yourself to get sucked into the "go along to get along" mind-set when you know it to be wrong.

Retooling

It's not easy creating an organization of people who understand that though they are expected to take the rules seriously they are also expected to break them occasionally and eventually to tamper with them.
—Marine colonel quoted in *Corps Business*[2]

Retooling is a concept that is vital to a healthy organization. It has to do with how you develop, nurture, evaluate, apply, change, and reevaluate policy and procedures throughout their life cycle. Changes in technology, construction, standards, equipment, personnel, knowledge base, environment, politics, the world situation, leadership, and many other internal and external influences drive which policy, procedure, guideline, or rule is applicable at any given time. If we do not constantly evaluate our goals and objectives with regards to running the business and running emergency incidents, we will never realize when and how to change to retain optimal effectiveness.

From a liability standpoint, we cannot afford to allow policy, procedure, guidelines, and rules (PPGR) to remain stagnant. Think about the implications of working from a 10-year-old incident management policy, a 10-year-old tactical guidelines procedure, or a 15-year-old vehicle maintenance policy, for example. I can assure you that if something goes wrong, if someone gets hurt or killed, or if there is extensive property loss, someone will ask what PPGR led you to that decision-making process.

Finally, firefighters are not dumb, despite what you might think. They look at policy, and they know what they can and cannot get away with. If you have a written policy or procedure and you are doing something different, then rest assured that when you demand accountability, they will pull out the policy and say, "Nope, it says so right here."

Retooling yields multiple benefits for the organization and the members. Most important, it keeps your members safe by ensuring that you are using state-of-the-art strategy and tactics to address the critical emergency issues that we deal with. It keeps you aligned with best-practices changes in ordinances, statutes, and consensus standards and eliminates old methods of doing business that are no longer applicable, safe, or prudent to use.

Next, retooling ensures that your organization covers the policy area from a liability standpoint. There is nothing worse than saying you are doing one thing but having something in writing, usually signed by the fire chief, department head, or manager, that is in direct conflict with what you did. It's a surefire way to get into legal trouble in court cases, on receiving Freedom of Information Act requests, or when dealing with discipline and human resources issues.

Finally, but no less important, retooling gets rid of silly policy. One job that organizations and their officers are not very good at is eliminating goofy, hairsplitting rules on the books. Celebrate getting rid of these rules: advertise it to the organization, by holding a ceremony to burn or shred a copy of the rulebook. Acknowledge that this is no longer needed, and thereby accentuate the growth of your members, who no longer have to be held by the yoke of silliness. Later in this chapter, we revisit the notion, introduced in chapter 3, of transitioning from a rules-based to a values-based organization.

There are lots of techniques and methods of retooling. Learn from outsiders; don't isolate your department by figuring that you have it all squared away already. Learning from outsiders saves you time and effort that can be focused on other matters. For example, in case you have to write a new SOP, realize that 90% of them have already been written. There are Web sites galore where you can download them; then, review and make the changes to fit your specific needs, without reinventing the wheel. Take the time to have guest instructors present new ideas and new methods of accomplishing business—and don't be afraid to try them. Don't get locked up in the dogma of the past, how it has always been done; instead, ask "How can we do this better?" and strive to accomplish that.

Make sure you have a stakeholder SOP and policy review committee. Their first task should be to identify a time line that each policy will have assigned to it for review and change, if change is needed. High-risk/low-frequency policies are much higher maintenance and require shorter review times (perhaps every two or three years) as compared to uniform or grooming standards. The desire is to have a diverse group from the organization, acting as an advocate for the change, that sits down and reviews the proposed changes, gets input from the appropriate stakeholders within the organization, formats the suggestions, and presents the final proposal to the senior staff. Suggestions for change should be sent out to the field, specifying a time frame in which feedback is invited. Even though policy creates so much upset among our individual members, when given the opportunity to provide input and feedback, you will be lucky if you get a response from 20% of the department—and that is being optimistic. Someone has to take the initiative to spreadsheet the review times, so that you don't miss them.

Have a method to get interim policy into the field quickly without being halted by formal roadblocks. This is critical when something is discovered during an incident that had previously not been identified and is a health or safety risk. Sometimes rules change overnight because of changes in ordinances, laws, or statutes, and we must be able to retool our doctrine in short order. Sometimes we make mistakes when we develop policy, and either we cannot admit it or we "fail to be able to get out of the room," as Mr. Ligachev put it. When we see something silly or realize that we made a mistake about something—Fu Manchu mustaches, for example, or some other minor trifle—have a method to get it changed immediately, and follow it up with a formal change later. You would be surprised what this can do for morale. Sometimes people come up with excellent ideas for the enhancement of the organization, and being able to get it out quickly is important. Make these a different color, so they stand out.

Change when change is necessary is also a good retooling policy. Try to streamline your rules, policies, and procedures to just those that make sense and don't require 52 volumes to sort through to make a decision. Spend time not only changing policy but also changing perspective and retooling understanding. Take the time during your training and in-service to review and update people's understanding of your PPGR and what the intent of these is.

Finally, understand that you cannot make a rule or policy to cover every situation. You have to take the time to teach people how to think, not what to think. If you saw all the spur-of-the-moment initiative that is taken by small units on the fire ground that is technically outside policy or procedure but works—and works effectively—you would have a stroke. This is about instilling values in people and allowing them to think on their feet and is the next subject of our discussion.

Rules versus values systems

Earlier in this book, we spent quite a bit of time discussing values and how they apply to organizations. One of the most difficult conversions that organizations—and thus supervisors—make is the transition from a rules-based organization to a values-based organization. The reason this is so difficult at both the human and organizational level is twofold. First, it's a break with the way in which we have done business for hundreds of years. Second, and even more difficult, it requires our members, our supervisors, and our bosses to think and act differently.

The vast majority of fire service organizations have been rules driven for as long as they have been in existence. The bookshelves are filled with policy manuals in which nearly every potential scenario is outlined and documented in great detail. Often there are different manuals for different sections of the organization. What is most hilarious about these books is that when you ask the frontline firefighters or officers about the contents, they can put a name to a given rule and tell you the ancient story about how it came to pass.

Rules-driven organizations are designed to limit the choices that their members can make in favor of consistent decision making. Every organization needs rules, even a values-driven organization. There are certain rules that we must have, and outlining rules certainly makes for a clear understanding of expectations and enables consistent and equitable enforcement. The pitfall, however, is that many rules-driven organizations put such a high value on following the rules that it takes on an importance well above the benefits; in many cases, management loses sight of the larger goal of protecting lives and property and serving the customer, as it instead focuses on administering the rules.

Values-driven systems look at organizational life differently. As we discussed earlier, values-based decision making operates around a set of core values. An understanding of these organizational values is provided through the education of all of our members, and they are a fundamental part of how we operate on a daily basis.

Values-driven organizations are focused, even rabid, about achieving a successful outcome and making decisions based on the values system. Thus, the values system focuses on outcomes instead of the process—a massive change from the way in which rules systems operate.

The fundamental difference is that in a rules-based organization, there is only one method to reach a given outcome, while in a values-based system, there are many paths to a given outcome. To many old timers and to some officers, this is a scary proposition, because they see it as a threat to consistency and application; it certainly subjects the chosen method to reach an outcome to scrutiny, since there are many possible paths to the same outcome. It is quite discomforting to let go and trust your people to make the right decisions, especially when you have not educated, trained, and taught them how to think, rather than what to think.

VBF&R has a mission statement that includes "making citizens and visitors feel safe anywhere, anytime." This is a values statement. The question this poses is, How can you write a rule to ensure that people feel safe anywhere, anytime, and how can you teach members to apply that rule? Think back to the story, from chapter 1, of Mrs. Smith broken down, on the side of the road. What does a rule provide when it comes to making her feel safe anywhere, anytime? Could we write a rule that bars civilians from riding in the fire trucks, thus limiting our ability to deal with this based on a values-based filter? Absolutely, but we would not be accomplishing what we said we were going to do in our mission statement. If we tell our officers, "Here are our values—figure out the path to accomplish them," we open up many paths that our members can explore and apply to reach a successful outcome. What we want is a successful outcome to a given problem that is aligned with our values, takes care of the customer, is safe, and strives for excellence.

When you trust your members to seek and obtain the best possible outcome for your internal and external customers, while not compromising the organizational values, the need for extensive PPGR loses its importance and statue. Values-driven organizations create an innovative and creative decision-making environment for their members. They is much more flexible in their ability to deliver service and reach outcomes in a timely manner. Most important, the organization thrives on human initiative and ingenuity driven by a set of commonly held values.

You may think that this puts you out of the woods with regards to rules, enforcement, and accountability—that moving to a values-based organization puts the insane in charge of the asylum and that senior staff might as well start herding cats if they plan on having any control. Rules-driven and values-driven organizations are different, and the leadership/management styles required in order to be successful may be at opposite poles. This is because no organization is purely one or the other; you must blend the best of both worlds to form a true values-based organization.

Take heart, values-driven organizations needs PPGR also. The striking difference in adoption, understanding, and enforcement involves the issue of timeless conduct and behavior. Simple behavioral norms—like "be nice, don't drink, don't fight, be safe, and don't steal"—set the norm for behavior. While PPGR in a values-based organization are as few as possible, they apply to everyone all of the time and are strictly enforced.

Inheriting Policy Violations

Despite your best efforts to create a values-driven organization—or to enforce or understand the rules—you will, without a doubt, during your career as a supervisor, inherit (either by design or chance) a policy violation. How you deal with this will be based on your understanding both of your duty and responsibility and of the rules and values of the organization. There are also some basic personal rules that you can apply to make this flow easier.

First, never take it personally; it's always business. A rule or policy violation hardly ever represents a personal attack on you. Usually, either someone has made a mistake—or has a supervisor that has allowed it to become the accepted norm—or someone has decided to push the system, based on their values or lack thereof.

Understand that you are an officer for the organization, not on A, B, or C shift or staff. Your authority transcends shifts and divisions: you have the authority and responsibility to correct problems that you encounter or that are handed to you. You cannot look away simply because the individual does not report to you all the time.

Professional courtesy to your peers is important, even when you know they have allowed the violation to occur. Make sure you give the immediate supervisor of a violator the opportunity to know about the event. In some cases, you may be able to immediately refer the problem to him or her, while in other circumstances, you will have to deal with it immediately and then notify the supervisor. One way or the other, make sure that he or she is aware of the decisions you made and that you saw the violation.

When you're angry, don't do anything that makes you feel good. If someone decides to challenge your authority or bucks your order, remember that it is business; remember the Universal Three; and finally, remember to give people a second chance. We examine in detail how to do this shortly.

Explain to people what the perceived problem is and what you expect as a solution. Be prepared for feedback (good and bad) from the offending party, and take the time to listen before you restate your expectations.

Keep it at the lowest level. Always attempt to solve problems and correct deficiencies within your circle of influence. Moving outside that circle, as does become necessary at times, brings new layers of decision making and often makes the event more emotional and complicated. However, it is not always possible to keep it at the lowest level, especially when people want to be asinine about it—but try.

Use policy violations as moments of learning. Use them to teach people what is expected and how they can meet those expectations. Also use them to project what you stand for, what you believe in, and what you will and won't tolerate.

Understand that you will not always be supervising your regular crew. You may have personnel reporting to you on relief, you may work a trade, or you may be assigned somewhere else on a temporary basis; regardless, you are responsible and accountable when you arrive at work. It's different when you go into someone else's house than when they come into your house: it often becomes a tightrope walk when you are the visitor.

Relief for Firefighter Jones

Today Firefighter Bob Jones has been sent on relief to your station. He will work for you for the next 24 hours, but he is not permanently assigned to you. This happens hundreds of times a day in fire departments all across the country. For this scenario we are going to keep it simple and perhaps a bit exaggerated, because we are interested in the application. There are many paths this event might follow, so let's examine them and see how we fare.

When Firefighter Jones arrives and checks in, you notice he has a long Fu Manchu mustache, in violation of the policy. (Can you tell I am obsessed with this policy?) The possible paths range from the extremes of ignoring it (not an option in this book) and tackling him and shaving it yourself (again not a smart move). What we are looking for is to apply the rules just covered so as to reach a successful outcome.

You call over Firefighter Jones and say "Your mustache is way out of policy guidelines. I need you to trim it." He grumbles, "Yes, sir," and goes off to trim the mustache. Simple enough, you may say, but let's see how the rules apply here. First, you didn't take it as a personal affront. Second, you approached him professionally and explained what you wanted done, and he gave you feedback and obeyed by trimming the mustache (cursing you under his breath all the way to the bathroom).

Case solved? Not quite. You need to make sure take care of the professional courtesy part. So you telephone Captain Smertz, his supervisor, letting him know that Bob showed up on relief with a grooming issue (as if he did not already know) that you have addressed. Regardless, you have accomplished several things here. First, you have let him know what you did to his direct subordinate, and even more important, you have subtly told him, "Don't ever send your policy violations to me again." Finally, you have solved this at the lowest level (with no battalion chief involvement), kept it between supervisors, and done your duty as an officer.

Now, what other options might we explore? We could always send Firefighter Jones back to his station with directions to not come back until his grooming is appropriate. That does not solve the issue and does not keep it at the lowest level, though. The minute you are short staffed, the battalion chief is going to get involved, inquiring about what happened to the relief. Next, Captain Smertz will either send you someone else (which does not address the issue) or have Firefighter Jones cut his mustache and return, with the caveat that "I wouldn't have made you do this, but Captain Sargent has a case of rectal-cranial inversion on this." Don't expect a supervisor who doesn't do his duty instantly to do it because of you; they will find a scapegoat to blame.

Let's see how one last scenario plays out. You approach Firefighter Jones and ask him to trim his mustache. He replies, "Well, Captain Smertz lets me have it, so I don't see what the big deal is." You reply, "I understand, but you are way out of policy guidelines. I need to you cut your mustache," to which he replies, "No."

So the game is afoot—you are being directly challenged. Of course, you could whack him right then if you wanted to, but remember to keep the humanity in your leadership, to give a second chance, and to keep it at the lowest level. So how do you accomplish this? You direct for Firefighter Jones to follow you to the office and to sit down, you open up the policy manual to the section on mustaches, and you ask him if he is familiar with the policy.

Understanding and Enforcing Policy

You then repeat your expectation and give him one last chance to go trim his mustache. If he refuses this time, you should by all means ring him up! At this point, you take him off the truck (notice I did not say relieve from duty, and I will tell you why in a minute) and inform him that he is now not only in violation of the policy but also insubordinate and failing to comply with a direct order. In short order, you advise the battalion chief or next-level supervisor and his supervisor of the situation.

Here is where I take issue with this situation. Firefighter Jones was given every opportunity to comply, even a second chance. Yet he challenges authority by refusing a direct order! I did not work to promote just to have some member tell me *no* when I am enforcing department policy. So I would try to get him on all the charges I could, knowing that at least something will stick. I would still do this in a businesslike manner, making sure I knew the process, protecting his rights as an employee, and following the steps to the letter to make sure I don't lose because of a technicality.

Earlier I said how important it is for officers to know policy and its intent. The reality is that the kind of event we are describing here is a high-risk/low-frequency human resources event. Because it does not happen often enough for us to know the policy as thoroughly as we should, we need to make sure the chain of command is involved right away. Moreover, from experience, I advise you to be careful about how you relieve this individual from duty, to ensure that you give the process, investigation, and the subsequent outcome time to reach its conclusion. You have time, so use it!

Everyone's rules, regulations, human resources policies, collective bargaining guidelines, and union protection or the Firefighter's Bill of Rights are different. In VBF&R, a captain can suspend for 24 hours, a battalion officer for 48 hours, and a division chief for 72 hours; the fire chief can suspend for two weeks or more.

We had an employee who created problems when I was a division chief. The incident occurred around 2300 hours at night, and I directed the captain to suspend him for the remainder of the shift and to have the employee report to the fire chief's office the next morning. After the investigation was complete, the fire chief determined that a two-week suspension was in order, but human resources overturned it, stating that he had already been suspended and that you could not put an employee in double jeopardy. They were right; I had screwed up and been caught on a technicality. I should have sent the employee home on annual leave, instructing him not to return to work until the investigation was complete. Then, once our investigation had been completed, we could have given him his annual leave back and suspended him for two weeks, rather than nine hours.

There are several lessons to derive from this story:
- Use all the time you are allotted for the investigation
- Don't dole out any corrective action until after the investigation, to ensure it is both warranted and applicable
- Know the policy and its impact
- Understand what your options are to immediately solve the problem at hand before a full investigation is completed.

In this example, the immediate problem was that the employee cannot remain at work when he is in violation of a policy and has been insubordinate to an officer. How do we remove him from the work space in a fair and equitable manner to solve the immediate problem and then undertake the investigation to solve the long-term issue(s)?

Conclusion

Understanding (including intent), enforcing, administering, and retooling PPGR are some of the most complicated and emotional events that an officer will have to undertake during his or her career. Understanding the aspects of policy development, influence, rules, values, duty, and how to react to inappropriate situations takes education, a few mistakes, and experience to feel comfortable with it.

Notes

1. Covey, Stephen R. *The 7 Habits of Highly Effective People*. New York: Free Press, 2004.

Chapter 8
Evaluating and Compensating People

Minimum qualifications become maximum expectations.

—James O. Page

The Culture of Forming Evaluations

Evaluating people is a supervisory responsibility that has its joy and sorrows. The good news is that we usually hire great people, who put their heart and soul into the job and the team. They want to do well, they strive to accomplish good things, and for the most part they do; from time to time, everyone makes mistakes, but it's never very bad. There are times, though, when we hire or inherit the wrong people for the job; they sneak through the system, or someone who was supposed to document their performance did a poor job. For whatever reason, we still have employees who don't belong. Nevertheless, our life is mostly very good, as we are surrounded by great, dedicated people who really care about the human beings they work with and for.

There are those who could argue quite convincingly that you do not need evaluations—that given a true values-based system there are other methods to recognize and broadcast how members of your organization are accomplishing the desired outcomes. To some extent, I do not disagree with these people, and I would love to believe and work in a system where everyone is so motivated, committed, and professional that we just continually slap them on the back and say, "Great job!" In time, we may reach that state, but it's unlikely in the government structure. I would also love to find a municipal government that really does not care what it gets for its budget money, which is as likely as my being given a lifetime supply of free gasoline, free alcohol, and a million dollars a day deposited in my bank account.

Evaluations are formed by drawing from the history of the organization and are put together by the human resources (HR) department, with input, both positive and negative, from management and labor. The history of the organization may also dictate that an organization does not form evaluations at all; my concern with this is twofold: First, unless the HR and legal departments buy into it, there is no way to get rid of problem employees without proper documentation—as has been my experience as a career firefighter and officer. Second, there is no way to reward great players without some form of documentation and evaluation of their work. Moreover, even when outstanding work is documented, most municipal systems are established such as to reward the norm—that is to say, everyone gets the same reward, regardless of effort and results.

Evaluating and rewarding people in municipal or federal government is an entirely different ball game than in private corporations. If someone meets the minimum qualifications in government, there is nothing that can be done to force improvement. By contrast, as those readers who have worked in private enterprise can attest, if you don't produce in that environment, you will face a variety of consequences, from the loss of a bonus to the loss of a job.

The Myth of Merit Raises

There is a UPS commercial in which Rusty Wallace, the NASCAR driver, goes up to the UPS truck driver and exclaims, "I want to drive the big brown truck." To this, the driver replies, "I'm sorry sir, you can't drive the big brown truck." For those of us unfamiliar with the culture of UPS, that would make no sense; the comedy is that Rusty, despite being a professional stock car driver, does not qualify under UPS evaluations to drive the big brown truck. He would have to work his way there by spending time in other jobs and proving himself.

There have been countless studies on what motivates people to work harder, and one of the resounding themes is that money is not a motivator, at least not in the long term. There are many types of monetary reward systems practiced in private business that municipal government will not or cannot adopt. For the most part, municipal government adopts *step raises*, or what some would call merit raises. "Step" is a much better term, as raises based on merit are really a myth.

Closer examination of this time-in-service, graded pay system reveals that what some organizations call a merit raise system is in fact nothing more than an entitlement system. How do we come to a starting salary base? Do we just make it up and say, hey, that sounds good? Regardless of whether you are in a right-to-work or a collective-bargaining system, the baseline for salary is established by doing a regional study of departments of comparable size, capability, and responsibility. Even union and management, when they go to the table to bargain salaries, know what that baseline is and that they can negotiate 5% or 10% up or down; thus, starting salary ranges are always known.

Suppose that every year you work, you get a percentage increase of your salary. At year one, you get X%, and at year two, you get X%—until you reach the top of your scale. This does not count cost-of-living raises or benefits; rather, it is simply what shows up on your paycheck. Although some might call this a merit raise, it has nothing to do with merit and everything to do with meeting minimum qualifications in a municipal system: Meet the minimum qualifications, and you will get your raise; quality and effort have nothing to do with it, because in this system, the person who "sits the bar" gets the same as the person who excels.

Why is there a *top out*, a point at which you will no longer get these raises unless you promote? It's a business decision that finance and management calculate. Eventually, it becomes cheaper to hire someone new at a lower salary, rather than continue to pay you more for the same work. When you have learned all you are going to learn about your particular job, you have reached the top-out point. If you want to make more money, then either move up and promote or sit and be happy.

In an ideal world, we would have what they have in many private organizations and corporations—namely, pay for performance. This concept entails paying better performers more than moderate performers. The problem in municipal government is establishing a system that can measure performance and equate it to a given level. In private industry, that's easy to quantify through quotas, sales levels, calls taken, profit earned, and so forth. But how does one measure that in a municipal government that has a history of mediocrity in just about everything it does? Even while an individual organization such as the fire service could develop a system to measure only their people, all the other departments not receiving equitable pay incentives would then cry foul.

Pay for performance also takes into account acquired specialty skills that are needed for job performance above and beyond basic requirements. Examples in the fire and rescue services include hazmat team, technical rescue team, SWAT team, helicopter pilot, crew chief, coxswain, and boat crew.

Along with this comes the problem of money and evaluation. First, to have a true pay-for-performance program, you need a pool of money from which to pay everyone who meets the criteria. That means that a set amount of money in the pool would never work, because there may be times too many people excel. Governments hate that and refuse to enter into a system with those financial criteria.

The next problem is developing a system that captures, defines, and qualitatively measures performance. There are three critical issues here:

1. An evaluation system must define qualitative measurements that can be documented to measure levels of performance. A simple number system or quantitative measurement will never suffice, and the reason for this is our number-two problem.
2. A number system lends itself too much to the good-old-boy system of giving people something they didn't earn without justification based on hard data.
3. Supervisors must commit to and be trained on completing the evaluations in such a manner and with sufficient qualitative data to justify the raise.

The point is that even though it has been proven that money is not a motivator (if I gave you $50,000, there is no guarantee that I would get a return of $50,000 more in production from you), most government organizations lack any structure or measurement tool that lends itself to paying better performers more and poorer performers less.

There are a couple of other issues that officers need to understand about financial reward and evaluations in government organizations. The first is the issue of *compression*. In many locations, people are hired at a specific starting salary. Perhaps in the next two years, that starting salary increases owing to regional studies, inflation, and other factors. Thus, new hires may well make the same as someone who has already been employed for two years. If current employees do not receive pay adjustments upward within their range, as well as adjustment to the new staring salary (as usually occurs), they might even find themselves paid less than the new employee. As this is repeated over time, the result is an entire class of employees who have been compressed, or are making only a few dollars more than employees with much less time. This is further complicated in municipal government because it may take a commitment of many millions of dollars to correct.

The second issue is the aspect of steps within the pay range. This system allows an employee to move upward by steps in a pay range and then on to the next pay range. This decreases the chance of compression by moving people along pay grades based on seniority. Unfortunately, it still does not address the question of pay for performance.

This brings us to the concept of the bar. This is what municipal government and HR departments consider as minimum requirements.

The bar

Just about every job classification in municipal government has a set of minimum requirements. Evaluations are based on an employee's meeting the minimum requirements. Fail to meet the minimum requirements, and there are consequences; exceed the minimum requirements, and we say, "Great job!" Here, at least in municipal government, is a reason to have evaluations: You must be able to ask whether members are meeting the minimum standards. It's unfortunate that we have sunk so low as to require only the minimum to stay employed, but this is the culture in and legal position of most municipal governments.

To survive in municipal government, all you need to do is sit the bar. Fall below the bar, in either actions or job performance, and suffer the consequences—such as retraining, discipline, financial impacts, or other organizational direction. Rise above the bar, where we really want you to spend most of your time, and in most instances, there is no fiscal reward. Thus, it is incumbent on us to find methods of rewarding people who live well above the bar.

Don't get me wrong: the people who sit the bar are not bad employees; they simply meet the minimum. They don't cause us much problem; they show up on time, in the right uniforms; and they do their jobs. However, they are not going to give you anything extra. Those who sit the bar may consider the fire department their second job and perform accordingly. Most of them are okay firefighters and fair EMTs, but unless they have specific talents that can be applied, they never really excel or give anything extra. You can have a few of these people on your shift, but the entire shift cannot be composed of minimum performers.

The good news is that, despite municipal government's best efforts to make the fire and rescue services into minimum-performance jobs, most people we hire won't have anything to do with that concept. Most excel in one aspect or another, while only a small percentage of our members sit the bar.

The bad news is that some do fall below the bar, although they represent an even smaller percentage of our members. In fact, most of us have fallen below the bar once in awhile in our careers; for this, we got a reprimand, or a talking to, or something else, and we came back above the bar and stayed there. The problem is not occasional, once-or-twice-in-a-career below-the-bar performance, since as a supervisor, you recognize it, correct it, and move on. However, when you have an individual who is continually below the bar, because of performance, actions, values, or rule breaking, then a problem exists.

What Do We Know about Evaluations?

The key values of supervisory evaluation are demonstrated by Colin Powell's performance principles:[1]

1. Make performance and change top organizational priorities
2. Define the new game, and expect everyone to play it
3. Make sure that your best performers are more satisfied than your poor performers
4. Get rid of nonperformers
5. Consider the possibility that if nobody's upset, you may not be pushing hard enough

Organizations need to ensure that individual, team, and department performance is a top priority. In addition, to succeed, you will need to evaluate the world around you and make the changes necessary for optimum performance. In the end, it's never about who you are, the color of your skin, your sex, or any other surface trait; it's about performance. As a company officer, you should evaluate performance from several vantages.

Rules, directions, strategic plans, procedures, policy, and operations change. When this happens, it is critical for everyone to be educated about the changes. Once granted an understanding of the changes, everyone should be expected to play by the new rules.

Although we have previously examined the fiscal limitations of working for a municipal government with regards to rewards for performance, there are other methods to ensure that the best performers are satisfied, such as bonus programs, choice assignments, and training opportunities. Even saying thank you, awarding plaques or medals, or announcing individual successes in the organization's newsletter accomplish this goal with little expenditure.

Getting rid of nonperformers in a municipal service is a tedious, drawn-out affair. This requires time, patience, and thorough documentation; also, the company officer or supervisor, as well as the team, must accept that it will be a miserable time. This is so important that later in this chapter, a section is devoted to fringe employees.

In any progressive organization that requires performance, someone will eventually get ticked off that you are pushing too hard. However, if everyone were content to sit on their laurels, you would not be driving your shift or your organization in the proper direction.

Realize that evaluations are a tool developed by the organization to look at the minimum performance criteria. Their value is proportional to the time taken to develop them and the criteria assigned to make qualitative judgments. Just as important is the necessity to train your officers on how to evaluate your members. This includes understanding the evaluation system itself, how to score people, what type of documentation is required, how to counsel people throughout the year in anticipation of the evaluation, how to rate superior performers, and how to rate poor performers. When done properly, evaluations require time and effort of the company officer. Poor or no training only lessens the viability of the evaluation. Finally, consistent standards are critical, to eliminate the remaining good-old-boy evaluations out there that reward the fishing buddies.

Evaluations are ultimately about helping employees to reach their potential and perform their best. Done right, evaluations—both verbal and written—are tools for growing and learning. When done poorly, though, we wind up doing exactly what does not motivate employees. The following pitfalls demonstrate this.

Incentive pay plans do not work. There are reasons for this, but it is primarily because incentive pay plans put only so much money to pay out. Whereas in private industry, it's easy to say, "Sell this much, and you will get this much," in public safety in general—and the fire service in particular—it's much more difficult to set a benchmark to judge who gets incentive money and who doesn't. How many fires did we prevent by doing inspections? How many lives did we save by putting in smoke detectors? In other words, where are the data to form a basis for providing incentives?

Rewards are advertised as methods to spark performance may rupture relationships. Competition between peers can lead to unanticipated actions, such as a decrease in motivations or even giving up, or negotiations and unethical behavior. Unless you have enough rewards for everyone who performs exceptionally, or beyond whatever benchmark you establish, rewards don't typically work.

Finally, rewards may eliminate the risk taking that drives innovation. People may be afraid to take risks because, if they fail, they will not be rewarded. Perhaps we ought to develop a celebration for the most innovative idea that failed during the year.

Beginning the Process

The evaluation process begins long before the initiation of any paperwork or official documentation. It begins with an educational endeavor by the organization to ensure that all members understand how the system works. The most important aspect, from a values and culture standpoint, is to ensure that everyone in the organization—and in their particular work unit—clearly understands the expectations of both the organization and their supervisor. Without a clear understanding and discussion about expectations, how would one ever know what to strive for and what performance will be measured against?

Every time we undertake an operation, a customer service contact, a project, or any other task that is necessary to the organization, we need to apply it as a learning point. Moments of learning are critical components of evaluations: they provide feedback to the participants regarding how well their performance compared with what was expected, whether they met or exceeded the expectations; they give the opportunity for reverse feedback in the form of questions; and they allow individuals to make adjustments to their performance if necessary.

Nurturing growth in people, being a talent scout, and using moments of learning are all concepts that have already been discussed. Evaluations can be enhanced by using the talents of team members to mentor those who need it. Teaching is learning twice, and mentoring develops a set of skills in both the mentor and recipient that enhance their understanding of performance, open their eyes to new capabilities, and provide a fertile ground for communicating change as it occurs.

Despite our best efforts and our desire to perform, we have all tried and failed at times. One of our expectations should be that our members become innovators, questioning the status quo and seeking new and better methods to accomplish or improve on what we are doing. When we challenge people to do this, when we explicitly state this as part of our culture, we had better be prepared to see some failures and reward them nevertheless, as attempts at innovation.

The GE Engineer

A General Electric engineer approached his boss, the vice president, with an innovative idea. He indicated that he believed he could develop a new capacitor that would revolutionize the capacitor industry and make GE lots of money.

After listening intently, the VP asked how much the project would cost and how long it would take. The engineer replied that it would cost about a million dollars and take his team about a year in development. His logic, numbers, and reputation were regarded as reliable, and the VP approved the project and the funding.

A year later, the engineer walked into the VP's office and placed an envelope on his desk. When asked what was in the envelope, the engineer said, "It's my resignation letter. The design didn't work, and we couldn't develop the capacitor."

Picking up the envelope, the VP scanned the letter, promptly tore it up, and threw it in the trash can. "I just spent a million dollars on your education; if you think I am going to let you go, you have lost your mind. Now get back to work!"

Nothing else was said.

Human-level contact is the key to beginning the evaluation. Constant feedback is important. Meeting with your team members individually at least once a quarter to review accomplishments, as well as potential areas for improvement, is imperative. This also provides an excellent mirror to determine how you are doing as a supervisor.

Good or bad, no employee should ever be surprised by the written evaluation they get at year's end. Everyone should know their performance standards and how they are doing throughout the year. The written evaluation should be simply a cumulative document collecting what you have discussed, worked on, and expected throughout the evaluation process. When employees are surprised, it means either that the supervisor has not been doing his or her job or that the employee refuses to listen or understand.

Using the Door and the Desk

There are several tools that supervisors use when doing evaluations, but none more telling or powerful than the door and the desk. Each of these particular devices, which inhabit all of our offices, is used improperly every day.

The door is a symbol in the fire station. When meeting with your members, leaving the office door open signals that everything is okay; closing the office door leads to whispers of, "Uh-oh, what's up?" It may just be for privacy, but to the firefighters, when someone has been called in and the door closed, alarm bells go off. They run to the phone, call Station 4, relating in no uncertain terms that Chase has Dave in the office and is hanging him by his feet! They knew something was wrong and that Dave was involved; they just don't know what exactly it is yet. In actuality, I might be thanking Dave for covering last shift, I might be asking him how his son's broken arm is, or I might be telling him I need him to do something for me—or perhaps I *am* scolding him for not meeting an expectation. Regardless of the conversation, when you close the door, firefighters infer that something sinister is taking place.

Without a doubt, there are times when privacy and protocol require the use of a closed door. As a supervisor, you should understand the message that is sent when the door is closed. Use it as necessary.

The desk is another powerful tool. When you are meeting with someone about performance, good or bad, 99% of the time you should come out from behind the desk and either pull your chair up beside his or hers or sit on a couch if there is one in the office. The desk is a barrier, and its use conveys a clear message of subordinate and supervisor. When you bring someone in the office and seat them across from your desk, and you sit behind the desk, the message it sends is, "Listen to what I am about to say, because I am not going to repeat it. Sit up straight and take notes." The desk should be reserved for those events where you want it perfectly clear that the time has come for a serious, no-bull conversation regarding performance and expectations.

Positive Influence for Positive Employees

Most of our evaluations will be good evaluations. This is because the vast majority of the people we hire are dedicated and committed to what they do, and at the very least, they meet the minimum qualifications. It should not, however, be construed that positive evaluations are not an important part of your job as a supervisor. Even when a positive evaluation does not mean additional money to your members, you should still take the time to document performance and give feedback. Positive evaluations should be used as an effective communications tool: to reinforce the performance of your members, to set goals for the next year, to create a process whereby an employee can let you know about career goals and aspirations, and to provide a two-way conversation about each other's performance.

Evaluations are an excellent opportunity for you as the company officer to seek feedback on your own performance. If you want to know how a leader is doing, don't ask his or her supervisor; instead, ask the people he or she is supposed to be leading. This was driven home to me quite poignantly when, as the special operations battalion officer, I asked my nine captains each to list three things I was doing right and three things I was doing wrong or needed to improve upon. When I received their lists, I was happy to see I was meeting many of their expectations. However, a consistent message was that I could be "economical with the truth" at times, or in firefighter terms, I lied. I was stunned, hurt, and ego busted. What, I asked myself, was I doing to make them believe I was lying to them? I racked my brains, and finally it came to me. When the news was good, I was giving it to them straight, but when the news was not so good, I was holding back. Inevitably, they discovered the facts about the bad news, so it was stupid not to give it to them straight when they were adult enough to hear it.

Did this hurt? You bet, but it taught me two valuable lessons. The first lesson was to take a hard look in the mirror, and by doing so, I was able to change my performance and behavior for the better. The second, lesson I learned was about commitment and that was to *always undercommit and overdeliver; never overcommit and underdeliver, because you will always come up short.*

Make sure you recognize positive evaluations an occasion to celebrate the success of your members. Find some way to say, "Thanks for meeting the organization's needs and my needs."

Myths and Truths about Discipline and Employees

Before we talk about problem employees, we need to dispel some myths and affirm some truths about evaluations and discipline.

Discipline is designed to correct behavior, not to punish people. Punishment may seem like the logical outcome, but discipline is really about adult choices and adult consequences. The people we hire are expected to make life-and-death decisions as part of their jobs. It's not too much to ask for them to make rational, adult decisions and understand organizational parameters. Discipline should always be directed at identifying a problem and correcting a behavior, thereby enhancing performance and reaching the desired organizational outcome.

Discipline is always progressive. Don't knock someone out of the park just because they make a mistake. Failure is a possibility, and as a supervisor, you should consider the circumstances and facts surrounding a given event. Negligence and dereliction of duty are usually apparent if we do a proper investigation. We talk about decision making in greater detail in chapter 11. Discipline is based on past performance (how much they have within a given time frame) and is specifically guided by departmental policy and rules. Some disciplinary action is not flexible, and there are specific and consistent rules that you must follow, like it or not. If the policy defines AWOL as 15 minutes or more late to the assigned work station and in the course of fact finding, you discover the employee was late because he or she didn't get out of bed in time, then that meets the definition of

AWOL. If the policy says the first violation warrants a written reprimand and this is the first violation, you don't have any choice but to issue the reprimand; if the policy says the second violation in 12 months warrants a one-day suspension, you have no choice but to issue the suspension. In other words, decision making may be constrained to merely gathering the facts and enforcing the policy.

There is an old HR saying: "Supervisors need to spend more time with their less talented, poorer employees." The theory is that if you spend more time with them, your teaching and mentoring can raise them above the bar. However, that theory has definite limits—and once these limits are reached, the concept is of little or no use.

Suppose you have four members on your shift. Three of them excel and require very little supervision. Then you have Dan, who is not a problem but just sits the bar. He shows up, does the minimum to get by, and does what he's told, but never giving anything extra. Wow, you say, I must not be doing my job as a supervisor, so you spend time helping him to excel. As a result, Dan moves just a little above the bar, but the minute you direct your supervision elsewhere, he goes right back to sitting the bar. You say, Obviously I am at fault for not exercising all my supervisory skills, so off you go to work with Dan again. Sure enough, he gets a little better, but once more, the minute you stop, he goes back to doing the minimum.

At this point, stop. It's like trading stocks. While you are investing your time in someone who is giving you a 1% return on an 80% effort, you could be investing your time with your better employees and getting an 80% return on a 10% effort. After a few tries, accept that Dan is a minimum employee, and spend your time with your better employees. Still expect Dan to work hard, make it clear that he must continue to meet the minimums, and say that you would certainly like to see him do more—just don't waste the valuable commodity of your supervisory time on a losing investment. If you want to see excellence, spend your time around excellence; if you want to see mediocrity, spend your time around mediocrity.

Don't get upset if discipline gets overturned. There is a process in place to protect both management and employees. You should want the process to work for both sides. If you as a supervisor make a technical error in the disciplinary process and it gets thrown out or overturned, use it as a learning experience. If you are high in the chain of command and overturn an officer's disciplinary action(s), it is critical that you discuss with the officer the reasons why, so that he or she understands what happened. Leaving a vacuum of information causes supervisors to take the attitude, "I did everything I could, but those people upstairs just don't care."

Most discipline has a time frame. In other words, most discipline is only on the record for a specific amount of time. When that time goes by, let it go, don't hold it close to you or the employee. If there are no further occurrences, then you and the employee have done what you needed to do. If you issued discipline during an evaluation cycle, it should go on the evaluation. Likewise, if the employee has corrected the performance issue and there have been no other violations, then you should note that as well.

Fringe Employees

Now and again, you will get a fringe employee. Hopefully, you will not hire any by mistake, but they can also be inherited from past supervisors. Fringe employees are those whose performance and behavior are continually below the bar despite the best efforts to help them. The issue may be constant tardiness or causing agitation in the station owing to interpersonal behavioral and/or anger management problems, among a myriad other problems.

I remember one young firefighter who was AWOL a couple of times; during vehicle checkoff, the officer had to leave the bay, only to find him with the pulse oximeter on a part of his anatomy it was not designed for! I cannot tell you what the saturations were, but I can assure you it created a problem.

No one likes to get the fringe employee. These people make life so miserable that organizations have established some methods that have historically been used to deal with these kinds of people—all of them wrong! You see, any organization that does not address this problem head-on is wrong. Furthermore, there are supervisors who refuse to do what is necessary to either fix or fire this kind of employee. Often this neglect has the tacit approval of the organization and the chain of command, because it's too much of a hassle to deal with the problem employee. We will talk more about accountability later, but to be brief, when you put your head in the sand and say, "I hope this gets better," it never does—it just gets worse. Ignoring this kind of problem, rather than dealing with it up front, will cost you much more later on.

The first option usually considered is to simply transfer fringe employees, keeping them moving so that no single supervisor or shift has to deal with them for too long. This way, no one is responsible for supervising and documenting the behavior. In many cases, supervisors just let these employees meet minimum requirements, because they were lazy, scared, or didn't get any help.

The second commonly used option is to send fringe employees to the "fixer"—that one company officer who we send our problem children to. Let me tell you, though, that officer quickly gets awfully tired of that solution. Moreover, by putting the problem individual with the fixer and walking away, all you will have done is to ignore the problem.

In your entire career, next to the death of a firefighter, you will never have a more stressful event than dealing with and evaluating a fringe employee. It will consume your time and your shift's time; it will detract from more important matters; and if you do not get the support you need from the organization, it will make you so bitter that you will swear that you will never tolerate such an employee again.

Hillary Rodham Clinton has said at least one smart thing, and that is, "It takes a village to raise a child." Well, with fringe employees, "it takes an organization to fire an idiot." Dealing with fringe employees is not a single-supervisor event. It requires support and guidance from the chain of command, help from HR, consultation with and direction from the legal department, and continual support from the shift. It requires that the organization get behind the officer who has been assigned or inherited this unenviable task.

Organizations have historically dumped problem employees with other supervisors and then assumed the problem has been taken care of. In reality, what this does is sets a supervisor in a life raft, adrift alone in the sea. If you do not handle this situation in the right manner, though, you will end up with a bitter supervisor, a worn-out shift, and a fringe employee who stays on the job because the documentation is not comprehensive enough.

The purpose of dealing with fringe employees is to correct their behavior and performance and thereby salvage them. It has been my experience, however, that these employees are usually not salvageable. Thus, the fix-it-or-fire-it mentality needs to be geared toward two specific goals: first, changing the behavior; and second, documenting everything we do.

Rules for dealing with fringe employees

The very first thing you need to do is to ensure that you understand the process and know the rights of the employee. You can bet your stars that they know their rights and know the policy better than you will ever know it, since they live and die by technicalities and policy. The bottom line is that everything you do when you get to this point has some legal, HR, or technical implication with regards to documentation, time frames, and process. In all organizations, a process and a set of labor and management rights exist that you must familiarize yourself with, to play the game right and win. In Virginia, we have the Firefighter/EMT Bill of Rights by which we must abide. If a grievance is filed, certain steps must be taken and time lines must be followed, to protect both parties. The chain of command and other departments within the organization can help you to understand all aspects of the process, including the technicalities.

The next step is to determine where the official HR file is. You can keep reports, records, and documentation in many different places, but in legal terms there is only one official HR file for the employee. You need to find out where that resides, because every piece of documentation that you create needs to end up there; if it does not end up there, it will be as though it never existed. In the system I once worked in, the official file resided at the HR office for the city. The file kept by fire administration and the file kept in my office meant nothing when it came down to brass tacks. Thus, it is critical to find out where this is, so that everything you do is copied into it.

When you get to this sad stage in which you are in the fix-it-or-fire-it mode, if it's important enough for a sit-down, it's important enough for write-down. If it's not in writing, I guarantee you that the employee will claim that it never happened. It's a shame, but realize that these people must be documented with as much care as if you were writing history for all humankind.

Documentation in the right form is critical to success. Because it will be used by you, for you, and against you, it must be timely and within the guidelines established by your organization and process. Any documentation that you undertake to record poor performance—or even when you are simply writing a reprimand—must include specific pieces. When dealing with the fringe employee, make sure all of your i's are dotted and your t's are crossed. The following are the minimum standards that you want to adhere to when documenting behavior:

1. *State what was done wrong.* Make sure you clearly state what the violation was, when it occurred, and how it was discovered.
2. *What rule(s) or policy(s) did this violate?* Extract verbatim the policy, procedure, guideline, or rule that was violated. It is a good idea to cut and paste the direct verbiage from the manual into your document. If there are multiple violations, list all of them separately.
3. *What should have occurred?* Clearly state what the employee should have done with regards to the policy that was violated. List each policy and each expectation separately.
4. *What will happen if this occurs again?* This comes directly from your policy and is based on several factors, including how many previous violations the employee has for this or other events. Is this particular policy a cumulative policy—that is,

do subsequent violations occurring within a given time frame incur specific penalties? This can be indicated by a general statement, such as "Future violations of this kind or additional policy violations may result in loss of pay, suspension, demotion, or termination." These criteria come directly from your organization.

5. *What is expected in the future?* State the organizational expectations for future behavior by this employee, clearly identifying what is and is not acceptable.

6. *What will you and the employee do to make this happen?* This is a **work plan**. Notice this is in boldface, because if you don't do this, you will have just wasted your time: if the matter ever goes to court or a grievance panel, they will ask what you did to reeducate and assist the employee in reaching goals. This can be indicated by a statement—in conjunction with assignment to a course to beef up the skills in question—such as "Firefighter Sargent will report to the Fire Training Center and attend a driver-operator program from March 1 to March 15, at which time his proficiency shall be tested in accordance with standard." The plan that you put together could lay out company training, in addition to what you are already doing to assist the employee. It might include referral to employee assistance programs (EAP) for anger management—or, for a matter involving a physical fitness issue an employee is experiencing, a referral to a health and safety officer for an upper-body strength program. The bottom line is that you must develop a plan, in writing, to be disseminated among the employee, you, and the chain of command, that will be placed in the employee's official HR file, so that you can show you made every effort to provide the necessary KSAs for the employee to be successful.

7. *Specific statement regarding right to appeal/grieve.* Find out what your system requires and put in a statement that tells the employee what his or her rights are regarding this action.

8. *Signature page.* All documents need to be signed by all parties as accepted, seen, and delivered.

This process is repeated every time you document an event. Check with your chain of command, HR department, and legal staff to see if your system requires more than the minimum standards outlined here.

Documentation and Evaluations

So there you have it: we have collected a bunch of documentation all year long as we talked with our employees about their progress. Now what happens to all this stuff—good and bad, but mostly good—that has been collected over the year? Well, at some point in time, you are going to do the employee evaluation. That means you are going to have to take the time to pull up the form provided by the organization, making any necessary comments and additions, and then sit down to review the amended form with the employee, getting it signed and moving it on to the HR file.

Let's talk about this in sections. First, everything you documented needs to end up referenced in the evaluation. It's critical that you do so, because the evaluation is a milestone and a cumulative event in most HR systems. What that means is that once the evaluation has been completed, filled out, signed sent to the official HR file, no other documentation exists on that employee. Thus, if you fail to reference and identify all the documentation that you did over the year and fail to place it in the evaluation, you will have wasted your time.

Suppose you had a fringe employee who received five reprimands over the year and you were going to mark the evaluation in such a manner that it would have identified him as not meeting minimum requirements. In the process of doing the evaluation, you do not reference any of the reprimands, but simply say that he or she did not meet the requirements. You review it, get the employee to sign it, and out the door it goes. Well, guess what? In most cases, it will get kicked back, and you will have to change the evaluation because you did not provide the supporting data for your conclusion. Moreover, if not listed in the evaluation, all of the reprimands will have just disappeared, because once signed, the evaluation is the only official HR document.

Every piece of paper you complete, good or bad, over the year should be considered an event. At the end of the evaluation period, all of those events need to show up in the cumulative document we call an evaluation.

Now suppose you know you are going to have a bad evaluation (not meeting minimum requirements) on an employee and suppose that there will be some sort of consequences for that bad evaluation, such as loss of a promotion or of a pay raise. As a good supervisor, you have documented, worked with, and advised the employee and your chain of command throughout the year, so there should be no surprises. Nevertheless, you can anticipate that fringe employees will play the game—grieve, threaten to sue, and otherwise make your life difficult—so what can you do?

First begin the evaluation early on, at least 90 days before its due date; you can always add details and revise toward the end. At this early stage, also let your chain of command for feedback, to verify that the documentation so far is tight, meets the criteria, and is appropriate. Then, ask permission to take the evaluation and go to HR to let them look at it; since they tend to sympathize with the employee, you will get an idea of how your evaluation will be received. Next, get permission from your chain of command to have the legal department review it to make sure it's correct from a policy/legal standpoint. Finally, take all this feedback and use it to make the adjustments and as a learning tool; you will be surprised to find out about what you can and cannot do during this process.

Once the employee gets the evaluation and either blows up or calmly calls an attorney or union representative, you can rest assured that everyone who is going to be on your side is already aware of the situation. No surprises await, so you are all together, in the same life raft.

The organizational key is for the entire chain of command to support the initiative of maintaining documentation and doing thorough evaluations on all employees. The shift must pick up more of the workload, as you spend time documenting and attempting to assist the fringe employee. The shift must be prepared to do extra drills in support of the work plan that you develop, and someone on the shift may have to step up as a mentor.

Hopefully, you will never be faced with such a stressful situation, but if you are, you need to be prepared. I have had to deal with these situations as a supervisor, and as a division chief, I have had to assign some of these problems to my officers.

A word about personnel files

Personnel files are the connection between the organization, the supervisor, and the employee when it comes to a lifelong employment history. While each state has laws regarding what can and cannot be in a personnel file, who can and cannot gain access to the file, and exactly what contents can be accessed at all. Following are some general rules, but there is certainly no substitution for specific legal counsel from your legal and HR departments, to clarify how these rules relate to your specific organization.

Some organizations will choose to separate the personnel file into sections, while others might not. What is important is that under the Freedom of Information Act, it's possible that the HR file could be requested if some form of litigation takes place between employee and employer. Most records personnel file records fall into one of three categories:

- *The main personnel file.* This usually includes documents that show the history of the employee with the organization. Employees

usually have the right to inspect these documents, which are for day-to-day administration. The main file may include the following:
- Employee's application and resume
- Employment letter
- Payroll items
- Compensation change orders
- Employer-assigned duties and job descriptions
- Notices of commendation, counseling, discipline, and/or termination
- Attendance records
- Grievances

- *Documents the employee, but not a supervisor, may access.*
  To protect employees from potential discrimination, supervisors in your system may not have access to the following information:
  - Medical and health-related information
  - Health and other benefit claims, usually excluding general health care enrollment forms
  - Workers' compensation claims
  - Wage garnishment information
  - Immigration forms

- *Documents an employee may usually not inspect.*
  With the exception of the attorney/client letters, supervisors can usually have access to the contents of this file, but usually only after the need to know has been established:
  - Letters of reference
  - Documents related to the investigation of a possible criminal offense
  - Attorney/client privileged communications concerning an employee and litigation-related materials

Finally, a word about records retention is in order. Most organizations should have a records management policy that supervisors should have access to in case they have specific questions. Here are some general rules for records retention:

- Pension records are typically retained permanently
- Employee medical records must be retained for 30 years
- Occupational safety log summaries must be retained for five years
- Payroll records, rate schedules, employment applications, and other individual employee information must be retained for six years

- Unemployment, Social Security, and tax/withholding records must be retained for four years
- Family and Medical Leave Act records must be retained for three years
- Immigration forms must be retained for two years from the date of hire
- Employment applications of those not hired must be retained for two years

Some learning points on documentation and attitude

Use the same guidelines to evaluate everyone. Notice I did not say evaluate everyone the same; there is a subtle but important difference. Your job is to use the same guidelines and policy for everyone, but not everyone is the same. Discipline is progressive, as is reward. Someone who is consistently doing great things for the organization deserves to get more than someone who is not; someone who has received three reprimands in one year gets a different level of punishment than someone who has received only one.

As a supervisor, you cannot have a zero-defects mentality. First, human beings are not zero-defect machines; all of us make mistakes. Second, if you want people who take risks in trying new things and evaluating new methods of doing business, then you can expect some failures. Don't expect everything to go right all the time.

The key question you should ask right out of the gate when you start to investigate an incident is, "Was someone negligent?" Negligence is different than making a mistake or having an accident. There is a big difference. For example, negligence is putting your portable radio on the ground and having a truck run over and destroy it; an accident is having your radio in your turnout coat and, during a fire, having it slip out of the radio pocket and fall through the window, destroying it.

Repeated mistakes of the same nature are a problem and need to be handled accordingly. Someone who continually violates the same policy or comes back to the same negligent act over and over is a problem. One definition of insanity is doing the same thing over and over again and expecting different results.

A guideline that I have developed for when you should never accept mistakes that are in any way due to poor technical application or negligence is what I call the 5% time. That's the time during which we are engaged in emergency operations on the fire ground or rescue scene. Don't get me wrong: I have been known to push Phenergan on a highly nauseated patient without getting the physician's order by accident, thinking it was a standing order.

It's unfortunate that we do sometimes hire or inherit people who don't belong in the organization. As stated previously, document everything and always follow and administer policy. Keep a journal, especially when you are involved in a long-term evaluation process to fix or fire an employee. Journals cannot be used for discipline, but if you have to go to a personnel board or to court after the employee has been fired, you are not going to remember the events in sequence over a two-year period. You can bet that the employee is keeping a book on you, so you had better do the same. Understand, however, that journals are "discoverable" evidence if the legal system does become involved, so keep the entries professional and restrict them to the specific incident, as a reminder of exactly what happened.

The Hoover

I was a young captain at Fire Station 3, a busy company stuck back in the alley, aptly named the Alley Kats. Enter a rookie firefighter nicknamed Hoover. Hoover was a big kid, well above the six-foot mark, but his faculties did not match his size. As legend would have it, the name Hoover stemmed from a specific event when this young man was a volunteer firefighter at Fire Station 2.

There was a Krispy Kreme doughnut store down the boulevard from the station. In the early-morning hours, when the doughnuts were just being made, the station would send someone to pick up three or four dozen, steaming-hot, still-rising doughnuts fresh from the oven. It was rumored that as a young volunteer, Hoover had consumed two dozen fresh, hot doughnuts by collapsing them into a single ball and swallowing them in several bites. It was also rumored that the rising process had not finished and as a result, the doughnuts expanded in his stomach, pressed against the inferior vena cava, shut his blood supply down enough to cause a syncopal episode of some duration that required an ambulance ride to the hospital to get his stomach pumped. I cannot confirm the physiology behind this theory, but people who were in the department before I was hired swear they were present for this event.

Hoover came to us as a rookie firefighter with both volunteer fire and EMS background, but from day one, there was something about his skills and abilities that led us all to believe we had our work cut out for us. He never seemed to get things quite right; he was slow; and on EMS incidents, especially when there was blood, he was difficult to find. We had several counseling sessions about what was expected, and he often asked me whether I thought he was in the right job. I replied that we would keep on working on it and that if I came to him and said, "I think you should consider different employment," then that would be the clue that I didn't think he would make it. It was a long first three months, with multiple counseling sessions and several pieces of paper (reprimands, work direction, and expectation letters), in an effort to get him on track.

As fate would have it, we departed one afternoon for a motorcycle accident at one of our busy intersections. Big Frank was driving, Hoover and Gary were in the jump seats, and yours truly was up front. As we rounded the corner just behind the medic unit, you could see a damaged car and the motorcycle, and the motorcyclist was lying in the street with that telltale bright red bloodstream emanating from the helmet and meandering down the road like a small river. This guy's dead, I thought to myself, and with horror I watched the medics exit the unit and start CPR. It was like slow motion, and I was inside the cab yelling, *No!*

We exited and surveyed the scene. The motorcycle had run the light and struck the vehicle in its left rear door. Interestingly enough, the motorcyclist then ejected from the bike and went through the car's side window, through the back seat, past a baby in a car seat, and out the opposite window, like a human bullet, landing squarely on the his head after using it as a battering ram!

So off we go to help—paramedic Captain Sargent to the rescue! We joined in, getting two suction machines going with catheters in the motorcyclist's mouth, like a dentist office except we were pulling more blood than the machines would handle. As in the old days, we called the hospital to stop resuscitative efforts, but to my befuddlement, the doctor insisted, "No, fly him to the trauma center."

If you are going to play, then play, I always say, so I figure an airway is in order if for no other reason than to make it look good. Big Frank hands me the laryngoscope, the suctions are going, and I cannot get the tongue moved around to see past the pool of blood. Hoover is no where to be seen, but I am past that now. "Frank," I yell, "hand me the Magill forceps," figuring I can move the tongue enough at least see the vocal cords or some bubbles. I reach in and, lo and behold, when I snag the tongue and pull, the entire soft pallet, tongue, teeth, and oropharynx comes out in my hand. I am holding this in the air like a trophy; it is hanging from the Magills like a large bloody fish, about the length of my forearm.

As I glance to my left, I see the young woman who was driving the car look at me with eyes wide and pass out on the grass. Frank can hardly contain himself: "What you gonna do now Mr. Paramedic Man?" I look around, stuff the entire mess back in the mouth—need I add, not anatomically placed—look at the medics, and declare, "Bag him!"

Soon the helicopter arrives, the patient transfer is made, looks of disbelief are exchanged, and the flight paramedic is ticked off because blood is sloshing back and forth across the floor of the entire rear of his helicopter. They take off, we go to clean up the scene, and I ask Gary, "Where's Hoover," to which he replies, "I don't know."

As I round the rear of the engine, here sits the wayward son, on the rear step, shaking, mumbling repeatedly to himself, "I can't do this." We console him on the way back to the station, and off he goes to six weeks of EAP. I think to myself, This is probably not the job for him. I document the incident, periodically check on Hoover's progress, and await his return.

On his first shift back, Hoover was sitting at the kitchen table drinking coffee and reading the paper. Little did I know that Gary had gone over to Be-Lo's grocery store and purchased a pig's brain. While Hoover sat there peacefully, Gary entered the kitchen and yelled, "Heads up!" casting down the pig's brain onto the table; like a hockey puck on ice, it slid under the paper and landed squarely in Hoovers lap. Before I could even gasp, the paper dropped, Hoover was shaking and once again mumbling, "I can't do this." Another six weeks of EAP followed. In today's fire service, for pulling such a prank, we would more than likely have been looking for a job ourselves, but those days were kinder and gentler.

When we finally got Hoover back from his second EAP, he and I had a long talk about what the job requires. I suggested that it might be in his best interest to find a job he was comfortable with and that it would be to his advantage to take the initiative to resign, rather than have me do all the paperwork and documentation on his job performance. Unfortunately—for me and him—he had friends in the department who convinced him that mean old Captain Sargent was out for him and not to resign.

As a result, I spent the next eight months documenting, developing, and implementing work plans and evaluating Hoover's performance. In the end, he saw it coming and resigned anyway. I liked Hoover. He was a fun guy, and I am sure he will find his calling. Unfortunately, he was just not cut out for the fire and rescue services.

Sometimes we hire or inherit people who are not right for this job. That does not make them bad people; it just means they made a wrong career choice. Some are smart enough to recognize it and move on. Others require suicide by supervisor and make it miserable for everyone. You may never have to face this kind of situation in your career, but if you do, you had better have the leadership and management tools to deal with it, as well as a supportive and educated chain of command and a strong will.

Notes

1. Harari, Oren. *The Leadership Secrets of Colin Powell*. New York: McGraw-Hill, 2002.

Chapter 9
Prejudice, Diversity, and Sexual Harassment

> *One day our descendants will think it incredible that we paid so much attention to things like the amount of melanin in our skin or the shape of our eyes or our gender instead of the unique identities of each of us as complex human beings.*
>
> —*Franklin Thomas*

It Is What It Is

This chapter is intended to raise your awareness of your responsibilities as supervisors, not to provide you with the legal or policy positions that often drive these very volatile issues at the local or regional level. I am not an attorney, and courts and people can sometimes be fickle. Therefore, the best direction with regards to the topics covered in this chapter may come from a ruling by the legal or the human resources department. Here, we will outline federal law in terms of the requirements it establishes and how they affect your decision making.

This is a hard chapter to write, since my personal experiences have made me fully cognizant of the varying opinions that exist. My intent is not about righting the wrongs by explaining how people and institutions use race, diversity, and a range of other issues to further their own personal agendas. Neither is it about discussing how certain racial or ethnic groups can have exclusive organizations and not be called racist, while other ethnic or racial groups would be crucified for doing the same. It's not about quotas, because I do not have enough data to argue intelligently for or against quotas, and I am positive that I could never defend quotas personally if we were to hire anyone of any color or background who was unable to do the job. It's not about preferential hiring or testing procedures that are skewed so as to make it either easier or harder for someone to enter the service. I cannot explain these things without stirring up emotions in people, myself included, on one side or the other of the aisle. I prefer to simply say these are outside my circle of influence, and they are what they are. I am smart enough to know they exist, wise enough to know where and when and around whom to express my opinion on any of these given topics, and savvy enough to understand that as a supervisor I have a responsibility.

On a personal note, I prefer to view people as people—saying, Here is what is required to do the job, and hiring people based on their ability to perform. I try very hard to see past the amount of melanin in one's skin, religious or culinary tendencies, mode of dress, or anything else that makes us different from one another. I prefer to focus on actions instead of surface traits and make my decisions based on that. With the way I dress, I have no business looking at surface traits.

This chapter is not about interpretation and process by law or legal agencies and why they make the decisions they do. As I write this, for example, the Department of Justice has written to the police departments of Virginia Beach and Chesapeake about an investigation into whether their entrance examinations are discriminatory against African Americans and Hispanics. The hiring practices of the Virginia Beach police and fire departments were called into question in 2004 when a study showed that it didn't have the same proportion of minority employees as other city agencies. Some—such as the Department of Justice—would say that alone was proof of government bias, but that conclusion would confuse unmet hiring goals with pernicious intent.

The problem according to government lawyers was that 59% of the black applicants and 66% of the Hispanic applicants passed a standardized math test given to new recruits, compared with 85% for whites. Applicants had to have a score of 70% to pass. This test is administered all over the country and includes questions that wouldn't be out of place in a middle school exam. I am going to show you some of the math questions so you can see what I mean. The only prejudice here is against anyone, regardless of ethnic or religious background, who cannot add, subtract, multiply, or divide. Here are some of the questions:

1. On Tuesday, Officer Jones worked the 3 p.m.–11 p.m. shift. At 10:55 p.m., he was called to the scene of an accident, where he remained until 1:30 a.m. How long past his regular shift did Officer Jones work?
   a. 55 minutes
   b. 1 hour 50 minutes
   c. 2 hours
   d. 2 hours 30 minutes
   e. 3 hours 5 minutes

2. An appliance store is burglarized and 21 clock radios are taken, with a total value of $1,050 stolen. What is the average value of the radios?
   a. $10.50
   b. $50.00
   c. $55.00
   d. $105.00
   e. $22.05

Why do I take your time with this? In the first place, you need to understand that sometimes lawyers, organizations, and individuals need a remedial course in common sense. Furthermore, sometimes the direction, rulings, or positions of organizations or individuals is simply illogical.

Importantly, recognize what you can and cannot influence. Your influence is at the supervisory level of your given unit. You should use this influence to create a work environment free from the aspects that lead to these issues, a place where everyone, regardless of race, creed, or color, can have a place to work and thrive and maintain human dignity.

Finally—and no less important—this chapter is about keeping out of trouble. If you want to land in deep water immediately, let sexual harassment, prejudice, or ethnically or racially biased actions infiltrate your work space. Don't try to rationalize harassment, and don't try to make sense of bias; realize these are part of the world and understand what your roles and responsibilities are in the work environment.

Sexual harassment, prejudice, diversity, and pornography all are difficult subjects that many supervisors would prefer to have go away. It's an unfortunate fact of life in the fire service that dealing with these is incumbent in the job.

The Changing Face of the Fire Service

For years the fire service was predominantly composed of Anglo-Saxon white males—and in some communities, specific immigrant populations, such as Irish, Italians, or Germans. In those days, the attitude was that people of color, different ethnic backgrounds, and different religions, as well as women, need not apply or volunteer. Our world has changed; we have begun to break down the barriers that have historically kept the fire service from becoming a diverse work environment.

This book is all about change. If you have received only one message so far, it should be that the only thing certain about change is change. When I first joined the fire service, a bumper sticker prevalent on many private vehicles read, "VBF&R. Don't send a girl to do a man's job." We can all agree that those days are over, as our workforce and the attitude that we will accept as rational human beings have changed.

Historically, at one time or another, every fire station in America, even in Salt Lake City, had a collection of pornography. As a chief justice once said about pornography, "I can't define it, but I know it when I see it." It might have been the magazine rack or cabinet in the bathroom, the videos or 16-millimeter films shown at night around the fire station, or the pictures hung up in plain sight. With the advent of computers, the Internet, and cable and satellite television, pornography can be accessed at the touch of a button. Today, however, the attitude toward the display of pornography has changed.

Men and women are different, thankfully. Viva la difference! The biological differences are accompanied by different perceptions about each other and about our capabilities to do a given job. However, the differences have led to the creation, not just in the fire service, of a *glass ceiling*. A woman's ability to compete physically with a man is not at issue; rather, it's about being able to perform the skills required as a person, and there many men on the job are in much worse shape than the women we hire. For example, far fewer women have had heart attacks and strokes on the job as compared to the men!

By the year 2010, females will represent twice as many workers as males in the overall workforce. Youth in the workplace is declining and was already well below 20% in 1990. As of the year 2000, women made up 47% of the U.S. workforce. Approximately 5,200 women work as full-time career firefighters and officers, representing just over 2% of the total.

In 1998, there were 27,000 African American and 9,000 Hispanic career firefighters, representing 11.8% and 3.9% of the total, respectively. African American women comprise about 10% of female career firefighters and officers. Detroit now has more than 20 African American women firefighters, including District Chief Charlene Graham, who was promoted in 1996. The District of Columbia Fire Department, which has been one of the nation's leaders in hiring black women, employs more than 50 as firefighters and an even larger number in EMS. The Oakland Fire Department currently employs more than 15 black women, out of a total of 491.

In New York City, fewer than 6% of FDNY's 11,000 firefighters are men of color, and women represent 0.3% of the total. Compare this to the overall population of New York City, which is 30% Hispanic, 25–30% African American, 10% Asian, and 51% women.

The following urban fire departments (having more than 75 career personnel) have the highest percentages of women firefighters:

- Madison, Wisconsin: 14.8%
- Boulder, Colorado: 14%
- Clay County, Florida: 13.8%
- San Francisco: 11.7%
- Montgomery County, Maryland: 10.2%

However, several large urban departments have no women at all.[1]

I think it's clear that our workplace does not simply require tolerance of diversity—it demands it. If the fire service is going to be successful in the years to come, we need to embrace change.

Diversity in the Workplace

So what is this issue of diversity about? I don't think anyone would disagree that we need qualified and capable people to do the job. Therefore, it's not about skin color or sex, it's about performance. That is to say, if you can perform the job, you should be given the opportunity.

Although there have been in the past and always will be people who use their ethnic background or their sex as a means to get and maintain employment, they do not represent the majority of minorities. I don't even like the word "minorities"; I prefer Americans. People should be judged based on their performance. The minimum performance criteria should determine who is competent—and thus who should be given a shot at the job.

Every member of society has strength and weaknesses, both physical and emotional. We know that firefighting and EMS require upper-body strength, a strong back, analytical skills, mechanical skills, cardiovascular health, muscular endurance, and a wide range of other physical and mental skills, as well as emotional stability. Everyone who comes to us has the ability to apply and strengthen their given skills.

I cannot tell you how women, blacks, Hispanics, Filipinos, or any other minority firefighters feel about their work space and treatment in any fire department other than my own. Conditions, working environments, culture, and history vary from place to place. Nevertheless, I can relate, from a secondhand perspective, what the minority firefighters at VBF&R told me about how they felt in the workplace and about what makes a good working relationship between people of different genders, ethnic, or racial backgrounds. They told me in no uncertain terms that we need to see "a relationship that involves the entire person." We need to look past all the obviously different surface traits, to look at the total person. Judge people on their ability to perform as members of the team and do what is needed.

Take the time to get to know people—not just minority firefighters—for who they really are, including background, upbringing, schooling, and influences. What influences one person may not influence another. For example, Martin Luther King, Jr., was a great man, and did some wonderful things for this country and civil rights; still, he did not have the influence on me, as a white kid growing up in middle-class suburbia, as he had on my black friends and black firefighters I have known, as well as on their parents and grandparents.

The minority firefighters at VBF&R told me to think about our relationship in terms of a shared history. Past actions of people and organizations have an impact on how people's perceptions of the world are formed. We all have to realize that we cannot change history, but those who do not study it are doomed to repeat it. That goes beyond American History: build a shared history between your team, your organization, and your shift, and think about the stories (history) of experiences, incidents, and events that are related by both old and new members. These events bind us in ways that we don't realize at all, creating a shared history that gives us a common point of reference.

The minority firefighters also indicated that we need to have relationships that are collaborative rather than competitive. Never did any of these firefighters indicate that it was important for you to eat dinner with their family, understood their religion, or embrace their culture as your own. In my organization, I never saw anyone try to force anything on anyone else. We were forturnate, however, because I have read of and visited other locations where this was not true. Each member brings to the table talents, education, experience, and capabilities that greatly enhance our collaborative effort, and the following are just a few examples: Having a Hispanic firefighter fluent in the Spanish can be a life saver—when we went to certain neighborhoods, having a racially diverse shift helped us effectively deal with the event. Women firefighters are inevitably much more adept at dealing with trauma in women, especially when it is physical or sexual. Our relationships need to collaborative and focused on the outcomes we want for the organization.

Finally, these young men and women reminded me that you have to show people that you value them. Your words, actions, and the interest that you show are the first steps in accomplishing that. In the team, there needs to be a strong sense that everyone respects each other and values their individual talents.

You can go to every seminar you want on diversity; you can listen to people preach about how it should be and why it is how it is now; and you can listen and you can watch—and you will still learn very little except about the legal aspects of the issue, because everywhere is different. The diversity aspects of working for the San Francisco Fire Department are entirely different from Salt Lake City, New York City, Miami, or Virginia Beach. There is no pixie dust that you can sprinkle to create an environment that promotes mutual understanding. Instead, it requires human skills and understanding—from human beings, dealing with human beings, in a human business—to develop a diverse and harmonious work environment. Ultimately, if you want to find out how minority firefighters feel and why, then you have to ask.

Equal Employment Opportunity: A Primer

All organizations are required to have an Equal Employment Opportunity/ Affirmative Action (EEO/AA) policy and someone with the expertise to evaluate complaints and manage them appropriately. EEO/AA polices are designed after the federal law, to promote equal employment opportunities by providing a internal resolution of sexual harassment complaints and/or discrimination complaints based on race, color, national origin, sex, age, religion, or disability.

The regulations of the U.S. Equal Employment Opportunity Commission (EEOC) are published annually by the U.S. Government Printing Office, at Title 29 of the Code of Federal Regulations (available online). There are currently 44 specific regulations, ranging from provisions on age discrimination to the Americans with Disability Act, that fall into the equal employment category. It would be impossible to examine all of them here, but listed below are some of the specific areas in which EEO applies:

1600	Employee responsibilities and conduct
1601	Procedural regulations
1602	Record-keeping and reporting requirements under Title VII and the ADA
1603	Procedures for previously exempt state and local government employee complaints of employment discrimination under section 321 of the Government Employee Rights Act of 1991
1604	Guidelines on discrimination because of sex
1605	Guidelines on discrimination because of religion
1606	Guidelines on discrimination because of national origin
1607	Uniform guidelines on employee selection procedures (1978)
1608	Affirmative action appropriate under Title VII of the Civil Rights Act of 1964, as amended
1610	Availability of records
1611	Privacy Act regulations
1612	Government in the Sunshine Act regulations
1614	Federal sector EEO
1615	Enforcement of nondiscrimination on the basis of handicap in programs or activities conducted by the EEOC.
1620	The Equal Pay Act

1621	Procedures—the Equal Pay Act
1625	Age Discrimination in Employment Act
1626	Procedures—the Age Discrimination in Employment Act
1627	Records to be made or kept relating to age, notices to be posted, administrative exemptions
1630	Regulations to implement the equal employment provisions of the Americans with Disabilities Act (1630.16 and Appendix to Part 1630: Interpretive Guidance on Title I of the Americans with Disabilities Act)
1640	Procedures for coordinating the investigation of complaints or charges of employment discrimination based on disability subject to the Americans with Disabilities Act and section 504 of the Rehabilitation Act of 1973
1641	Procedures for complaints/charges of employment discrimination based on disability filed against employers holding government contracts or subcontracts
1650	Debt collection
1690	Procedures on interagency coordination of EEO issuances
1691	Procedures for complaints of employment discrimination filed against recipients of Federal financial assistance

As you can see, the provisions are extensive. Thus, a degree of expertise is required in order to manage all aspects of a program of this magnitude. Following are basic guidelines regarding policy and procedure.

Every policy begins with definitions, ranging from discrimination to sexual harassment. For example, since we are going to specifically talk about sexual harassment, here is a definition of discrimination:

> **Discrimination**: The demonstration of bias, whether intended or not, against an employee with respect to the terms and conditions of his/her employment on the basis of the race, color, national origin, sex, age, religion or disability. Such bias may be demonstrated by the actions of another employee, of a non-employee, or by the application of City, departmental, divisional or other policy, practice or procedure to an employee or group of employees.[2]

Next you must have a process in place for dealing with those complaints. This includes, but is not limited to, right to file of the complainant, right of the respondent (the person named in the complaint), administration of the policy, confidentiality, informal and formal procedures for dealing with the complaints, time frames, criteria and method to investigate the complaint, and conclusion and notification guidelines.

Hopefully, you will never be involved in an EEO complaint, since they are emotional events. The best approach, though not always possible, is to be educated as to your responsibilities and the events that trigger such complaints.

Sexual Harassment: What It Is and Is Not

Sexual harassment affects everyone in the workplace. What we believe to be sexual harassment may not fit the category; conversely, we may be involved in actions that we believe to be harmless but are in fact sexual harassment. As in any system, there is always the potential for abuse, for individuals to file claims that either have no merit or do not meet the criteria. Checks and balances have been designed into the system to identify such abuses and the counter with the necessary action.

As a supervisor, your job is to ensure that the workplace is free of sexual harassment. Human resources and legal departments, as well as chain of command, should be available to assist you in this endeavor. Your first task will be to make understood the expectations regarding behavior around the fire station. Without understanding what sexual harassment is, though, it would be hard to lay out concrete behavioral expectations for your members.

Sexual harassment is unwelcome and unwanted treatment. Furthermore, it's harmful to employees and employers and affects victims' physical and mental health, as well as their ability to function in the workplace. Finally, numerous court rulings have established that sexual harassment is illegal.

There are two forms of sexual harassment: *quid pro quo* and *hostile work environment*. Quid pro quo means something for something. It is a kind of sexual harassment that involves threats or rewards. Threats include firing, blocking promotion, transferring, or giving a bad evaluation if the person does not go along with a sexual request; rewards include hiring, promoting, or giving a raise in exchange for a sexual favor from the person. Quid pro quo is usually a supervisor/subordinate act.

For example, suppose a chief officer approaches a female firefighter who has been asking for a transfer to the investigations division and indicates he can get her transferred if she sleeps with him or goes out with him. Or suppose a female supervisor promises a firefighter a glowing evaluation if he sleeps with her. Both of these examples, while very blatant, are quid pro quo sexual harassment.

Hostile work environment covers *regular and repeated actions*, or things displayed around the workplace that unreasonably interfere with job performance or create an intimidating, hostile, or offensive work environment. Hostile work environment includes sexual pictures, calendars, graffiti, or objects, as well as offensive language, jokes, gestures, or comments.

While quid pro quo is pretty straightforward to interpret, hostile work environment is often subject to the eye of the beholder; what may be offensive to one person may not be to another. Also, most court cases have required the behavior to be regular and ongoing in order to be considered sexual harassment. For example, a female firefighter must listen to remarks about her physical characteristics whenever she is in the bay doing vehicle maintenance, or a male firefighter is offended by a beefcake photograph put up on the outside of a locker by a female firefighter.

Recognizing sexual harassment may be difficult. Deciding what actions are okay and what actions are not can often be confusing. The best I can offer are some guidelines:

- Sexual harassment can be related to power—forcing someone else to put up with or do something he or she doesn't want to do
- Sexual harassment can be physical contact such as touching, holding, grabbing, kissing, hugging, "accidental" collisions, or other unwanted physical contact. In the worst case scenario and transitioning into violence in the workplace it could be rape or physical assault
- Hostile work environment can be verbal, such as offensive jokes or offensive language, threats, comments, or suggestions of a sexual nature
- Hostile work environment can also be nonverbal, such as staring at a person's body, leaning over someone's desk, making offensive gestures or motions, or circulating letters, e-mails, pictures, or cartoons that are sexually oriented

A lot is left open to individual interpretation. For that reason, you need a litmus test for sexual harassment. This can be performed by asking yourself two simple questions: First, would I want my spouse, parents, children, sister, or brother to have to see or listen to something like this? Next, if a citizen, councilperson, or some other external customer walked in and saw or heard this, would I be in trouble?

As a supervisor, how can you tell if someone, even yourself, might be perceived as harassing? This is not meant as condemning these actions in every context, but rather as a method for you to realize that someone could potentially perceive that harassment is taking place:

- Do you often touch people in your dealings with others?
- Do you use nicknames such as honey and sweetheart?

Prejudice, Diversity, and Sexual Harassment

- Do you tell jokes that the opposite sex, people with disabilities, or individuals of different ethnic or racial groups are the brunt of?
- Have you posted pictures of scantily clad men or women in the work space?
- Do you have a history of becoming romantically or sexually involved with colleagues in the workplace?
- Have you ever been called a ladies' man or a man chaser in the workplace?
- Have coworkers ever told you to keep your hands to yourself?
- Do you laugh or otherwise participate when colleagues tell jokes that degrade a group of people?
- Do you use sexual innuendos in your normal conversation?
- Do you ask coworkers or lower-ranking workers to take care of personal needs for you, such as getting coffee or running personal errands?
- Do you often compliment colleagues on what they wear or how they look?
- Have you ever responded, "I didn't mean anything by it," on hearing someone was offended by something you said or did?

That covers just about everything in the firehouse that was never meant to harm anyone, right? What I am pointing out here, which has been extracted from some very robust sexual harassment training programs, is that numerous innocuous actions in the fire station could nevertheless be perceived as sexual harassment. You cannot simply dismiss the concern of people who are inclined to view these actions as sexual harassment as being too sensitive, because as a supervisor, you will be held responsible for the actions of your personnel—and your own inactions.

What the courts have said

To determine whether an act is or is not sexual harassment, the courts use a set of behavioral standards to evaluate the claim of sexual harassment and universally apply these criteria when making decisions.

1. The first behavioral standard is *severity*. This refers to an offensive act that is extreme. There exists a wide continuum of inappropriate behavior, ranging from jokes to assault. Severe behavior represents the worst type of offense and is usually sexual assault or battery. When faced with an accusation, the EEO office or the courts will assess the severity. When inappropriate behavior is severe, only one occurence may be sufficient to qualify as harassment.

2. The next behavioral standard that is applied is *pervasiveness*. This refers to the frequency of an inappropriate behavior. An isolated remark or occasional slip of the tongue does not qualify as harassment. If less severe acts are continuously repeated, however, they can become harassment.
3. Next is the *reasonable-person* standard. This means that the courts attempt to view the harassment from the perspective of society, instead of what is normally accepted in the workplace. For example, even though rough language and racial slurs may be the norm among a group of employees, that would not be a reasonable defense against a charge of racial harassment.
4. Finally, the courts will ask whether the act was *unwelcome*. The conduct under complaint must not have been encouraged, solicited, or invited by the victim. A sexual advance, a request for sexual favors, or sexual conduct must be unwelcome to qualify as sexual harassment.

Supervisory actions

The supervisor should take a proactive approach, ensuring that all employees understand the expectations of the policy and the organization. Everyone who works for you needs to understand what constitutes acceptable behavior. Finally, as a supervisor, you need to set the standard by your actions. I know plenty of fire captains who cannot seem to stop dating firefighters in the same station or battalion, and in almost every instance, it ended badly. It's never a good idea to fish off the company pier.

So what do you do when you get a complaint that alleges sexual harassment? First, sit down with the policy and look at it, since you have probably never seen it or had to use it. The next step is to consult the chain of command and seek guidance and advice; it's possible that someone has been through this before, and you don't want to make the same mistakes.

When investigating a compliant, go through the following process:

- First ask who was involved. Determine whether it was an individual, a group, or a nondepartmental member.
- Then, ask what happened. What was the act that led to the complaint?
- Next ask where this happened. Was it in the station, away from the station, at another work site, on at an incident?
- Finally ask when it happened. Find out the exact date, including time, month, and year. It sounds foolish to go to such detail, but sometimes complaints are made long after the event.

Avoid asking why. You do not want to give the appearance of placing blame on any party. Your job is to investigate the facts, so don't try to determine why. Anyway, once you get the facts, the reason(s) should become clear without your having to ask.

When you meet with the employees involved, make sure they understand that you may have to ask some intimate questions. Explain the importance and meaning to them, to get better cooperation with your investigation.

Finally, document, document, document. Put in writing everything that you are doing. At the end of the investigation, provide a written report to all parties involved with your findings and recommendations.

Pornography in the Firehouse

Go to the Internet and search for "firehouse pornography," and you can read any number of articles about firefighters being fired for possessing pornography, articles about firefighters in jail for sending or possessing child pornography, and articles on disciplinary action for using departmental computers to access pornography. Therefore, a definition of both pornography and obscenity seems in order.

The Encyclopedia of Ethics provides a good definition in this context, defining pornography as "the sexually explicit depiction of persons, in words or images, created with the primary, proximate aim, and reasonable hope, of eliciting significant sexual arousal on the part of the consumer of such materials."

The most important development in the United States with regards to censorship and the Internet has been the Communications Decency Act (CDA). The CDA was voted overwhelmingly into law in 1996 and made it a criminal offense to send "indecent material by the Internet into others computers."[3] The law was attached to the Telecommunications Reform Act of 1996 and passed by Congress on February 1 of the same year. It was signed by President Clinton the following week. On the same day the bill was signed, the American Civil Liberties Union (ACLU) filed suit in Philadelphia on the grounds that that the statute banned speech protected by the First Amendment and subjected the Internet to restrictions that were out of line with regulations faced by other media.[4] After an injunction suspending the enactment of the law was passed in the U.S. District Court, the case of *Reno v. ACLU* proceeded to the U.S. Supreme Court. On June 26, 1997, the Supreme Court voted unanimously that the CDA was a violation of the First Amendment.[5]

The 1973 landmark case *Miller v. California, supra* (as modified by two subsequent cases) established a three-pronged test for determining whether a work (any material or performance) is obscene and therefore unprotected by the First Amendment. To be obscene, a judge and/or a jury must determine that *all* of the following obtain:

1. That the average person, applying contemporary community standards, would find that the work, taken as a whole, appeals to the prurient interest
2. That the work depicts or describes in a patently offensive way, as measured by contemporary community standards, sexual conduct specifically defined by the applicable law
3. That a reasonable person would find that the work, taken as a whole, lacks serious literary, artistic, political, and scientific value

Examples of hardcore sexual conduct that an obscenity law could include for regulation under the second prong of the test are patently offensive representations or descriptions of

- Ultimate sexual acts—normal or perverted, actual or simulated
- Masturbation, lewd exhibition of the genitals, excretory functions, and sadomasochistic abuse

There has always been pornography in the fire station, so what's the big deal about pornography in the fire station? Well, be that as it may, a little education will help you to make an informed decision.

The prohibition on pornography in fire and rescue services has been tested in the legal system and come up short. In *Johnson v. County of Los Angeles Fire Department*, the U.S. District Court ruled that "private possession, reading, and consensual sharing of *Playboy* are protected by the First Amendment." The 1994 decision described fire stations in Los Angeles County as offering "sufficient privacy for a firefighter to read a magazine, either in his private bunk area or in the relaxation area without exposing the contents to an unwitting onlooker."

The lesson here is that organizations and supervisors who control pornography to ensure that a hostile work environment is not created first need to determine whether the consumer of pornography can look in private without offending others. Fire stations with separate bunk rooms, for example, would meet this requirement, so long as pornography was removed at the end of the shift. However, posters, calendars, and magazines in public view or in the dayroom would be prohibited.

We have already spoken about sexual harassment, and by now you should have acquired a reasonable expectation that pornography may be offensive to someone in the workplace. However, the question arises of whether a zero-tolerance policy for pornography should be implemented. Although you may have a policy on what you can and cannot access with station computers, that does not equate to a zero-tolerance policy on pornography in the fire station.

To manage this effectively, an organization must develop a policy that limits the use of company equipment, vehicles, and supplies to official business only. This policy would effectively prohibit pornographic images and materials that someone might find offensive from being downloaded or viewed on workplace computers, copied on organizationally owned equipment, posted on a bulletin board in the station, and disseminated by interoffice mail. Additionally, the policy would prohibit the viewing of pornography on television and would prohibit illicit behavior in the ambulance, fire truck, or station.

Without the support a policy of this more comprehensive nature, it is difficult to defend a zero-tolerance policy should an employee decide to download pornography at work. Such a zero-tolerance policy was tested when a 20-year veteran of the St. Petersburg, Florida Fire Department was fired when he was linked to pornography on the station computer. In defending—and winning—the case, attorney Robert McKee said that a zero-tolerance policy could not apply to computer-based images while at the same time permitting adult material on the fire station television. Thus, if you have satellite television that is not blocked, you do not have a zero-tolerance policy; if you have cable television that is not blocked and allows premium channels, you do not have a zero-tolerance policy.

Questions regarding personal space and access can also be raised. Speak to your chain of command and legal department about these, because my answers may not apply to every situation:

- *Can I use my own computer?* Usually not, because you are using city, county, or municipal telephones or wireless to tie into the system, thereby making it their system.
- *Can I have* Big Sweaty Fireman's Magazine *in my locker?* That depends on whose locker it is and what policy is in place on pornography in the station.
- *Who determines what pornography is?* Pornography is usually determined by community standards. What you can see in San Francisco or Vancouver is not the same as in Salt Lake City. The organization may also have a policy that outlines their interpretation, which might be much stricter than the community's interpretation.

Okay, enough about pornography. We all know that it exists, that it's an eight-billion-dollar-a-year business, and how to access it. As long as it's legal, whatever you do on your off time is your business. As a fire department supervisor, you had better be explicitly clear on what is acceptable at work.

Showstoppers in the Firehouse

How can you lose your job or put it in jeopardy in two minutes or less? By ignoring the lessons in this chapter, to begin with! Allow sexual pranks, language, or visuals to get out of hand in the station, and you just might find yourself saying, "Would you like fries with that?" Allow ethnic or racial slurs to become a normal, everyday part of descriptive conversation around the fire station, and you might wind up in front of a judge. Let any of the things we have discussed here become pervasive and normal for your workplace, and you will have real troubles beyond your wildest dreams.

Modern society is a very intolerant place when it comes to sexual harassment, pornography, and ethnic/racial discrimination. The quiet dogmas of the past, the things we accepted whether they were right or wrong, are long gone: these are bandwagons that will ride you out of town.

Several very good monthly newsletters provide updates on court cases and legal issues dealing with all aspects of fire, EMS, and human resources. As further reading, I would suggest *HR Monthly*, *Best Practices in Emergency Services*, and *Legal Briefings for Fire Chief*, among others. They offer excellent guidance for times when we have to ask these hard questions and make difficult decisions.

Notes

1. Women in the Fire Service, National Fire Protection Association, National Multicultural Institute, and President's Initiative on Race
2. City of Virginia Beach Policy 4.03 EEO Discrimination Complaints
3. Wilkins 1997.
4. Ibid.
5. Ibid.

Chapter 10
Anger and Violence in the Workplace

> *Violence is not merely killing another. It is violence when we use a sharp word, when we make a gesture to brush away a person, when we obey because there is fear. So violence isn't merely organized butchery in the name of God, in the name of society or country. Violence is much more subtle, much deeper, and we are inquiring into the very depths of violence.*
>
> —*Jiddu Krishnamurti*

What Is Violence in the Workplace?

Violence in the workplace is a serious safety and health issue. Its most extreme form, homicide, is the third-leading cause of fatal occupational injury in the United States. According to the Bureau of Labor Statistics Census of Fatal Occupational Injuries (CFOI), there were 551 workplace homicides in 2004 in the United States, out of a total of 5,703 fatal work injuries.

—OSHA

Patrick Henry Sherrill's name probably means little today, but in 1986, it was seared into the collective public consciousness. Probably more than any single person, Sherrill was responsible for making the general population aware of a kind of rage that had been increasing but had, for the most part, been overlooked. Sherrill, a postal employee, brought the issue of violence in the workplace into the media spotlight. Reprimanded by his supervisor and told he could expect a poor performance review, Sherrill fumed over the perceived injustices and soon thereafter walked into the Edmond, Oklahoma, post office and fatally shot 14 coworkers, wounded 6 others, and committed suicide.

NIOSH has found that an average of 20 workers are murdered each week in the United States. In addition, an estimated one million workers—18,000 per week—are victims of nonfatal workplace assaults each year.[2]

Outside the fire service, homicide is the second-leading cause of death on the job, second only to motor vehicle accidents. Among females, homicide is the leading cause of death. However, men are at a risk three times higher of becoming victims of workplace homicide than women. Additionally, 76% of all workplace homicides are committed with a firearm.

The fire service is not immune from these kinds of events. There have been murders of chief officers and staff in headquarters, there have been domestic-strife–related killings and woundings in the fire station. Most recently a firefighter in New York City was charged with aggravated assault when he struck another firefighter in the head with a chair in the kitchen, fracturing his skull and rendering him unconscious.

Workplace violence for the fire service may occur in the station, or it may be perpetrated against us while responding to or working incidents. Assaults, shootings, and stabbings are no stranger to fire and EMS personnel working the streets. For example, in Kansas City, a female firefighter-paramedic was shot during the response to an incident that escalated into a full-blown gun battle involving police.

Coupled with the violence is the human aspect of anger. We have all witnessed or been a party to an angry outburst between peers or between firefighters and customers. Although not as devastating as the acts of violence, displays of anger can still be cause for concern.

Progressive departments have already-established policies addressing violence in the workplace policies. OSHA has taken a mandatory step toward outlining a program for all fire and EMS agencies. Progressive agencies should already have a program in place and active.

Nearly one year ago, volunteer EMTs on an ambulance in Arlington, Virginia, were attacked at knifepoint by a psychiatric patient who had already been searched and cleared as "safe" by police. Thankfully, no one was physically injured.

Several months ago, paramedics from Montgomery County, Maryland, were called to an apartment to assist an unconscious person. Without any warning, the patient jumped up and began assaulting the paramedics. As they fled with the patient in pursuit, one of the medics tripped, and the patient slit the medic's throat with a knife. Fortunately, the injury was not fatal.

Stories of violence against firefighters and EMTs are becoming more common. Owing to the increasing levels of violence seen by health care workers across the country, OSHA has created a new standard on workplace violence, which may be issued as early as December 2008. Standard 3148, Guidelines for Preventing Workplace Violence for Health Care and Social Service Workers, will apply to all types of health care workers, including EMTs and firefighters who perform EMS first-responder functions.

The new standard will require employers (fire and EMS departments) to

- Create a written hazard prevention and reporting program
- Perform an analysis of workplace violence hazards
- Provide employees with training on recognizing and defusing violent situations or behaviors and establish systems to prevent violence
- Provide employees with equipment necessary to protect personnel who may be placed in violent situations
- Create a treatment program for those subjected to violence

The NIOSH definition of workplace violence is a great place to start:

> Workplace violence is a physical assault, threatening behavior or verbal abuse occurring in the work setting. It includes but is not limited to beating, stabbings, suicides, shootings, rapes, near suicides, psychological trauma such as threats, obscene phone calls, an intimidating presence, and harassment of any nature such as being followed, sworn at or shouted at.
>
> The Workplace may be any location, either permanent or temporary, where an employee performs work-related duty. This includes, but is not limited to, the buildings and surrounding perimeters, including the parking lots, field locations, clients' homes, and traveling to and from work assignments.

That is a pretty overarching definition, and it would benefit us to differentiate between anger and violence. Anger is defined by Webster's as "an intense emotional state induced by displeasure. Anger, the most general term, names the reaction but in itself conveys nothing about intensity or justification or manifestation of the emotional state." By contrast, violence can be defined more broadly: "exertion of physical force so as to injure or abuse" or "intense, turbulent, or furious and often destructive action or force."[1]

For the purpose of discussion, realize that an incident in which anger is displayed may not lead to violence. Alternatively, an act of anger may progress to become an act of violence, or a cumulative buildup of anger may result in a terminal violent outburst.

As supervisors, we must not tolerate anger or behavioral outbursts by our members while at work, either in the fire station or out serving the public. The potential ramifications should be obvious, from loss of public trust to lawsuits.

People at work may display anger because of life stress such as personal relationships, financial problems, chemical use or dependency, a feeling of injustice, or personal disappointments. It is up to the company officer to recognize the emotion and then seek a solution, by keeping the reasons in mind. This chapter is about recognition of the signs of anger that may lead to workplace violence and application of tools to defuse the situation.

Managing Anger and Disappointment

It's a fact of life that things will at times make you angry. Several times during my career, I have been disappointed or flat out angry with someone because of a decision or action, and sometimes I spoke or acted inappropriately. It is human nature to let emotions take control of us occasionally, but in our business and in our work environment, we cannot afford to let this jeopardize our operations.

Our personal lives and those of our coworkers are complicated, to say the least. Like all segments of the population but to a greater extent, we encounter marriage problems, financial problems, substance-abuse problems, and everyday stress. These kinds of life pressures sometimes spill over into our workplace.

Everyone has bad physical or emotional days. One of your daily tasks as an officer is to get a read on your members as they come to work. It is not a complex process: you know the people who you work with, and you know their demeanors, so you should be able to pick out changes in personality and attitude. The value of a 15-minute shift meeting in the morning to discuss business of the day and to get a read on your people should not be undervalued. When you see people who are having an emotionally down day or appear to not be themselves, pull them aside early and in private and ask if everything is okay. It's your responsibility then to adjust accordingly, sending these people home sick if need be or making a change in riding positions. The point is to take care of your people and take care of the team. Those whose heads are not in the game can be dangerous to you, the team, and the customer, not to mention themselves.

When matters escalate to shouting in the firehouse that is a clue that everything is not all right. Anger expressed verbally and at high volume is not appropriate for the fire station or its surroundings. We have all seen employees who initiate shouting matches about issues ranging from a dirty jump seat to not being properly relieved at shift turnover. Most such events are not because of spur-of-the-moment issues; there is usually some background insult between these individuals that has elevated to the point of conflict.

When you witness this, you have to defuse and then get contact. An immediate closed-door meeting is critical to reach a quick resolution. If the conflict involves members of other shifts, the other shift supervisor must be engaged as well. You job is to establish the expectation that this behavior is inappropriate, find out what is causing the conflict, and resolve the conflict so it does not occur again.

Understand that some people may require prescriptions to manage their emotions. While not widespread, a chemical imbalance can create aggressive behavior; this can be managed effectively with medication. I tell you this from experience, because I have seen several very good firefighters who took an entirely different approach to their fellow workers once they sought medical attention, were diagnosed, and received drug treatment.

Others will need anger management. It's unfortunate but true that some people we hire have an antisocial approach to life that is not screened out in our hiring process. That kind of outlook can be very destructive and is not conducive to life in the fire station or service to the public. Most organizations have EAP to address these kinds of events, but you cannot accept this and say, "That's just the way Chase is," because a continually angry employee will say and do inappropriate things to other members and/or customers. When you see repetitive behavior of this nature, counsel the employee and recommend EAP. This is a values-driven and action-driven decision that you cannot afford to ignore; if you do, I guarantee that at some later date, there will be an escalation against a member of your organization or a customer, and you will have many more problems on your hands than if you had simply addressed it early on.

Good old Uncle Bruno (Alan Brunacini) once told me, "Chase, when you're angry, don't do anything that makes you feel good." These are wise words for working with people. I just wish his advice had come about 10 years earlier! As a supervisor, don't make the same mistake as I have of taking action when you are mad, angry, or disappointed. Don't make a telephone call, type an e-mail, go face-to-face with someone, or talk bad about someone when you are angry. Take a couple of shifts to cool down and think about the event. When we are angry, our emotions bypass our brain, and things get said (or done) that we later regret; however, once it's out in the open, it is impossible to take back what you have said. This requires tremendous self-control willpower, but it's a critical aspect of growing up as a supervisor.

Getting bad news about a variety of issues can lead to disappointment and anger. You will not always get what you want, despite your best efforts.

There are matters beyond your circle of influence that affect how successful you are in certain areas of your career; politics, money, personal agendas, alliances, enemies, and timing, to name a few, will play a role in decision making at times. True leaders take bad news with a stoic approach that indicates (despite churning stomach, rising blood pressure, and increasing pulse rate), "I expected that anyway." It's a skill that you have to learn, and it's difficult to pull off. It takes practice and discipline, but it is worth learning how to do it well. I am not sure I ever mastered this skill!

In situations that have the potential to cause disappointment among your members, alter the method in which you present things. In his book *Leadership*, Rudolph Giuliani makes a very wise observation to always "underpromise and overdeliver."[3] This reduces the potential for building up people's hopes only to dash them later, with a scaled-down assessment. If you undercommit by speculating on a lesser outcome, then when the decision or product comes in at a higher level, people will be excited. If you overcommit and underdeliver, though, people tend to get disappointed, and if this happens repeatedly, the impact on your and the organization's credibility is significant. Disappointment can lead to anger and hostility, which you do not want.

As long as we work with people there will be emotions. Anger and disappointment are two of the emotions that we must recognize and learn how to effectively manage from both a personal perspective and within our work unit.

Recognizing the Potential for Violence

Violence in the workplace doesn't come only from angry employees. It can erupt from a number of sources and for a number of reasons. Before we move on to explore warning signs and actions, consider the following types of critical incidents:

- *Random.* The attacker does not know the victim(s) and probably doesn't work for the organization. This occurs most frequently to fire and rescue personnel during response or in hospital emergency rooms. Good street-level training programs are an excellent method to give employees the tools they need to anticipate and survive these kinds of events.
- *Authority directed.* The attacker targets an authority figure who can be either an actual, representative of the bad news, or a perceived source of problems. This can happen in response to disciplinary actions or as a result of terminations.

- *Vengeful*. The attacker targets people or an organization who are perceived to have done some wrong. The attacker may or may not know the victims. Such an attacker may be an employee or an external customer.
- *Domestic*. The attacker targets a coworker who he or she has some connection to outside work. Commonly, the attacker targets a former romantic partner or spouse. The victims of these attacks are overwhelmingly women; however, a recent instance of domestic violence led to the death of both a male paramedic and a female paramedic in the station.
- *Argument driven*. The attacker escalates the event and responds to an immediate conflict with violence.
- *Felony*. Although all of the previous events have the potential to be a felony, this represents in particular the murder or maiming of someone while a crime is committed—for example, a robbery or rape.

Most homicides in the sales and service industry fall into the felony category. Incidents in the fire and rescue services are usually random or the result of a domestic situation in which we become embroiled.

Most feared and, of course, most publicized are the random violent incidents by angry, disturbed, or disgruntled employees. In January 2006, during the writing of this chapter, a postal worker in California walked into the post office and killed five coworkers. That this type of violence is so nonsensically out of proportion to any possible grievance is what makes it so frightening.

In my career, I have witnessed workplace violence firsthand on several occasions. I remember one day at the station when an employee simply blew his top, took a Halligan bar, and begin to attack and dismantle the Coke machine. Thankfully, his rage was directed at an inanimate object. I have seen members throw punches in the station as a result of an argument that has escalated. I was party to a termination when an employee with a history of behavioral problems assaulted a patient on the gurney; unfortunately, past supervisors had refused to accept and deal with his repeated acts of anger and violence in the fire station, so it had taken a new officer, myself, to solve the problem, which had been compounded by years of action with no documentation.

Warning signs, coupled with your intimate knowledge of your employees, should be the baseline for your decision-making. Although some of what we will outline in this section may seem to come from way out in left field, I can tell you that no one ever expects a violent act in their work space—just like we never expect a firefighter to get killed on our shift. Choosing to ignore personal changes or warning signs because of your disbelief that this could be happening and not intervening early enough could make the difference between interceding in an event and allowing the event to occur.

There is a wide range of warning signs that a supervisor should look for to anticipate a violent or hostile act. Changes in personal behavior and a continual paranoia that the system is unfair should be reason enough for concern. Even while some warning signs may be so extreme that no confusion is possible, others warning signs may be more subtle and thus must be taken in context, by examining the circumstances and the individual employee's actions and reactions. Consider the following as early-warning signs of a potentially violent employee:

- *A history of anger management problems or violent behavior.* Although we should not be hiring these people in the first place, we often inherit them; also, people change based on emotional and personal events in their lives.
- *An obsession with weapons.* Any discussion of weapons as a method of solving a personal problem indicates that rational thought is not taking place. This is different from simply owning and using firearms. If owning an excessive number of firearms and enjoying shooting was an indicator, I would be locked up!
- *Making verbal threats.* Any direct or veiled threats need to be considered as real and immediately dealt with. We will address these in further detail in the next section.
- *Intimidating coworkers or supervisors by bullying, threats, and physical intimidation.* No one should use physical size or an aggressive nature to harass or intimidate another employee.
- *Obsessive involvement with the job with no apparent outside interest.* Although you might say this is 80% of the fire service—and I would hope that you are wrong—being a professional and being committed are different. The potential impact of having no other life than the job is that when something happens to disappoint the worker, change the work environment, or disrupt the normal work process, such an individual may feel as if their entire life has been turned upside down.
- *A loner who has no involvement with coworkers.* Again, this may well describe several members of your organization, but must still be considered as part of a total evaluation. After all, the case may be that they have little contact with coworkers but have a robust family life. The people you are looking for have no one inside or

outside the job. These kinds of personnel should not be hired in the first place, but sometimes people slip through the cracks.
- *Getting angry at criticism.* Any member could get angry, depending on the emotional aspects of the day, but escalating the event and chronic disputes with coworkers or supervisors are clear warning signs.
- *Verbalizing a hope that something bad will happen to someone on a continual basis.* We have all made off-the-cuff comments like "The best thing that could happen to him is to be hit by a meteor," but repeated and directed comments about an individual are cause for concern.
- *Expressions of desperation over recent personal problems.* Sometimes our problems drag us into a pit of despair. When this is expressed, it needs to be taken seriously, especially if it is out of character for the employee. Some people seem to be depressed throughout their entire life, but even these people should be evaluated when they begin to discuss extreme measures of desperation.
- *Fascination with other acts of workplace violence.* We have all seen the movies where the killer tacks up newspaper articles about his or her murders or of similar events. Expressing an interest in workplace violence should be suspect.
- *Continually pushing the limit of normal conduct.* This is typical fringe employee behavior and needs to be evaluated. Included here are pathological finger pointers, who blame others for their problems or actions. Being allowed to push the limits and exhibit unacceptable behaviors can escalate to a wide range of problems, including anger and violence.

All of these potential warning signs need to be evaluated in the context of the bigger picture, taking into account their frequency, the nature of the event, and the surrounding environment. The key as a supervisor is to take a several-pronged approach:

- Realize that anger, hostility, and ultimately violence in the workplace are very real aspects of the job.
- Know your people to be able to spot behaviors that are out of character and intervene early.
- Do not accept the actions of those few individuals who continually exhibit the aggressive posturing that often leads to anger and hostility in the workplace.

Evaluating Threats and Taking Action

Should one of your members make a threat of violence or hostility or should an outside customer do the same, specific actions are required of you. First, be educated about your organizational policy and cognizant of the steps to be taken when a threat is encountered. The guidelines presented here are generic and should in no way be interpreted as telling you to do something different from what your policy outlines.

First—and most important—take all threats seriously. While many of these may be expressions of anger made in the heat of the moment, they are still inappropriate for the workplace. Make sure you let your members know that reporting a threat is not an act of sabotage or that they are tattletales for doing so; on the contrary, it is expected in the workplace.

Your next step should be to confront, in a timely manner and in an appropriate setting, the member who made the threat, inquire as to why the threat was made, explain that it is unacceptable behavior, and note the problem and counseling in a formal file. In many cases, this may be all that is necessary to resolve the issue. An apology and explanation to the party that received the threat (perceived or real) is in order, together with an assurance to you, the supervisor, that the problem has been addressed sufficiently and will not recur.

As an officer, your job entails peeling back the layers of the onion to get to the root of the problem. Then you can intervene, examining the seriousness of the event and finding and recommending a solution. Solutions may range from professional help to a simple apology, but must be reached quickly.

Your job is also to offer assistance when it is needed. To do this, you need to make an honest assessment of the incident, by getting all the facts from all the parties involved and putting the pieces of the puzzle together. Only then can you offer solutions and assistance that will really remedy the issue at hand.

Never be afraid or embarrassed to offer an employee professional help, in the form of EAP, professional financial guidance, or an occupational health evaluation to determine fitness for duty. Many of the root causes of violent events are well outside our expertise, and that is why these professional support and counseling systems are in place.

If policy allows, it may become necessary for you to make a formal referral requiring that the employee seek help. Of course, it is always better to get voluntary compliance. The VBF&R policy allows mandatory referrals for known substance-abuse and violence-in-the-workplace violations. Know what your policy says, so that will know what options are or are not available to you as an officer.

When a threat is made, supervisors have a *duty to warn*. This is meant to protect you, the organization, and the employees against potential violence and deflects the liability that arises when a threat is ignored.

Every threat must be reported in some manner. Your policy should outline the steps required for reporting, based on the type of threat. The source of the threat is immaterial. If is a customer or a vendor, inform them that the threat will be reported. When reporting a threat, you have to ask yourself, "Is the employee venting anger?" If so, make sure to confront, instruct, and consider some form of referral.

I want to stress once more that you need to educate all employees that passing on a threat is not snitching, but is expected. Be aware of the legal and liability implications, at both the personal and the organizational levels, of non-response when there is a history of threats.

Finally, get the chain of command involved and seek assistance from them. You may never have managed an incident like this before in your career, and you may feel uncomfortable or unsure of how to proceed. The secret is to deal with it head-on and immediately. You took the job as an officer—now do your duty!

Defusing Violent Situations

No one wants to believe that a violent situation can occur in their work environment. It does, however, happen so we need to be prepared to deal to the best of our ability.

As soon as you sense the situation could turn violent, get the police on the way. This applies to in-station violence and on-scene violence. Once it escalates to a point where there is the potential for physical harm, I realize that everyone will do what is necessary to protect themselves; nevertheless, things sometimes occur that are out of our circle of control. It is important to understand your escape routes and safe havens in both station and operational environments. Don't try to take on a firearm, an edged—or even a blunt weapon. You will usually end up on the losing end.

There are two rules that apply to violent situations. The first is to keep your cool. Don't display anger, fear, or anxiety. Talk in a calm voice, lower and slower than normal, and don't be patronizing. To successfully understand the mindset of the person at this particular time, you must listen and attempt to communicate.

The second rule is to open up dialogue by asking what's wrong or simply listening, without comment or judgment, to the aggrieved party air his or her grievance. People generally get violent after a triggering event. Giving them a chance to talk—or yell—may calm them down. Listen attentively, maintaining eye contact, while they explain their problems. Closely watch the hands, since a violent act usually requires the use of hands to deploy a weapon. Asking what's wrong allows the aggrieved party to offer a solution, and the dialogue this allows either buys you time or begins to defuse and de-escalate the incident. Giving the aggrieved party the opportunity to solve the problem that triggered his or her anger not only will calm him or her down but also may lead to a solution that you can both live with.

Finally, help the individual to save face. If an employees are at stake, find a method of discipline that will preserve their dignity and won't humiliate them. Of course, being dragged out in handcuffs can be humiliating, but if it escalates to that point, you have no choice!

Although at times, violence can be irrational and thus have no logical solution, more often there are a range of possible remedies. Research has shown that employees or customers involved in an angry or violent action seek a particular remedy based on the gravity of the event:

- Upset employees may want to be taken seriously. If it has escalated to a violent action or outburst of anger, you had better be taking them seriously. Some people want to be shown respect that they feel they have been denied. By following the steps outlined previously, you may be able to establish a level of respect that lets you reach a solution.
- Sometimes upset employees want to be listened to. Break down the communications barrier by being attentive and seeking solutions from the aggrieved.
- In some instances, employees want compensation or restitution or to right a wrong. Don't make promises you can't keep, but listen to the employee and initiate discussion to understand what they want.

Finally, these people want immediate action. The steps you take in dealing with the violent event may provide the first indicator to the upset employee or customer that you are indeed taking action to resolve their grievance.

Conclusion

No supervisor likes to deal with events that get emotionally out of hand, especially when they involve his or her own members. These kinds of events, while limited in overall scope, do represent a real threat to the safety and health of you and your members. Furthermore, failure to act and respond appropriately can lead to personal and organizational liability and legal issues that are best avoided.

Notes

1. *Merriam-Webster's Collegiate Dictionary*, 11th ed. Springfield, MA: Merriam-Webster, 2003.
2. Sentera Health Care, Dealing with the Aggressive Employee (manual)
3. Giuliani, Rudolph W. *Leadership*, New York: Hyperion, 2002

Chapter 11
Decision Making
—It's Not Tarot Cards

> *It's choice—not chance—that determines your destiny.*
>
> —Jean Nidetch

Neurology and Decision Making

Decision making starts in and is in many instances controlled by fine balance maintained by the brain. A recent article in *Harvard Business Review* reports that this balance has been determined using advanced scans. A human "has a dog brain, basically with a human cortex stuck on top, a veneer of civilization," says author Gardiner Morse. He continues,

> This cortex is an evolutionary recent invention that plans, deliberates and decides. But not a second goes by that our ancient dog brains aren't conferring with our modern cortexes to influence choices—for better or for worse—without us even knowing it.[1]

Patients with brain injuries who have lost the section of brain that deals with the emotional portions of decision making have been shown to be unable to make decisions in a balanced and timely manner. Thus, it is critical that the brain be able to carry on a conversation between emotion and reason in order to make good decisions.

Sometimes emotion can badly distort our judgment. Members of our organizations have ruined their careers because of drinking and driving or domestic abuse. As the brain evolves during growth, there is a developmental process that interlinks the modern and primitive parts of the brain in order to make conscious decisions based on risks and rewards. Why do toddlers crawl up onto shelves where they can fall? Why do teenagers make decisions that seemingly lack willpower? The answer, from a purely scientific standpoint, is that their brains have not completed the wiring required for the parts of the brain to talk to each other. What makes us different from animals is that once our brains are fully wired and functioning, we can look to the horizon and evaluate what consequences might come from a decision to chase immediate gratification; likewise, we can get immediate pleasure from the prospect of some future gratification.

Part of the brain is geared toward the thrill of the hunt. This drive for reward by the brain is a primary source of bad judgment. Sex, chocolate, money, music, attractive faces, and hot cars, among other things, arouse the reward-system portion of the brain. This reward system relies on communications, between parts of the brain, that are enabled by the reactions of chemicals that scientists refer to as neurotransmitters. Dopamine is the primary neurotransmitter for the facilitation and regulation of pleasure. Dopamine is produced in the animal part of our brain and regulates the brain's appetite for rewards and its parallel evaluation of how well the reward meets the expectations. Also regulated in this portion of the brain are emotions like fear, loathing, and revenge, as well as other aspects of our animal brain.

The reality is that our brains are wired to communicate at certain levels and create both analytical and intuitive connections in making decisions. We can learn to control these aspects of our brain to some extent, but as we have seen, the brain and its chemicals are responsible for setting the stage on which we make decisions.

The Pitfalls of the Decision-Making Process

This entire book is about decision making. It's about all the skills, that officers need in order to begin their careers and make decisions. Regardless whether it is an emergency decision or a decision about discipline, planning, personal career moves, or training, both leadership and management skills are necessary. Throughout your career as a fire officer, you are going to be required to make all sorts of decisions. Many decisions you make could be life-or-death choices, while others may be prioritization to make sure you get the main thing accomplished. There are perils in decision making that new and old officers learn about as they go. Unfortunately, not too many officer programs teach decision-making techniques. Most of the time, we learn to make decisions through trial and error—and hopefully based on some training and experience.

This section is about identifying the pitfalls and how to avoid them—or in other words, about the tools available to you as an officer making decisions. Some very good scientific studies on decision making have been done. *Sources of Power, How People Make Decisions*, a book by Gary Klein, is an excellent starting point to search for data on how people in general make decisions. Here, we will focus on what is important to officers.

The more training and experience you obtain the better—or, at least in theory, the better your decision-making skills should become. As an officer, you should obtain all the training and experience you can get, even if it does not apply directly to the job you have at a particular moment. The more often you experience and see something, the more times you will be able to examine options and apply them, discovering what works and what does not. The organization must provide the necessary classroom education to its officers before they have to apply the decision-making process.

To get the level of training and experience that is required, several things need to happen. First, from an organizational perspective you have to exercise caution in placing someone in a staff job during his or her first promotion. My feeling has always been that the first assignment as an officer needs to be in emergency services, where one can learn to make decisions about critical aspects of the job. If 90% of our workforce is in operations, it makes sense to spend time teaching those KSAs first. Moreover, it's very likely that a staff officer may wind up in command of a fire or assisting at a major incident, and those decision and command skills cannot be learned behind a desk. When we put someone in a staff job right after promotion and leave her there, sometimes she gets promoted to the next-highest level and ends up back in emergency services at a command level while having never spent the necessary time on the apparatus, seeing and experiencing firsthand emergency events.

The downside of such an organizational decision is purported to be that since we put new personnel in emergency services (operations) positions and since there are only so many positions available, some of the officers in the field have to move to staff jobs. We sell this job to people on the merits of the 24/48 basis—one on/two off—or whatever novel schedule you follow. We need to begin to educate people from the time they walk in the door that they don't have a vested right to shift work and that, in fact, they may not do shift work their entire careers. Stop feeding the people you hire information that is not true: let them know early that the organization will put them where they are needed, and try to align that with their desires whenever possible. That brings me to my next point about learning how to make decisions.

To become a good decision maker, you need to mix it up. As an officer, you should expect to spend time in a staff position, since those tasks require a different kind of decision making, political savvy, and a wide range of other skills that you don't get to practice at the operations level. The decision to spend time in a staff position is ideally one made by the individual; however, if officers are not educated or mature enough to realize this and to make the decision for themselves, then the organization needs to make it for them. Organizations should establish this as a value and let people know this is an expectation. There is a seat for every butt in the organization.

The next factor that drives decision-making is the old PPGR (policy, procedures, guidelines, and rules) that every organization has. Sometimes you will be constrained or guided in your decisions by these Jersey barriers: Remember, "stay between the lines; the lines are our friends."

In a values-driven organization, decisions allow you to choose among many roads to reach the same outcome. Since we already spent time on this subject, I won't beat it to death here, but values certainly are another factor that drives decision-making.

At times, decisions should be based on human compassion. Decision-making cannot always be an accountant's operation, with columns for credits and debits. Based on circumstance and facts, you may need to throw the rules out the window and focus on doing the right thing, rather than doing things right.

Sometimes decision making is forced on you, and you are told what decision to make. This may be because the responsibility for making the decision in the first place was shirked (see chap. 5, under "Keeping the Kittens in the Box")—ignored or passed along to someone else. Alternatively, may be because of political or fiscal considerations from outside the circle of influence.

Decision making affects lives on the fire ground and organizational direction. At the strategic level, either on the fire ground or from a planning perspective, decision making is usually global in nature and takes care to look at a number of choices that need to be made to reach a long-term goal or objective. Tactical decision making applies to those groups of tasks needed in order to accomplish the strategic objective; in the simplest possible terms, individual tasks correspond to the individual decisions made in support of tactical direction.

As human beings, we tend to either to spend too much time evaluating a decision and end up in paralysis by analysis or to make snap judgments that end up taking us somewhere we don't need to go. There needs to be a balance between gathering data and ultimately making the decisions. I have always been a "70% guy": if I have 70% of the data, I will make a decision. It may not always be the best one, but we will get a decision and move on. With the exception of the fire ground, we can always add or subtract to get back on track if we find we needed a little more data.

Colin Powell's Decision-Making Guidelines

I have paraphrased these decision-making guidelines of Colin Powell here because I think they are universal and brilliant.[2] They provide officers with a universal method of staying out of trouble, maintaining credibility, and making decisions in a timely manner.

- *Don't get stampeded by first reports.* This applies to all decision-making, especially when you are angry or in critical situations like working the fire ground. The initial information is often incomplete and/or inaccurate and will begin to develop more clearly as time progresses. One of the real skills for fire ground commanders is to have the KSAs and savvy to use all their senses, understanding that the reports are not complete and nevertheless making and applying an appropriate strategic plan. When dealing with human resources and personnel issues, take your time. You never get all sides of the story right out of the gate. I guarantee that if you try to make a decision based on first reports, you are going to be wrong. Since plan A may not have been developed and implemented on the basis of accurate information, always have a plan B.
- *Don't let your judgment run ahead of your facts.* In particular, when dealing with other human beings, we let personal judgment get the best of the facts. Don't allow your personal experience with a given individual or group of individuals cloud your judgment and preset a decision. Make sure you assemble all the facts you need in order to make an accurate and informed decision.
- *Even with the supposed facts in your hands, question them if they don't add up.* Even when you investigate and collect all the information, it may not add up. Great detectives excel at recognizing when something is missing or just does not make sense. Perhaps you have been deceived or there is an economy to the truth being provided; perhaps you missed something in gathering the data; perhaps you did not ask why five times. Regardless of the reason, if you examine the information and it just does not add up, go back to the drawing board and find out what is missing.
- *Something deeper and wiser than bits of data informs our instincts.* You may have to make a decision that you know is right, regardless of what the data tell you. Opportunities present themselves that suggest an unconventional approach to a conventional problem. When dealing with personnel and making deposits in the emotional bank account, you may have to put more weight on the side of the human being. On the fire ground, you may weigh what the book says against what your experience tells you. The bottom line is not to disregard your instincts in favor or pure data.

- *Get the facts out as soon as possible, even when new facts contradict the old. Untidy truth is better than smooth lies that unravel in the end anyway.* I can't tell you the number of times when my department's leadership decided that it was better to hold back information than to produce it in advance of the event—the captain's job evaluation by human resources, the EMS program, and so on. If you think that people don't put together the facts quicker than you can establish them—or fill the vacuum with something other than facts—then you must be living on another planet! Let the troops know what you know when you know it. Yes, the situation might change, and yes, you might have to go back and say, "Well, that was not exactly correct," but that is nevertheless far superior to allowing someone else to fill the information void with something that is simply not true.

I have discovered that most great leadership ideas are not fads and have been around for quite some time. Had someone told me these principles when I was a young officer learning to make decisions, my life would have been much less complicated—with much less "Yes, Chief," "No, Chief," and "I don't know, Chief."

Gut Decision Making—and Why It Is Important

As in other businesses, there are often times in our business when it is necessary to make decisions from the gut. Decisions of this nature are emotional, requiring fortitude and nerve. Gut decisions require courage and confidence in one's ability and decision making, both of which are admirable traits for a leader. Usually, gut decision-making is used in times of crisis, when there is no time to weigh arguments or calculate probability of the potential outcomes. At times, we make these decisions on the basis of our own experience when the situation is without precedent and there is little concrete evidence pointing to a likely outcome. At other times, decisions are made in defiance of the evidence.

A number of studies have been undertaken to investigate gut decision making. In a study conducted at the Harvard Business School, Jagdish Parikh showed that respondents used both intuitive (gut and experience) and analytical (data) bases to make decisions. However, Parikh credited about 80% of their successes to instinct.[3]

There are also studies suggesting that using gut instinct as the basis for decision making is a dangerous game because our brains are prone to many natural mistakes. Historical data support the assertion that many bad business decisions were made from the gut, and even very intelligent people will say, "Don't trust your gut."

I disagree with this research at every level for the following reasons. First, few decision makers are going to ignore good information when they receive it. The balance between gut and brain is not an overriding chemical reaction that makes wise people with experience and training simply ignore what they have in front of them. However, when the information is not available or you cannot get it, you have to make a decision based on instinct. In reality, most people use a combination of gut and information to make decisions; research has shown that intellect informs both intuition and analytical thinking and that people's instincts are often quite good.

In the book *The Fifth Discipline*, Peter Senge succinctly expresses the balance between intellect and analytical skills:

> People with high levels of personal mastery . . . cannot afford to choose between reason and intuition, or head and heart, any more than they would choose to walk on one leg, or see with one eye. . . . A blink after all is easier when you use both eyes and so is a long penetrating stare.[4]

The other aspect of using your gut is based on history, experience, and thousands of years of neurological inbreeding in human beings. In our business and even when we are traveling, we have always said, "If it does not feel right, it more than likely is not." How many times when you have been in a bar have you felt that a confrontation was building up, even without extensive verbal exchanges? Sometimes you can feel the karma or energy in a place change, and you should not ignore this—it has been built into your genetic code, from when humans were hunters. How many times have you been on an aggressive fire attack when something just did not feel right and you decided to change what you were doing? How many soldiers are alive today because something just did not feel right and they changed their direction, patch, or approach? On the fire ground, instinct based on experience and informed by the senses can be used to make decisions and save lives.

Completed Staff Work

VBF&R and other organizations have adopted the completed staff work model to develop a process by which senior staff can evaluate ideas and determine their feasibility based on data and merit. Originally, this model was developed by the city as a management tool that they then required all departments to follow.

The completed staff work model has its pros and cons. The cons of completed staff work are that it can be misused by managers to stagnate excellent ideas under the guise of "more work needed" or "not enough data" to make a decision. On the one hand, the model can be used to slow a process down to the point of frustration when time and action are necessary commodities; on the other hand, I have witnessed pet projects given approval because of political connections, despite lacking the slightest application of the completed staff work model.

Completed staff work entails the study of a problem and presentation of a solution, by a staff member, in such form that all that remains to be done on the part of the boss is to indicate approval or disapproval of the *completed action*. The words "completed action" are emphasized because the more difficult the problem is, the greater is the tendency to present the problem to the boss in piecemeal fashion.

It is your duty as a staff member to work out the details. You should not consult your boss, no matter how perplexing the matter. You may—and should—consult other staff members. When presented to the boss for approval or disapproval, the product, whether it involves the prescription of a new policy or alteration of an established one, must be in finished form.

The impulse of the inexperienced staff member is to ask the boss what to do, and this is compounded when the problem is difficult. This instinct owes to a feeling of mental frustration. It is easy to ask the boss what to do, and it is easy for the boss to answer. Resist this impulse, however: You will succumb to it only if you do not know your job.

It is your job to advise your boss on what to do, not to ask your boss what you ought to do. The boss needs answers, not questions. Your job is to study and write, restudy and rewrite, until you have evolved a single proposed action—the best one of all that were considered. Your boss merely approves or disapproves. Unfortunately, it's rarely that easy, as your boss may want to take the proposal before the entire senior staff and have it reviewed, which means that a number of people—as many as you have on senior staff—will have a shot at dismissing the proposal as incomplete.

Do not worry your boss with long explanations and memos. Writing a memo to your boss does not constitute completed staff work, whereas writing a memo for your boss to send to someone else does. Your views should be placed before the boss in finished form, so that the boss can make them his or her views simply by signing the document. In most instances, completed staff work results in a single document prepared for the signature of the boss, without accompanying comment, unless asked for. If the proper result is reached, a good boss should be able to recognize it at once.

The completed staff work model does not preclude making a rough draft. Still, the draft must represent a wholehearted effort. It must be complete in every respect, except that it lacks the requisite number of copies and need not be neat. A rough draft must not be used as an excuse to shift the burden of formulating the action to the boss.

Completed staff work may result in more work for the staff member, but it results in more freedom for the boss. This is as it should be. Further, it accomplishes two important aims:

1. The boss is protected from half-baked ideas, voluminous memos, and immature oral presentations.
2. The staff member who has a bona fide idea to sell more readily finds a market.

When you have finished your completed staff work, the final test is to ask, if you were the boss, would you be willing to sign the paper you have prepared and stake your professional reputation on its being right? If the answer is no, do it over, because it is not yet completed staff work.

I discuss this model because it is a major trend in large organizations. It is time consuming and is designed expressly for major changes in the organizational method of accomplishing business. For application in any other format and to be used in an organization such as the fire service, it is too time consuming.

Making Choices

Choice is one of the true blessings with which human beings have been endowed. In our personal and professional lives, we have choices to make; sometimes we choose wisely, and other times—well, we stumble. Experience is a great teacher when it comes to making choices. The dichotomy of that is that experience is sometimes based on bad judgment, since you have to make some mistakes before you can have enough experience to have good judgment. As leaders, we are faced with a continual source of information, which we apply to the choices we make or don't make, as the case may be. Experience and training are critical aspects in informing how we make choices.

Decision-making and choosing are skills that are learned over a period of time, but can, to some extent, also be taught. Choices and decisions are methods by which we evaluate what to do, when to do it, and how to do it. The foundation is our ability, when presented with many options, to nevertheless choose the right solution, action, or direction consistently.

As information, data, culture, and policy change, so does the filter we use to make choices. For example, there is currently a flood of information about combat and special operations medicine. Unlike medicine in the civilian world, special operations medicine cannot be applied using a purely classroom, protocol-driven methodology, or it will fail. Captain Frank Butler of the Medical Corps of the U.S. Navy, put it best at a recent conference: "It is very important that we consider scenario-based management plans advisory rather than directive in nature because it is unlikely that anyone will encounter a casualty scenario in future combat that exactly reproduces one of our workshop scenarios." General Peter J. Schoomaker, the former Commander in Chief of the U.S. Army Special Operations Command, emphasizes that we have to "train people how to think, not just what to think."

The point is that we must first provide training opportunities for our new officers that teach them how to think, how to reason, and how to make the choices that have the most effective outcome. To compound this matter, the organization needs to determine what role rules, regulations, and procedures should play in guiding the decision-making process. Does the organization want officers to strictly follow the rules, unwavering, or do they desire a balance between strictness and making decisions based on values, justified by a few rules? Do we want our officers to take the initative to make choices, or do we want them to be afraid that if they make the wrong choice once and a while, we are going to punish them?

As I have mentioned earlier, you know instinctively that you cannot begin to raise your kids when they are 23 years old; you must instead raise them from the day they are born. However, the fire service has done a terrible job of raising its kids, our new officers. For no apparent reason, we never figured out that the company officers of today are the battalion and division officers, deputies, and fire chiefs of tomorrow. The skills and decision-making training (choices) that we provide them today will guide their future actions, decisions, and choices.

When you tell give someone an order like "see it," "own it," "solve it," or "do it," there are constraints as to how far you expect them to take it. There are Jersey barriers that define the boundaries in which they must make decisions and render choices. If organizations do not provide an education on what tools to use to make these decisions, on how to evaluate options, on how to apply rules and policy, and finally, on what role organizational values play in decision making, we set our officers up for failure.

> *Organizations must accept responsibility for teaching officers how to think, not what to think. If we do not provide expectations on decision making and choices, how can we expect our officers—and, therefore, our organization—to be successful, safe, and progressive?*

As professionals and adults, we are given the authority to make adult decisions, and with that comes adult consequences. The take-home message is that when we make choices, we must be ready to stand behind them and to accept responsibility if the choices we made turn out to be wrong. Don't think for one moment that I have not made bad choices in my life. As a human being, it's inevitable, but as a fire ground leader, you want to make sure the choices that you make do not cost lives. When it comes to truly leading people who want to follow you, much of that willingness to follow will come from choices you made in the past and that they witnessed. People watch what you do, not what you say, and the choices you make drive what you do.

> *Being an officer means you have to make adult choices, and with that comes adult consequences. It's the same with the members you supervise. They must understand that the choices they make have consequences.*

Fire Ground Decision Making and Choices

The place where you will pay the highest price for decision making is certainly on the fire ground, because it is paid for primarily in lives and, as a secondary tax, in property. In chapter 13, we discuss identifying the main problem in terms of the external and internal restraints that have an impact on the fire ground, such as friction and uncertainty. In general, however, when you make decisions and choices on the fire and EMS grounds, consider the following points:

- *Choose to be aggressive enough early enough.* If you choose to be aggressive, which quality is certainly required of and admired in fire ground leaders, you had better do it early and put enough resources behind your aggressiveness to be successful. You had also better be sure that your aggressiveness is going to pay dividends in terms of lives saved or property protected and that the excise tax extracted will not include the lives of your members, lost for no good reason. If you cannot press and close with your enemy (fire or rescue scene) quickly and overwhelm it, you had better consider yourself outmaneuvered and outgunned and decide on another tactic.
- *Be aggressively defensive.* When being on the offense is no longer an option, there is nothing that says you cannot be defensively aggressive! There is something mighty impressive about 10 to 15 thousand gallons per minute from master streams, flooding a building and washing furniture, cats, dogs, stereo equipment, and structural members out the front door. There are two types of insurance: life and property. Always rely on the latter whenever you have a doubt about your ability to prevail offensively.
- *Bring it!* If you think you might need it, bring it! There is nothing more embarrassing or potentially more life threatening than needing a resource at the incident and not having it. Evaluate early and often; ask *what if* every five minutes; understand where you can get the resources you need; and call them early. I would much rather have a bunch of people and equipment backed up, waiting to be deployed, than stand there powerless, wishing that I had something to solve my anticipated or unanticipated problem.
- *Don't let your ego override your gold!* Accept that officers who are junior to you may very well have the solution or may have

more technical knowledge about a specific issue than you do. Listen and hear when they speak, and engage the plan by using the initiative of your members. The ultimate outcome is to be successful, and that means using every means at your disposal, including the collective intelligence you assemble, to accomplish that. Make the choice to allow initiative from your junior officers.

- *Have a backup.* As a leader, you had better have a plan A, a plan B, and a plan C—and that includes making the choices that leave you with the ability to make choices. If you do not make the right choice to exit a structure when it is no longer tenable, if you do not make the choice to properly stabilize the vehicle before you enter, or if you choose not to evaluate and estimate, you may quickly lose further opportunities to make a choice.

Emergency incidents require a series of ongoing choices and decisions. Miscalculations, not understanding the concepts of friction and uncertainty, and being tactically or strategically inept will take a heavy toll on you, your members, and the overall incident.

People Choices

When you live and work with people, as a supervisor, you have a plethora of choices to consider each day. Moreover, the people you work with will make choices every day. Over the past year, I have spent some time discussing good and bad leadership choices, as well as good and bad personal choices.

Don't be surprised when someone chooses to ignore policy and procedure and comes to you with that lost puppy look. When people make bad choices, they may adopt a victim mentality, claiming, "It's not may fault," or "I didn't know." If that is truly the case, then we had better also figure out why the person who was responsible for telling them and showing them didn't!

We have devoted much discussion to the importance of teaching members how to choose. Similarly, as a company officer, your job includes teaching your members how to make decisions and what the consequences associated with choices are. It's not just your job to teach, but to communicate what the organization expects from you and from your members with regards to choices.

All people make choices based on a wide range of input and output that they experience. Choices are made by means of a convoluted matrix of emotional input and data. Each time a decision is made—whether by the king or queen or by the newest member of the organization—the following areas all come into play at one point or another:

- Emotions
- Past experience
- Values
- Loyalty
- Politics
- Financial considerations

Emotional decisions

I guarantee that when you make choices on the basis of emotion, you are making the wrong choice. Every time I have made a choice out of anger, it has always—and I do mean always—come back to bite me hard. As an incident commander or as an officer, you cannot—and should not dare to—allow your emotions to interfere with decision making and choices during an emergency event. Letting your heart make decisions that your head needs to make is a surefire method to get someone killed or hurt. It's also a surefire method to lose control of the incident while you wallow in your emotions, rather than driving the incident, which may already be steeped with emotions because something bad happened.

When you find yourself getting ready to make choices based on your emotions in a non-emergency setting, stop! Take the time (which you have) to make the right choice after you calm down. In the emergency setting, you absolutely must learn to control your emotions, as well as your tone of voice, and focus on sound strategy and tactics, not on sorrow, anger, excitement, or disbelief. Nevertheless, emotions can still help you to make good decisions in some contexts, such as when you are in love.

Past experience

There is absolutely no better teacher than experience. In Sources of Power, Gary Klein has undertaken an extensive study of how firefighters make decisions.[5] What he discovered is that studies on decision making staged in laboratories and treating people as biased or unskilled are incorrect; by contrast, people develop a naturalistic decision-making process based on experience. Using experience to make choices is an excellent method, but in order for you to do that, either someone has to teach you about that experience or you have to see if for yourself.

Values

Values are the filter through which we make choices. We use them in our families to raise our children and guide our own actions and choices, and intelligent organizations use them to teach people how to make choices. Policy and procedure cannot cover every situation in which fire service personnel find themselves. When there is no direct guidance, we must base decisions on values. This is difficult for those members who always want something in writing to tell them what to do and when; it also opens the door for those members who want to challenge everything that is not explicitly covered in policy.

Loyalty

We make some decisions based on loyalty, or commitment to a person or cause. Although there is much to be said in favor of making a choice because of loyalty, being blindly loyal can result when emotion and past history are mixed. When we make choices based on loyalty, we stand on our principles in support of someone or something. When you find yourself making these kinds of choices, you must understand clearly their ramifications. Even when you are 100% correct, you may end up being dragged out and shot! I respect loyalty and value it highly, so my message is simply not to let loyalty sway you into making choices blindly.

Politics

These are usually quid pro quo choices, or something for something. Whether they are all good or all bad depends on which side of the aisle you are in when the decision or choice is made. The higher you go in an organization, the more often you will have to make political choices. Thus, if you don't want to make political decisions, then don't go very high in the organization or the union.

Fiscal considerations

When Alan Shepard was asked how he felt about going into space, he answered, "It's a very sobering feeling to be up in space and realize that one's safety factor was determined by the lowest bidder on a government contract." When money is involved, the choice that is made is not the correct choice but the cheapest choice. In many instances, our purchasing laws and rules and our organizations are set up to work this way, and, well, it is what it is. Just be able to justify choices that you make that take exception to this unfortunate fact of life.

Choices about Moving On

When a man retires and time is no longer a matter of urgent importance, his colleagues generally present him with a watch.

—R. C. Sherriff

In your career, you have to make choices about your position in the organization. The one piece of advice I can offer is that you choose to spend some time away from the operations/suppression area and, instead, in a staff job. Every officer needs to know how the business runs, so as to be more proficient in conducting the business of the business. This career choice will, if nothing else, give you an unprecedented appreciation for shift work and being a frontline service provider. Moreover, being in the actual business end of the job may give you more satisfaction and challenge than you ever received in the field. Regardless of the reason, make a conscious choice to spend some time in a 40-hour workweek to see what it's about. The motto "been there, done that" expresses succinctly one of the most efficient methods of making choices.

True leaders must also recognize when it's time to move on. The decision may be made for you, owing to circumstances and people, or you may make the decision for yourself, as a life change. Regardless of the reasons, be prepared for the day when you leave the organization. Early in your career, plan for this moment by entering into a savings program other than your retirement, by starting a college fund for your children early in life, and by creating a skill set for yourself above and beyond simply being a firefighter. Make yourself marketable for your second career! Importantly, saving and planning must start early, when it's the hardest to do, both financially and emotionally. Remember that choices made purely on financial and emotional considerations are always questionable, so face your fears and put something away at the very time when its hardest to do!

When one bases his life on principle, 99% of his decisions are already made.

—Source unknown

Notes

1. Morse, Gardiner. "Decisions and Desire." *Harvard Business Review*, January 2006.
2. Powell, Colin, with Joseph E. Persico. *My American Journey*. New York: Random House, 1995.
3. Buchanan, Leigh, and Andrew O'Connell. "A Brief History of Decision Making." *Harvard Business Review*, January 2006.
4. Senge, Peter. *The Fifth Discipline*. Currency Doubleday Press, 1990.
5. Klein, Gary A. *Sources of Power: How People Make Decisions*. Cambridge, MA: MIT Press, 1999.

Chapter 12
Accountability and Responsibility

The ancient Romans had a tradition: whenever one of their engineers constructed an arch, as the capstone was hoisted into place, the engineer assumed accountability for his work in the most profound way possible; he stood under the arch.

We must reject the idea that every time a law's broken, society is guilty rather than the lawbreaker. It is time to restore the American precept that each individual is accountable for his actions.

—*Ronald Reagan*

Firehouse to Fire Ground

When you say *yes* and pin on the badge and gold of an officer, you accept an entirely new level of accountability and responsibility. I have seen officers who never accepted this and thought they were simply firefighters with a different color shirt. The transition to first-line company officer is the most difficult, because you are fundamentally a working foreman. You live, eat, sleep, and work each day, doing many of the same tasks on the fire ground as before, especially in short-staffed companies, yet you are the leader and manager of the team; in other words, you are the "designated adult in house." You must lead your team, accept responsibility for your actions and those of your team and set the standard by which members of your shift operate.

First and foremost, you are responsible for the safety of your personnel both in and out of the station, in everyday work and emergency operations. You must instill in your members—especially your senior members—that safety is critical in order to return home each day. Also, as a risk manager, you are responsible for ensuring good practices, so that preventable on-the-job injuries are indeed prevented. Finally, you are managing liability for the organization while making sure your members do not place themselves in a position where the system might take advantage of an unfortunate event.

There are dozens of mundane or traditional events that occur in the fire station everyday that keep the organization running. Most of this is the everyday work of checking equipment and filing paperwork. Let's review some examples of the mundane that if ignored or forgotten, could cost your members or the organization. Most apparatus have a maintenance log, and most SCBA have a daily checkoff. However, with the exception of the SCBA, you don't check off the apparatus on a daily basis, and you trust your members to complete the logs. Trust is a critical element of leadership, but so is checking up periodically.

I remember specific instances when officers failed to regularly examine the checkoff books, and on inspection, many entries were missing. It may have been that the apparatus was out of service that day; it may have been that someone simply forgot to check off the SCBA; or it may have been that a relief crew was in. A wide range of possible causes exist, but the bottom line is that it was not checked off. Perhaps this would be no big deal, but perhaps not. Suppose that on the day in question, the SCBA was not checked off and someone gets hurt at a fire. During the investigation, the workers' compensation board discovers that the SCBA that Firefighter Jones was wearing that day had not been checked off; thus, they can claim there was no way of knowing whether the SCBA was working correctly before it "malfunctioned" during the fire! As a potential consequence, Firefighter Jones could be denied his benefits. Further, this one event could create a storm of other litigation, extending the time that Firefighter Jones has to go without money and health care while he fights the decision in court.

One day while I was a battalion officer, I was called to the city garage, having been told there was something I had to see. On arrival, I was led into the bay, where my frontline apparatus was up on the lift to reveal that the right inside dual was down to the steel belts! Now here is an apparatus that is supposed to be checked every shift; there is a process in place to check the tires, yet our frontline apparatus was driving around with a tire on steel belts, discovered only during a maintenance stop at the station. The log showed that the tire had been checked off day after day; nevertheless, steel belts were showing! What would have been the consequences if the vehicle was involved in a motor vehicle accident during a response and this was discovered on investigation? The responsibility for the accident would be shared by the driver-operator who checked off the vehicle and the officer in charge of the station on each shift. Perhaps it was a process problem related to the way in which we checked the tires, but any trained officer should know to get under the truck to look at the tires—and not just hit them with a mallet and say, "Good to go."

You are responsible and accountable for the training—or lack of training—that your crew receives. When you are on leave and you assign someone up front, you are responsible for any inappropriate actions by that person. It may be that you provided the proper training, but he or she made a mistake or a bad decision. As long as you have documented that you have been providing ongoing and realistic training, you will be okay. However, if you allow someone to ride up front in your absence (assuming you make this choice) who is neither trained nor competent to do the job, you will have to answer for the actions and choices of that unqualified person.

Accountability and Responsibility

Finally—and make no mistake about it—you are responsible for the decisions you make on the fire ground and during response. If you are (or your move-up person is) riding in the front of the cab and you allow the driver to run a red light and someone is hit, you—yes, you—and the driver are liable and responsible. You and no one else are in command of that vehicle. On the fire and rescue ground, you are responsible for the direction of your crew and for delegating tasks to complete the tactical objectives. It is up to you to choose how to accomplish that—reading the building inside and out, deciding where and when to apply water, planning how the search will be done, and so forth.

The reality of leading a fire station as a company officer is that you have to check up on personnel and equipment periodically, to ensure the quality and appropriateness of the operation. Furthermore, you are responsible for the decisions you make or allow to be made in and out of the stations.

The topic of planning is discussed in detail later (see chapter 16), but in brief, it is your responsibility to understand what the organization expects with regards to projects and tasks. You must recognize what the main thing is and plan effectively to get it done.

VBF&R has so-called *traditional programs* that every company must complete each year. These have included hydrant maintenance, building inspections, pre-incident planning, and target hazard refamiliarizations. At the beginning of each year, every company in the city is provided a list (of hydrants, occupancies, etc.) to complete. The lists are then divided up, and the officers know what they must do.

The organization used to put a time frame on the accomplishment of each task—for example, that hydrants would be maintained between July and November. The company officers, rightfully so, said, "Hey, let me manage when I do them as long as I get them done." Great idea! Therefore, the programs went to annual completion.

Now companies generate a monthly report that goes to the battalion chief and moves up the chain of command to show projects, progress, and problems at the company. For example, I noticed one of my companies was well behind on hydrants and had about 30 days left. I questioned the battalion chief, who replied, "They always leave their hydrants till the last month." Okay, I thought, his company, his decision. However, at the end of the year, the hydrants were not complete, and the excuse given was, "We had an increase in call volume and could not get them done." This is an unsatisfactory answer, since they had all year to get it done and made a conscious decision to wait until the last month. Poor planning on the part of an officer does not translate into an emergency on the part of his or her supervisor. We had a sit-down with the captain, which we used as a learning tool: we did not hammer him, instead making it clear that if the problem occurred again, the consequences would be more severe. To the company's credit, they corrected the problem and it never occurred again.

Every organization has reporting and record keeping that needs to be done. In the previous discussion of technical competencies, I suggested that it's important for a company officer to delegate these, as a training tool, and to make sure all members get a chance to manage each report and record, so they understand the process. It would be impossible and inappropriate to delegate everything that needs to be done and call it a training tool. However, the company officer is responsible for the timeliness and completeness of those reports.

As you can imagine, we could go on and on with examples of situations that have occurred—everything from water fights in the station, resulting in a broken femur, to horseplay on a ladder in the bay, resulting in a fall and a severe head injury. Personnel, apparatus, reporting, training, safety, policy implementation, and education—the list never ends, and it is your responsibility to ensure that all of it is done correctly, on time and safely.

Human failures at all levels of the organization occur for one of four reasons, which you should keep in mind throughout our discussion:

- *Standards failure.* Standards are unclear, impractical or nonexistent
- *Training failure.* Standards exist but are not known or ways to achieve them are not known, or ignored
- *Leadership failure.* Standards are known but are not enforced
- *Individual failure.* Standards are known but not followed

Why People Follow You: Understanding Trust

Few things help an individual more than to place responsibility upon him, and to let him know that you trust him.

—Booker T. Washington

The leaders who work most effectively, it seems to me, never say "I." And that's not because they have trained themselves not to say "I." They don't think "I." They think "we"; they think "team." They understand their job to be to make the team function. They accept responsibility and don't sidestep it, but "we" gets the credit. . . . This is what creates trust, what enables you to get the task done.
—*Peter Drucker*

You may ask, What does trust have to do with developing accountability and responsibility? But in the fire service, how can you do business if people don't trust you? I think everyone reading this would agree that trust plays a major part in our decision making when we decide to follow someone's instruction. The bottom line is this: At the company or organizational level and even in your personal life, you can't expect to accomplish much without trust. If you want people to learn to be accountable and take responsibility for their actions and if you are going to lead them, then they have to trust you and trust that you will be responsible and accountable.

How do you earn trust, when do you give trust and how much of it should you give? Trust is primarily a relationship-based currency that's either spent or banked, depending on your personal relationships, your actions, how you communicate your decisions, and of course, your ability to do your job in the operational setting, as well as in the administrative and human resources arenas. The next sections examine in further detail how trust is earned and given.

Relationship-based trust

You can't earn trust without establishing a relationship on some level. That's not to say you must personally know everyone, shake their hand, and eat dinner with them, but make no mistake, simple human interaction that allows people to see who you are is the most effective method of establishing trust.

We form relationships immediately on entering a work group. If you're a company officer, this relationship begins the minute you walk through the door of your new station or whenever a new member joins your team. If you're the chief, it begins the minute you show up for work. That's when the subordinates develop an impression about who you are, what you stand for, what you believe in, and how those characteristics blend with their professional needs and desires. However, these initial relationships are based largely on evaluations of surface traits—looks, body language, and initial conversation. Having only this first opportunity to decide whether to trust one another would be tragic, since surface traits don't expose the true person.

Forging personal relationships involves much more than showing up, shaking hands, and telling people what you stand for. Most of your personal relationships will manifest themselves through your actions.

As a subordinate, I consider a wide range of issues when I decide to trust you, my company officer. These issues include not only the personal contact I have with you but also the relationship I form with you by watching how you communicate with me and whether your actions support your words, the team, the organization, and me. Likewise, I consider whether you're willing to tell me when I'm off base and why you think so and whether you'll risk my making a bad decision so I can learn from it.

Building a lasting, trusting personal relationship is an ongoing endeavor that takes time and has its ups and downs. In our personal lives, many of us develop only a few true relationships or true friendships in which we have established unlimited trust. Consequently, we develop varying degrees of relationship trust.

For example, at the company level, we may trust one of our coworkers as a competent firefighter who we'd risk our lives with, but outside the workplace, we wouldn't trust that same coworker with our money or our kids because of his or her personal immaturity or an irresponsible lifestyle. At the organizational level, I might trust my direct supervisor, but because of their actions, I might fail to trust others in the chain of command.

(Because most human beings are smart enough to separate the wheat from the chaff, an untrustworthy leader who tries to protect him- or herself by throwing all his officers together in the they-don't-trust-us mix is transparent. Remember that principle and character are a clear glass. Everyone can see from one side to the other.)

As I've indicated already, trusting relationships are not just about human interaction. Many of us trust political or social figures we've never met. We form these trust relationships by watching, listening, and gauging the impact that their actions have on society. Building trust relationships with people you do not know very well or have not had the opportunity to spend time with is very difficult. Unless you pay attention to their words and compare them with their actions, you will have no basis for relationship-based trust. In the end, it's not really a matter of trust, since you really never trust someone you don't know well or don't work with. It may boil down to a matter of respect, if you can say they have always been good to their word.

The foundations of a trusting personal relationship are almost all behavioral and are discernible though indefinable to anyone who spends time around a given individual. To build more trusting relationships with your crew, consider the following behaviors:

- *Lead by walking around.* Never underestimate the power of a personal meeting, talk, or dinner. Candid discussions allow people to understand who you are and what you stand for. At the company level, quarterly meetings with each of your members to discuss their progress, including goals and accomplishments, are critical in developing relationships. Beware: If your words don't support your actions, people's trust in you will wane as they pay more attention to what you do and less to what you say.
- *Take the time to actively listen.* When people come to you with an issue or to use you as a sounding board, give them the time they deserve. Push away from the computer, don't answer the telephone, and engage in conversation. Listening alone is not enough—you must act, make something happen, and provide feedback. (I've learned this from personal experience: People want you to spend time with them.)
- *Build relationships early.* When you meet subordinates, take the opportunity to let them know what you stand for, what you believe is important, and what you expect from them and the organization. Also realize that they will watch very closely to see whether you follow up with your actions, once you have set out your expectations.
- *Be positive.* Attitude is infectious, so if you're the kind of person who constantly sends the message that "it can't be done," people will view you as a drain. At a recent business-plan meeting, after a lecture from the chief, the battalion officer facilitating the meeting looked at him and said, "Thanks for bringing us down." The entire staff sighed, since he had just verbalized what they were all thinking.

If you talk the talk, you had better walk the walk

Your mother used to tell you, "Show me—don't tell me." Your actions, your follow-up, and the direction you give your team say everything about who you are at a given point in your organizational life. I say "at a given point" because people may grow and mature, or they may fall prey to the pressure of making wrong choices, which pushes them below the line. Still, if you do fall below the line, the real mark of a trustworthy leader is how quickly you work to correct it.

To trust you, people must know what you stand for. This is exhibited in your expressed beliefs, but most important, in your level of integrity, or how you take action when presented with a specific issue. People want to know your core values and beliefs as they relate to running the team, battalion, or organization. You must stand by your convictions; the greatest failure you'll ever experience is the moment you decide, based on which way the political or social winds are blowing, to change what you stand for. Always be willing to seek compromise, but never give up on your core beliefs. When you do, people think you're selling out, giving up, or doing what's best for yourself in spite of everyone else.

In the fire service, we are surrounded by rules, polices, guidelines, and other processes typical of a paramilitary work environment. This creates organizational and personal conflict when, as leaders, we must adhere to rules we think are foolish or unnecessary. The quickest ways to lose trust are to inconsistently apply and enforce rules and to allow your personal feelings to dictate what you will and won't enforce. To avoid these pitfalls, consider the following guidelines:

- *Be consistent.* If you continually flip-flop on issues, policies, ideas, or procedures, you will send a message no one will understand. It's not about being popular, it's about being a leader, and that requires mature and consistent decision making. Don't be afraid to take up a member's cause that you feel is righteous, but don't be afraid to tell a member who is barking up the wrong tree to let it go. People remember the decisions you make and the reasons behind them. If you start making different decisions in the same situations, people will notice right away.
- *Undercommit and overdeliver.* Don't promise anything unless you're absolutely sure you can deliver, and don't confuse undercommitment with saying no. Undercommitment in no way communicates *no*. In fact, you should spend most of your time

saying *yes*. Commit yourself and your team to accomplishing what you need to do, but when you know it's going to be difficult, don't paint a rosy picture—paint a guarded, optimistic one. Look for alternative methods, funds, and processes to achieve your goals; if you do what has always been done, you'll get what you always got.

- *Don't surround yourself with yes-men or yes-women.* People watch how you make decisions and how you invest in others' ideas. Also, people can spot the teacher's pet very quickly. If you marginalize people and their ideas, you had better do so for a very good reason, because the message you are sending is, "Your value here is limited. Find somewhere else to work."

- *Follow through.* When you say you're going to do something, do it. If you say, "I'll write OSHA and get a ruling on mustaches," you had better do it and communicate the results. And when you tell people you will provide an answer or an explanation later, you had better follow up.

Communications—the key to long-term commitment

How we communicate on a personal level and an organizational level is critical to gaining and keeping trust. Electronic media have been both benefactor and detractor to modern communications. On the one hand, we can send information quickly from very remote locations. On the other hand, it eliminates the human interaction. Moreover—and destructive to the leadership process—it allows running, often angry discussions to occur between individuals or groups. When this happens, stop the electronic conversation immediately, seek out the individual, and speak face to face to solve the problem.

Electronic media have also created a "fog" of communications. We send out so much information that the really important details can get lost amid a mass of e-mails. How many officers do you know who instantly delete e-mails from a certain sender? Also, how many officers or members simply don't read their e-mail? In either case, no communication takes place.

You must communicate in some form when you decide you're going to provide direction to individuals, a team, or an entire organization. Most organizations and officers are remiss in realizing that before you implement, you must explain. Nature abhors a vacuum, so if you leave one out there, something will fill it. People expect to be provided explanations and reasoning. They don't have to agree with it or even like it, but for working adults, simply saying "Because I said so" does not fly. Here are some suggestions on how to develop trust through communications:

- *Communicate frequently, honestly, and openly.* Keeping people continual informed in a timely manner builds morale, confidence, and trust. The members of your shift and the organization must know what's occurring around them that might affect their ability to function. (If you think for one minute that information given during a staff or a senior staff meeting is kept secret for very long, you've lost your mind!)
- *Fill the vacuum before it fills itself.* One way to reduce your crew's trust in you is to implement a process or procedure and not explain why you did so until three or four days later. By that time, it's too late; the vacuum has been filled. People form their own perceptions when you fail to provide up-front explanations, and it's very difficult to go back and change those perceptions. The solution is to explain your reason for making the change, then let people digest it, and finally implement the change. Fill the vacuum with fact and direction.
- *Remember: A lie travels halfway around the world before the truth has a chance to put its shoes on.* Left to their own devices in the absence of open communications, members of your shift or organization will develop their own inferences as to reasons. And their reasons will be part truth, part perception, part rumor, part discussion with others, and part intrigue as to why "they did not tell us." Disseminate information early and often, even if it's not good news. When people decide to buy into a team, they buy in for good and bad, so don't be afraid to show them all your cards.

Don't fly if you don't know how

It makes little difference whether you're in operations or administration; members of your team must trust your abilities, but will build this trust only through watching your everyday actions in both non-emergency and emergency settings. Your expertise, or maturity, is a combination of competence and commitment. Competence means you have the KSAs to do your job and is reflected in your everyday decision making, your thought processes, your administrative reports, how you deal with human resources issues, and your actions on the fire and rescue ground. If you fail to maintain your KSAs, your team members won't trust you to make the right decisions at the right time.

Commitment involves choosing a course of action and sticking with it. When things get really tough on either the emergency or the administrative side, team members you to reaffirm that you'll see them through the situation. They want to know that regardless of the barriers and any naysayers, they can depend on you to continually ask what, how, why, and when.

In many organizations, people mistakenly promote unqualified personnel, which leads to functional incompetence. In the fire service, we expect everyone to be able to perform every job, but aside from being hired as a firefighter-paramedic (baseline requirements), not everyone is suited to be an officer or a chief or to hold an administrative position. Next, we tend to promote based on test scores and résumés, but we fail to consider past performance and organizational history. You might be the smartest kid on the block, but you might not have any people skills or you might have limited fire and rescue ground capability; nevertheless, we would still promote you and expect you to lead people. To avoid this pitfall, promote people to positions that draw on their talents.

A recent Harvard Business School study asked three questions of top executives. The first question was, "When you hire someone, how long does it take you to realize they are not right for the job?" The answer: 10 days. The second question was, "How long does it take you to remove them?" The answer: 10 years! The fire service is guilty of this as well, because we seldom have probation periods for our officers to evaluate their performance. The third question was, "If you could hire for KSAs or people skills, what would you hire for?" The consensus: people skills. Across the board, top executives believe you can teach someone technical skills, but you can't make them a people person. Never has this been truer than in the fire service, where the work is all about people.

That said, how do KSAs and people skills encourage and build trust? The world is changing and so should you—after all, this isn't your father's fire service! Everyone, especially officers, must stay abreast of the newest trends and science for fire and rescue. Beware of people who've never been outside the city for training and do business the same way they did 10 years ago. Research the following hot topics and ask, "How will this affect me on the fire and rescue ground?"

- *Reading smoke.* Are there indications that some interior attacks and rescue operations are anything but?
- *Ventilation.* Is the roof really the best place to be? The United States has one of the only fire services in the world that does vertical ventilation.
- *Special operations teams.* Now more than ever, advances in science and increasing threat demand dedicated and trained personnel.

- *Does ALS really work?* Are there better alternatives to patient care? Do helicopters really save lives, or are they just expensive toys?

Foster trust by demonstrating your abilities and commitment through the following means:

- *Let the crew run the ship.* Impress on your senior crewmembers the need to educate and teach junior officers with less experience, even if they're outranked. When you trust people to run the ship and they have a direct stake in the success (or failure) of the entire organization, performance will be impeccable, and trust will soar.
- *When you need help, reach out.* Just because you have the gold at any rank does not mean you have the KSAs to successfully perform the job. When you know very little about a job, admit it and seek help. I remember an officer on our technical rescue team who just could not bring himself to do this. Rather than admit he did not have the appropriate KSAs and seek out other firefighters to help him, he spent his time trying to slow them down. As a result, he failed, lost his crewmembers' respect, and ultimately left the company.
- *Seek knowledge and learning.* When you're assigned a job or task that you know very little about, dive in and learn all you can. Go out and get the formal education (hang the paper on the wall). Continually reaching out for intellectual knowledge and technical skills will pay for itself time and again. The trust you gain by learning from people—and the trust they give you when they see you are competent—is unbelievable.

Trust is as critical to leadership as oxygen is to life. It's an emotional concept that's exceptionally difficult to earn and too easily spent. By creating relationships and communicating effectively through a variety of means, you'll earn your members' trust.

Without trust, you'll never have more than a mediocre crew, team, or organization. It may well be that there are vast, untapped resources of skill, resourcefulness, drive, talent, and desire right under your nose, encased in your members and waiting to be released. Without trust, however, members will not spend that currency on you, and as a result, they'll never go the extra mile; until you earn their trust, they'll do only what it takes to get by from day to day.

Aspire to instill trust within your members. Understanding the importance of trust will enhance your potential and success as a leader and allows you to immediately recognize when you've fallen below the line; picking yourself up and getting back above the line creates even more trust.

Building trust presupposes an environment where people are accountable and responsible for their actions and the actions of the team. It is an all-encompassing fact that if you are not accountable and responsible for your actions, I will not trust you.

The Look-over-the-Fence Mentality

Don't for one minute think that every officer we have—or who you will work with or across from—is going to espouse these ideas. You have heard time and again about the differences between A shift, B shift, and C shift in the fire department. What occurs or is allowed on another shift fosters a look-over-the-fence mentality. When people see that policy or operational violations are tolerated on another shift, they ask why the violations were allowed, or else they ask, "If they can do that, why can't we?"

The first lesson is clear: as an officer, you cannot allow this mentality to infiltrate your shift. Establish your expectations for the team up front. Once everyone knows, they may still question you occasionally, but you can always bring them back to what your expectations have been from day one.

The other lesson is more difficult, and that is how to speak with people who are causing problems for you either because of their behavior or by allowing certain kinds of behavior in others. This is like the mother who goes across the street to tell little Johnny's mother that she is tired of little Johnny dropping his drawers in front of her daughter. When other officers allow behaviors to occur that are clearly wrong and have an impact on your members and teams, it's time to pay a professional courtesy call.

If the problem individual is an officer, the next level of the chain of command needs to get involved. We had a captain who would come to work and immediately put on his physical-training gear and his clogs—yes, clogs. If that was not enough, the guy wasn't in the best of shape anyway, and as an officer he is supposed to be setting the example. It was creating problems not just across shift but throughout the battalion as the word spread, personal sightings occurred, and the question "What policy are we going to enforce this week?" was posed. Finally, the battalion officer addressed the issue, and the physical-training gear/clog wearing ceased.

You cannot allow yourself to get sucked into a similar trap. First, address the problem officer to officer, and if professional courtesy does not work, then you have a decision to make: Do I ignore it and just tell my team members that we are going to do the right thing and not fall prey to this individual's example, or do I seek resolution through the chain of command by letting them know the impact that the problem behavior is having on my shift?

The Virus; or, My Stomach Hurts, I Don't Feel Well!

I saw a bumper sticker the other day on a pickup truck that said, "Being fat and lazy is no excuse to get a handicap sticker." Despite the lack of sensitivity of this bumper sticker, it has relevant social implications about a virus that is affecting America and the fire service. We have all been exposed to this virus, and in the case of the fire service, we have vaccines to prevent it!

In a world where 350-pound people sue McDonalds because they are fat and have heart disease, it's no wonder we have difficulty in quarantining the fire service from the virus. When 80-year old ladies spill hot coffee on their genitals and sue, claiming their sex drive has been taken away, it's a wonder this bug has not created a pandemic! The world is full of victims who have nothing better to do than look to someone or something else on which to blame their problems and who take great pleasure in being unable to extricate themselves from their miserable lots in life!

You may find it surprising but I do have sympathy for these people. The reason for this is that accountability should not be a bad word. the fire service has made it so by suggesting that we are only accountable when something goes wrong. In fact, accountability should be about the way we work and about how we live our organizational lives every day we come to the job. It's about giving 120% every day you work, regardless of the situation you find yourself in, who you work for, or how you feel about your situation.

As a case in point, I have not been very timely in delivering these chapters to my editor, who was fortunately very understanding. You might assume that lack of accountability had something to do with this, but you would be wrong. For one, I blame no one but myself for having not planned appropriately, scheduling too much to fit into the original time line. Moreover, no one is making excuses for me, and I am being accountable, by accepting responsibility when I fall below the line and taking the steps to bring myself back above the line. In this way, we work as a team, and when I let the team down, my editor picked up the slack.

Identifying the Disease

Lack of accountability leads to blame shifting, negative attitudes, retired on active duty (ROAD) status, chronic complaining, sloppy or incomplete work, laziness, and myriad other characteristic signs of a highly infectious organism that multiplies exponentially, sucking the life out of everything around it!

Company officers have a unique opportunity to see it, own it, solve it, and do it at the company level and even beyond. They have the opportunity to see the disease close up and to act much like the smallpox pioneers, who vaccinated people and set up quarantine rings to stop the spread of the disease. Likewise, organizations have the methodology to inoculate our new members and set an example for our vested members.

To diagnose this disease, you have to be able to recognize the signs. As you learned in EMT class, a set of baseline norms and abnormal observations are required in order to make a tentative diagnosis:

- *Wait-and-see attitude.* Never mind that the handwriting is on the wall—nope, just wait and see what happens. Organizations that lack initiative exhibit this sign, as do individuals and groups who know the outcome of a given action. This is the opposite of being proactive. When you wait to see if things will get better and do nothing to influence the outcome, most times things get worse! This is amplified on the fire ground, but applies to non-emergency settings as well. Similarly, company officers and organizations that believe "it won't happen here" reek of this—like a cytokine storm in a smallpox patient, you can smell it before you walk into the room!

- *Please tell me what to do.* Don't self motivate, don't get out there and work hard to get it done in anticipation of what needs to be done; instead, wait for someone to tell you what your job is. When you sit an employee down and say, "Hey, you need to improve," or "You're weak in this area," and a month later they come back and say, "I was getting mixed messages," get out the needle; the person is infected. If you don't know what your job is, then maybe you need a new job. Likewise, when people or organizations are not smart enough to see what the trends are or identify what needs to be done and do it without being told, the signs point clearly to the virus. Organizations that fail to anticipate questions "from across the street" (politicians) and fail to collect data and information to support their answers are classic examples.

- *It's not my job.* When you hear someone say this, what he or she is really saying is "Excuse my inaction"; "Allow me to redirect the blame to Firefighter [or Officer] So-and-So"; and "Allow me to avoid any responsibility." When you say it isn't your job, it clearly indicates that you know something has to be done but choose not to do it! How many times have you been shopping in a store or dining at a restaurant and been ignored or deliberately not assisted because it was not their job to do so. Again, the fire service is a team sport—that is, it requires for people to help other people and the organization when they recognize something needs to be done. We call that engagement!
- *Ignore and deny.* How many people have you met who simply choose to pretend that there is not a problem or remain unaware of how the problem affects them or others. We often call this incompetence. Incompetent people are unable to recognize that they are incompetent and therefore go on thinking that they have it all under control. Look around at industry to see what ignore and deny does for you—at Sheffield Steel, Bulova watches, American steel companies, Enron, Tyco, WorldCom, Detroit auto makers, and the list goes on. New York City can attest to the results of ignore and deny, because of the failure of previous police and fire commissioners to establish an integrated incident command system, and the problem persists today.
- *Finger-pointing.* In this classic move, members attempt to shift the blame to others, while denying responsibility for their own results. "It's not our fault that we backed the truck into the station. Our Captain never told us about the backing policy!" In the fire service, everything takes two weeks: Instead of getting to the root of the problem about why a vendor or someone cannot or won't deliver on time, we say "two weeks," meaning, "I really wish I could help, but its in someone else's hands."
- *Cover your tail.* How many elaborate stories have you heard as a company officer that have been contrived to explain why someone could not have possibly been responsible? Cover your tail is expressed in many ways—from excessive SOPs on commonsense items, to documenting everything under the sun, to sending backup e-mails to people to say, "See, I told you so." At the lower end of the scale are the "duck and hide" individuals, who

make every attempt to disassociate themselves from situations that could erupt into potential problems. In meetings, this is evidenced by silence on a much-contested issue; these individuals simply don't want to jump into the conflict.

Making Accountability the Normal Operating Procedure

Accountability has always been a proactive stance, even while we have sadly turned into a reactive stance. We hold you accountable for your actions after the fact—which is the right thing to do, but not the smartest. By contrast, every day you come to work, you should consider yourself accountable to do the very best you can, to function as a member of the team, to use your talents to the very best of your ability, and to anticipate what needs to be done and do it.

So, how can a company officer deal with an infectious virus, to halt it from spreading to the vulnerable members of your shift and the reminder of the organization? First, recognize the signs of the disease. This includes recognizing it in yourself; you must realize when you fall below the line and are infected. As human beings, we all fall below the line sometimes, but in and of itself, that is not the problem. If we identify that we are below the line and take steps—no matter how painful—to correct it, we go a long way toward persevering in our leadership and we set a necessary example. The difference between a winner and a loser is that the loser stays down when knocked down and the winner gets back up.

When you recognize the signs of this disease in others, you must take immediate steps to heal them. The recovery may be immediate, or in very stubborn cases, long-term care may be required, which people don't always survive. Let's look at some methods to recognize and treat your members:

- *Recognize the victim cycle.* Members who exhibit the behaviors we have described are sick with the virus. Some can be cured quickly; others will take a little longer to heal; and others still develop terminal cases. The victim cycle occurs when individuals refuse to take responsibility for personal behavior, actions, and work. Someone or something else is to blame for the way they feel, the way they are treated, and the miserable situation they are forced to work in! The fire service is not a really tough job most of the time: Show up on time and in the right uniform, do the right things, and take care of your team and your customers, and you are pretty much home free; it's certainly not brain surgery. You get a pillow, a blanket, and a gym and every day is Friday—find that on the open market! When you recognize the victim cycle, pull people aside and educate them about what they are doing, to help them get back above the line.

- *Ask yourself, Is this normal?* When you see someone under your supervision in the victim cycle, ask yourself these questions: Is this normal for this member? Is he of she a constant complainer or a continual victim? If this not normal for this member, has there been a drastic or sudden change in behavior and attitude? Chronic victims often just needs a good sit-down to establish expectations and direction or to have teamwork and standards of behavior explained. In some cases, they need to be told, "Shut up and stop spewing your negative viral attitude around here. I am not going to tolerate it!" It is fine if they want to be miserable away from work, but not in the firehouse. When someone has undergone a sudden change, ask the member what happened, what is going on—and there a numerous possible causes—that is fueling the victim cycle?

- *Develop a treatment plan.* Treatment plans vary by patient, depending on how badly the virus has infected the patient. The severe cases will require very firm direction and expectation, close monitoring, and tough love, and the most severe cases require quarantine away from the organization forever! At the company level, we accomplish this by setting expectations, followed up with counseling, feedback, and evaluations, and educating members so that they can fix themselves before we have to. Other cases simply require you to reach the root of the problem, provide them with the opportunity to vent, educate them about what they are doing to themselves, the team, and the organization through the behaviors described earlier. As a supervisor, your key job is to observe, diagnose, and fix.

Inoculations for Success

As mentioned earlier, we always hold people accountable when something goes wrong. By contrast, when things are going along just fine, to inoculate ourselves and our members, we need to ask the most important question: "Who is accountable for success?" It's easy to point the finger when the ship is sinking, but when you are cruising along and making good time, no one seems to care much about who is responsible. In the wonderful book *The Oz Principle*, Roger Connors, Tom Smith and Craig Hickman have given a wonderful definition of accountability:

> A personal choice to rise above one's circumstances and demonstrate the ownership necessary for achieving desired results—to see it, own it, solve it, and do it.[1]

Why is this definition so important? First, it suggests that accountability is not something that we demand of people when something goes wrong, but rather a way of living your organizational life. It stresses that what you do right now has the potential to yield better results, better teams, and a better organization. It frees us of the constraints that accountability is only after the fact and enables us and our members to recognize that we are responsible for our lot in life. Even when we are dealt a poor hand, or something goes wrong, or we make a mistake or get hurt, the real test of accountability is how quickly we recognize our situation and take steps to bring ourselves back above the line.

Second, this definition indicates that there is joint accountability from all team members for the success or failure of any given event, mission, or customer contact. Every member of your company has individual responsibilities, but each of them is also a member of the team. As such, each member wears two hats, one for him- or herself and one for the team. How we act, how we express ourselves, and how we are affected by positive or negative events all have an impact on the team and its attitude. Negative attitudes, finger pointing, and covering your rear end are viral behaviors that people can see, taste, and smell and that disrupt the ability of a team to function at its most effective level; in essence, the team gets sick.

Our success is interlaced, regardless of rank or position in the organization. Recall the story of Lieutenant Charles Plumb from chapter 1. In short, Charlie would tell you to never forget who packs your parachute. The fire service is a team from the top to the bottom, so on any given day anyone could be packing your parachute.

Finally accountability is about what the organization can do to help the company officer be a success. We can start by teaching basic organizational values, including the importance of teamwork to our recruit firefighters. We talk a lot about it, but very rarely do I see us requiring accountability from the group—for group success, rather than individual success. It's important that we let people know what kind of working environment we will accept. When you are a team, it pays to be a winner! Accountability is also about supporting our officers when they have a problem with a member who just cannot seem to shake the virus; sometimes, when you "can't change people, you have to change people!"

When you travel overseas, you first check to see what inoculations you would need against diseases you might encounter. In other words, you address head-on the microbiological rules associated with the environment in which you are going to live or work. Similarly, there are rules about organizations and people that you need to be aware of when dealing with accountability. You should learn to recognize lack of accountability from a distance and judge what impact it will have on you.

- *Life isn't fair.* Large organizations and bureaucracies are machines, and human constructs like fairness have no meaning to them. Only human beings understand these concepts, and even so, we often confuse fair with what we think it should mean. Especially at the macro level of a city, decisions are made that impact you and your crew, and there is nothing you can do about it. You can fall below the line and assign blame or point fingers, or else you can choose to say, "Okay, it is what it is. How do I work around it and make the best of it?" Many times, the rules or processes that organizations put in place and expect their members to adhere to are broken or disregarded almost immediately by senior members of the city or department. VBF&R just experienced this in the implementation of a new excellence program; city government ignored the process and did what they wanted to do without even a hint of attempting to follow the program. The result: many departments, including the Fire Department have to cede their technology staffs to the City Communications and Information Technology Department (COMIT). Never mind that COMIT will not do the same level of work for our department or others.

- *Not everyone in a leadership position is a leader.* Just because you reach a given level in an organization does not mean that you can lead, or even manage. Some people have no ability to develop relationships with the people they work with and consider them simply tools to get something done. Some people have no ability to see shades of gray, and the world is always either black or white. Some people surround themselves with members who are not accountable to anyone but themselves and their bosses. The real problem is that if you live in the organization long enough, this is all very transparent, no matter what level it occurs at, and it happens from company officer right on up.

- *You don't have to tolerate an infectious attitude at work.* This job is no longer merely about what kind of firefighter and EMS provider you are. There is so much more that we should be evaluating when we look at our members: How do they function as a team? What talents do they bring? Do they show up with a positive attitude and contribute? Are they enablers or restrainers? There are many more questions we could ask, but the bottom line is that, as an officer, I do not have to accept that you come to work everyday exhibiting the signs of the virus. At work, there are levels of expectation that everyone meets; off duty, you can be as miserable as you want.

Maintaining Accountability

People who lack accountability and responsibility consistently find ways to not solve a problem, complain that it's too difficult, find someone to blame, or come up with a reason it cannot be done. When you hear "We can't do that," or "That's not possible," that is below-the-line behavior. It may be that we cannot do it the conventional way or in the way we would prefer, or it may be that it will take much more time than we anticipated owing to budget, politics, or other road blocks. Nevertheless, there is nothing that cannot be done by motivated people who believe it can be done.

How do you foster an environment and practice behaviors that assist you in maintaining your personal and team accountability, when there are forces at work that push these traits aside? First, stay engaged with the issue or problem. When a task or project is difficult, people are inclined to give in and stop trying, to see whether things will get better on their own. I have had bosses who focused on what could not be done, rather than what could be done. Instead, explore creative and innovative alternatives to make a success out of a difficult project.

It may take several attempts to get the results that you want, especially when you are trying something new. VBF&R was very good at getting grant money, but there were several times when, for one reason or another, we did not get a grant. Had we just given up and stopped applying, we would not have been able to do many of the innovative projects for which we have become known.

One example I recall where we failed to stay engaged was with the mobile command post. With an influx of federal grants, the public safety agencies decided that we needed a functional mobile command post that could meet the needs and challenges of the 21st century. There were lots of obstacles in the way of what we wanted—in particular, the budget that was allocated for the project. We worked with manufacturers and finally found a very nice unit that was already produced and would meet our time frames. We did the show-and-tell with all of the department heads, worked out the fiscal aspects of the project, and brought in the unit for review.

Everything was in place, but we had one political stumbling block, and that was COMIT. They were determined from day one to to find every excuse why we could not purchase this particular unit. Everything in the unit, from supporting software and hardware to the type of digital clock, was a problem for them. In the end, with their political clout, they were able to crush the project.

COMIT's hidden agenda was that, by crushing the project, the money that had been designated for the mobile command post could be switched to their project, a new emergency communications center. As a result, we lost a phenomenal piece of equipment, which now costs twice as much as it did then. The irony of the entire story—which reiterates the point about staying engaged—is that afterward, senior staff would not support other alternatives, and the project died entirely. To this day, Virginia Beach, one of the largest cities in Virginia, has no functional mobile command post for major events, terrorist attacks, or operational incidents. We, the organization, failed to stay engaged and the project died.

Persistence is critical. To be accountable, you must ask yourself, "What else can I do?" Continually asking yourself this question allows you to develop alternative solutions to fix the issue as hand.

When we were forming a US&R task force, we lacked a training area. The members of the team worked on acquiring a chunk of land from the city landfill, on which to develop and build the first training site in the city. The technical rescue team, Fire Station 10, and Captain Mike Brown were instrumental in developing this training area.

After several years, the landfill kept expanding, the smell grew more intense, and it was becoming apparent that we could not use the location anymore. The team was at a loss for what to do, until Mike Brown, Rex Gurley (the site manager), Doug Kidwell, and others came up with a solution by asking, What else can I do to relocate the training center? They had to overcome some major obstacles put up by certain members of the organization who had no vision of the future. What exists today is one of the premier US&R training sites in the world and one of only three or four type I canine training facilities in the United States.

Being progressive requires you to think differently and to strive to understand perspectives other than your own. Albert Einstein once said, "The significant problems that we face cannot be solved at the same level of thinking we were at when we created them." You have to look outside the normal fire service mentality and look to other organizations, other sciences, and other models to see if they can be applied to accomplish our goals.

Continue to create linkages and forge relationships with as many of our internal and external customers and contacts as possible. This brings to the table critical new sources of information, funding, and direction. VBF&R has excelled at this in the past, especially when building special operations teams, but the philosophy has changed in recent years. In addition, I have always been impressed with the Virginia Beach Police Department's ability to create new linkages—funding part of the mounted police force through Friends of the Mounted Patrol, obtaining new Humvees on loan from local car dealers, and forming citizen advisory groups. By creating linkages with other groups, organizations can get equipment and complete projects that exceed their budget.

If you want to take responsibility for solutions, then you have to take the initiative. You have to assume full responsibility for discovering solutions that will ultimately deliver the desired results. Solutions like this come about only when you or your team take the initiative to explore other options, search for new methods, and even when you think you have done everything you can, ask, "What else can we do?". Always ask yourself how we can do this better and what we can do differently.

Staying conscious during these events is hard but critical. Never allow the program or process to get put on autopilot mode. Be attuned to anything that might offer a solution to your problem, because these tend to pop up in the strangest places. Remember the story of the monkeys in the cage (see chap. 7), and stay away from "we have always done it this way" mentality; that is the surest way to get yourself into a room that you have no idea how to get out of. Challenge every current assumption you have about the way something should or could work and get out of your comfort zone, to find unique solutions.

All of these techniques and mindsets offer a method to ensure that you act in a responsible manner and remain accountable for discovering solutions to the issues, problems, or processes. Because it is very difficult to stay this focused, it is rare to have an entire organization that operates in this mode.

Making Mistakes and Taking Your Lumps

Whenever you are put in the position of having to make decisions, you are going to make mistakes from time to time. You are inevitably going to choose the wrong path or, as I have done, make a boneheaded choice for which you must be held accountable. The secret, painful though it may be, is that when you make this mistake, take your lumps and get over it. People forgive mistakes that are understandable and acceptable. I could fill several pages with the mistakes I have made in decision making—sometimes on purpose, knowing full well the consequences. I had an unfortunate streak of asking forgiveness, rather than permission in my early career as an officer; eventually, I matured a little and moved on to make mistakes of a higher caliber and befitting a higher rank.

Many years ago, while VBF&R was transitioning to a predominantly career force but still had a rather strong volunteer fire service, the department inherited some apparatus from the volunteers. These apparatus were marked, "London Bridge Volunteer Fire Department" or "Davis Corner Volunteer Fire Department." As childish as I now realize that it was, many of the career firefighters (including some who came from the volunteer service) were upset that they had to drive trucks that said "Volunteer."

One day at Fire Station 3, my shift complained that they had to drive the tanker identified as volunteer, even though no volunteer had manned it in five years! "So take it off there," I said, knowing full well we had a policy that established a process to change anything on the truck. "What?" came the puzzled reply. I repeated, "Take it off there. Jim, you're an auto body guy, find the same color paint, sand it off, paint it, and make it look like it was never there." Thinking that they would never do it, I was just communicating, "I understand your issue," as the leader.

On the next shift, my intrepid firefighters had sanded off the word "Volunteer" and primed and painted the door of the truck. I swear it was beautiful job—you could not have expected a more professional job from a body shop. Now, everyone was happy, and I was not about to open up the circle of knowledge if I could help it.

As it happens, I worked for "Wild Bill" Adams, a retired marine and battalion chief who had instant flashbacks to boot camp and his days in the U.S. Marine Corps. Bill was a great fire ground battalion chief and always took care of his people, but he was also a strict disciplinarian and a policy guy. I will never forget the time when the guys at Fire Station 1 all bought clip-on earrings and wore them around the station. When Wild Bill came in, he went ballistic, screaming and threatening to fire all of them if they did not get rid of their "pierced ears." The guys and the company officer just watched him explode, and then said no. This was the absolute wrong thing to say, and as the verbal flow increased, they just smiled and said, "There is no policy saying we can't wear them." Determined that there was, Chief Adams poured over the SOPs only to discover that they were right. He stormed out of the station, went immediately to fire administration, from whom he demanded and received an SOP on jewelry. By now, the guys at

Station 1 had all thrown away the earrings and returned to their business. They had just been checking his blood pressure, so to speak, and by next shift, all was well among the family again.

Anyway, about two weeks after our chop shop paint job, Wild Bill walked into the bay, strolled past the tanker (tender for you West Coast folks), and froze. Slowly, he turned back and yelled, "Who the —— took off the 'Volunteer'!" I stepped forward and advised him that I had authorized it. After giving me a look of astonishment—shocked that someone actually admitted it—he calmly took me aside and asked me if I was aware of the policy—and whether I realized that I was not authorized to make this decision. I indicated in the affirmative, he wrote me a reprimand, we were still friends, the tanker never again had volunteer on it, and I was the hero to my guys. The moral of the story is not to go about breaking policy to be a hero—that was never the intent; rather the moral is that when you make a mistake, take your lumps and move on.

Believe me when I tell you that was not my biggest boneheaded decision. Sometimes we make mistakes and poor decisions either by accident or by design (it seemed like a good idea at the time). Admit your mistake and take responsibility for it; then, don't make that mistake again.

Some Leadership Traits for Accountability and Responsibility

The discussion of accountably and responsibility so far has probably been more complicated and in-depth than you might have expected. Primarily because the organization does not invest in officer training, we do a very poor job of providing the fundamental training and the proper culture to foster organization-wide accountability and responsibility. At the deepest level, there are a few fundamental aspects of accountability and responsibility that all leaders should understand to be successful.

A theme that runs throughout this book is that you must lead from the front. You have to be a motivated driver of issues, ideas, and character to demonstrate that you're accountable and responsible. People don't follow someone who sits back and does nothing, or continually says, "We can't do that," or has the virus of negativity. From administrative issues to the fire ground, your members need to know that you are out in front. Even when you are not physically in front, you need to be a cheerleader, the primary motivator, setting in motion those things that allow people to get things done successfully for you and the organization. You must be an enabler, not a restrainer.

Once more, it's not what you say; it's what you do that counts. Members pay very little attention to what you say and watch carefully what you do. If you tell me, "I am going to write to OSHA and get a ruling," you had better follow through on it. If you tell me, "When we build the new station, I am going to move the safety officer out of here and give your guys their space back," you had better do it. Fail this and fail the first-grade class on accountability and responsibility, on being good to your work, and on being trusted. Your character and commitment, what you stand for, will be judged by your actions, not your words.

Demand a boiling cauldron of an organization. You must expect and desire to be questioned in while continually seeking to improve your organization, your members, and yourself. Part of loyalty is speaking up and not going along with the status quo, and part of being an accountable and responsible leader is getting new perspectives. As a leader, you should remember that you don't have the market cornered on good ideas.

There is an excellent examination of accountability as it relates to people and organizations in *The Book of Five Rings for Executives*:

> *Empowerment is easy to talk about, but difficult to practice. Certain people can be empowered; others cannot. Unworkable systems will continue to be unworkable, even with empowered employees. Untrained and incompetent people remain so, even if placed in groups. Leaderless teams drift. Purposeless organization does not increase employee effectiveness. Useless activity wastes time and money. If people lack training, they cannot adapt. If organizations lack purpose, they will not endure. If leaders are absent, objectives will not be achieved.*[2]

Finally, especially as a frontline supervisor, technical competence, on both the fire ground and around the station, is your breath and heartbeat. Fail in these, and you will be dead in the eyes of your members. Your ability to take the initiative and institute training and programs shows that you are living up to the expectations and staying above the line.

Accountability is not a dirty word. For every action we take and every choice we make, there are consequences. Frontline officers must ask, "Who is responsible for success?" When the resounding voice comes back, "We all are," you will truly understand what accountability is really all about.

Do more than is required of you.
　　　　　　—George S. Patton

Notes

1. Conners, Roger, Tom Smith, and Craig Hickman. *The Oz Principle: Getting Results through Individual and Organizational Accountability.* Portfolio, 2004.
2. Krause, Donald G. *The Book of Five Rings for Executives.* Nicholas Brealey, 1998.

Chapter 13
Battlefield Firefighting

> *Everything in war is simple, but the simplest thing is difficult. The difficulties accumulate and end by producing a kind of friction that is inconceivable unless one has experienced war.*
>
> —Carl von Clausewitz

Ruining Your Career in 10 Minutes or Less

It's one thing to sit in a tactics and strategy class, take a command school, or work a new age fire ground simulator; it's another thing to do a live house burn or to work strategy, tactics, and tasks in a burn house or a natural gas trainer. All of these training tools are necessary for keeping skills sharp, but they can never replicate the real thing. Only through experience in the actual theater of operations can an officer learn the true aspects of a fire ground or significant rescue operation.

It's unfortunate that most of our tactics and strategy courses, along with our command courses, teach the same old things and never even touch on the critical aspects of working a fire ground. In 1989, Marine Corps Commandant General A. M. Gray distributed a remarkable 77-page document to every officer in the Corps; this doctrine was later published as a book, aptly named *Warfighting: The U.S. Marine Corps Book of Strategy*, and is a brilliantly written piece of work.1 When I first read *Warfighting* many years ago, I was stunned: The fire service had missed the mark in teaching people about what to expect on the fire ground. We had spent years teaching strategy, tactics, tasks, and incident command, but we had failed to train our officers on all of the intrinsic aspects of a fire and rescue operation. From *Warfighting* I have digested the critical aspects that apply to our world.

There are multiple jobs that an officer must learn well, but none is as important as being able to command a fire or rescue ground. Your reputation as an officer and a large segment of the trust you earn will be derived from your actions on the fire ground. Unfortunately, I have personally witnessed senior officers who would be better at cutting the grass than they are at running a fire ground; the worst were spastic or downright dangerous.

Make no mistake about it: The fire ground is combat, clear and simple. It is an environment that wants to win and take as many lives with it as it can; your job is to stop that from occurring. Want to ruin you career, at least with your peers? Then fail on the fire ground.

It's All Yours!

As a battalion chief, I was frequently credited with good fire ground strategy and tactics that were really other people's great decisions. Any battalion chief reading this will tell you that the fire or rescue scene success is determined almost exclusively by the initial decisions, communications, and commitment of the first-due company officers.

When company officers make correct choices with regards to strategy, tactics, apparatus placement, assignment of incoming units, clear and concise communications regarding the incident (painting a picture), and the safety of their crews, the battalion officer merely builds on the structure that is already in place to accomplish what needs to be done. In the end, the battalion officer gets patted on the back and congratulated on a great job!

We have all been to fire grounds where less-than-capable officers just didn't get it. As a result, apparatus did not get in the right place, orders did not get communicated, and subsequently, the entire event turned into chaos. Pulling everyone off an assignment and starting over is no small task. First, it's extremely difficult to reposition thousand-pound apparatus. Second, trying to collect firefighters committed to a poor operation is a lot like herding cats.

Company officers are the most influential people we have in our organizations. Their influence extends to the fire ground, EMS and rescue incidents, and any time people and apparatus are committed in order to solve a problem.

I assume that all of you would like to avoid the situation where your battalion officer is jumping up and down, unable to speak, red in the face from anger because you or someone else failed to make the right choices on an incident. In the following sections, I have broken down the discussion into manageable portions, so as to communicate the critical points for success in a readily understandable form.

Prioritization: Start with the End in Mind

Stephen Covey, author of *The 7 Habits of Highly Effective People* has a saying that is applicable not only to management and leadership but also to preparation for and engagement in emergency operations: One of your chief responsibilities is to ask *what if* every day you come to work as an officer. You have to be anticipating, planning for, and developing training for the incidents to which you might have to respond. From a pre-incident standpoint, you have to prepare your troops for the potential combat ahead and find a method for everyone to practice those skills. To accomplish this, you have to look for the resources to accomplish the job, plan how to do it, establish achievable outcomes that you want to see your personnel accomplish, and then put the plan into action.

On the fire ground, start with the end in mind. There is a pre-incident phase and an initial-response phase. The initial choices should be made with an achievable (assuming the correct tasks and techniques are applied) and identifiable outcome in mind. Even with the best intensions, a decision that is implemented using the wrong techniques in the wrong place will lead to failure. In addition, as the incident reaches its climax and is brought under control, the decisions that you make will be added to by someone else, who has to live with your first decisions and actions.

Beginning with the end in mind is like a construction project. For the entire project to succeed, the initial blueprints have to be meticulous: the foundation has to be laid properly, in accordance with the plan, or the rest of the project will be out of whack. This reminds me of a shed I built in my backyard. I laid the footers, sank and leveled the foundation, and reflected proudly on the fine work I had done. That was until I began the construction and found out I had added almost four inches to the total width by placing the 6 × 6 outside instead of inside. When you have preconstructed walls, that is sufficient to undermine the final product. Thankfully, my cousin has trade skills and helped me finalize the walls and roof. The difference on the fire ground is that there is no time to curse and to figure it out and fix it.

Identify the Main Problem

Each fire, EMS, or rescue incident has a main problem. The biggest mistake a first-arriving officer can make is failing to identify the main problem, which must be engaged immediately if the outcome of the event is going to be good. Classroom instruction and experience provide the necessary tools (most of the time) to identify the main problem when responding to an incident.

Identification of the main problem is critical, and failure to do so will always result in something bad happening. When you identify the main problem, you make the choices that establish your strategy and tactics, which then drive your tasks and commit people and resources to attack and rectify the main problem. By misidentifying the main problem, you will choose the wrong strategy and wrong tactics, thereby driving the wrong tasks—and when things go wrong at the task level, bad things happen.

Let's examine a basic example of using tools to identify the main problem. Suppose you arrive at a single-family dwelling with fire from the first floor auto-exposing to the second floor. Your water supply is established, and you have communicated direction to other units. So what is the main problem? Well you could say, it's a fire, Chase. Of course it's a fire, but as a company officer, you have been given tools to identify the main problem. One of the first tools you are supposed to apply is the size-up, which includes the walk-around. Lo and behold, when you go to the rear, you see Mr. and Mrs. Smith are hanging out a second-floor window, with heavy smoke around them. So, again, is the main problem a fire? No, it's the rescue of Mr. and Mrs. Smith. Had you put a line in service, under the assumption that the main problem was a fire, you would have blown the fire and hot gases up the stairway, roasting Mr. and Mrs. Smith in the rear bedroom!

When officers fail to identify the main problem, things go bad at the task level. We end up losing firefighters in a structure fire at 0200, losing victims who should have been found, and committing into structures that we should not have committed into. You must choose wisely, or else you will set the operation up for failure from the very beginning. Although it can be done (and is done by smart incident commanders), it is extremely difficult to catch up and regroup when things are going bad at the task level.

Communications, Command, and Control—First-Due Perspective

It is not my intension in this section to teach the proper terminology of incident command, how to repeat orders, and so forth. If you have not received that fundamental instruction at this point, you are way behind the power curve. This sections is about your actions in the first few minutes of an event, including the reasons why much that occurs is not as you saw it in your mind's eye.

Your very first job at any incident is to communicate clearly, effectively, and with as little emotion as possible. Yelling into the radio accomplishes as much as not saying anything at all. As the first-due company officer, you are the initial incident commander, as well as an artist: only you can physically see the fire or incident, so through your communications, you have to paint a verbal picture of the event so that other units understand what it looks like.

Once you paint the picture, it is your job to determine what you have been able to do up to this point, what you can accomplish with your crew, and what you need other crews coming in to accomplish. As simple as this sounds, it is often forgotten by company officers, who focus on engaging their own crew's initial strategy. This creates an environment where time becomes the enemy. When orders are not given or are not clearly understood, arriving crews will do one of two things: They will engage in what they believe needs to be done, which might not be what actually needs to be done, or they will await orders, wasting valuable time as the fire intensifies and/or the rescue grows more complicated.

If you are a second-arriving officer, it's your job to ensure you understand the communications and implement the tasks you have been given. Once you have completed this, you need to communicate the completion and find out what the next priority is.

The first-due perspective and direction are required in order to develop the incident in a way that leads to a successful outcome. There will be enough surprises and stumbling blocks to success that you do not need to add to them by not being able to communicate command and control the first few moments of the event.

Why Fire or Evolving Rescue Grounds Are Your Enemy

If you were going to war against an adversary capable of thinking and applying to the combat area specific resources, you would not think twice about the intelligence and human initiative that was about to confront you. In Warfighting, General A. M. Gray defines war as

> *A state of hostility that exists between or among nations, characterized by the use of military force. The essence of war is a violent clash between two hostile, independent, and irreconcilable wills, each trying to impose itself on the other. Thus the object of war is to impose our will on our enemy. The means to that end is the organized application or threat of violence by military force.*[2]

Although you might expect that something inhuman, like a fire or rescue scene, because it is not an intelligent, thinking being, would be incapable of either using hostile force or imposing its will on us, you would be wrong. The fire or rescue ground is a living and breathing entity, and being wrong about that will get you and your members in trouble First, given the opportunity—and compounded by lack of action—the event will grow worse, self-evolving until either we do something to stop the growth or it reaches its climax (burns up all the fuel, kills all the people it can kill, continues to collapse until it can't collapse anymore, etc.).

Second, these kinds of operations are violent and hostile events that have already killed or hurt people or are trying to kill or hurt people, your team members included. If you fail to recognize that these events represent acts of violence, you will find yourself stumbling into the area without the proper mindset, and that is how people get killed and hurt. The collapse of a retail structure, killing three firefighters, is a violent event; the flashover that traps and roasts to a cinder an entire crew is a violent event; and bailing out of a third-story window while on fire, to save your life, is a violent event. Likewise, attacking the fire, offensively or defensively, is a violent event; cutting apart a vehicle piece by piece while you hold the occupants together, stop the bleeding, and suction blood is a violent event; and reaching into the hole made by a gunshot and grabbing a bleeding artery is a violent event. Make no mistake about it, fire and rescue incidents are hostile events that require violence and speed of action to resolve them. If we do not apply violence and hostile action to stem the tide of the event, it will use violence and hostile action to grow, destroy property, and kill human beings.

The fire and rescue ground has a will of its own. Anyone who has ever seen fire jump across an open street, or a firestorm created by a large wild-land fire, or a building that is not shored begin to collapse knows exactly what I am talking about. If you let them, these events will impose their wills on you. Fire and rescue events seek a terminal end, and that "end" can come as a result of the event imposing its will on you, or through your efforts, tactics, strategy, you can impose your will on the event. Never let a fire rescue event command you like this. When a fire or rescue event reaches its own destination without your successful intervention, you usually lose. That means the loss of both blood and treasure most of the time.

As human beings we have a slight advantage in that we can apply our thinking to an action. The fire or rescue ground does not think, but given the opportunity, it can apply action on the basis of our inaction or wrong actions. However, what the fire and rescue ground has that is comparable to our ability to think—and that courses in tactics, strategy, and incident command do not teach—is the ability to create circumstances that can affect the outcome of the event.

On the basis of these premises, the fire or rescue ground is your enemy. It is trying to impose its will on you, using violence and hostile force. Given the chance, it will try to kill or maim you, before you can eliminate its resources to do so. Thus, you should familiarize yourself with the following characteristics of all fire and rescue operations.

The Enemy at Work

The enemy at times produces circumstances that directly affect our ability to reply. At other times, if we are not careful, engaged, and educated, the fire or rescue ground tricks us into letting it have its way. Keep in mind the following principles, to remain aware of what is happening, to anticipate what might occur, and to keep the situation from becoming critical.

Friction

The problem for most officers is that the fire ground is full of friction. Identifying a main problem and attacking it may seem simple, but you would be surprised how many times it is not done. The main problem must first be identified correctly, and as time goes on, it will change based on the actions that are not completed, updated, or followed up. Countless factors impinge on firefighting, making it an extremely difficult endeavor. Collectively, these factors are called *friction*, which Carl von Clausewitz has described as the "force that resists all action. It makes the simple difficult and the difficult seem impossible."

Friction on the fire ground comes in boundless forms and is always present. It's not possible to get rid of fire ground friction, but by understanding friction, it can be managed effectively. To clarify this, let's examine some forms of fire ground friction:

- *Mental friction.* Indecision over a course of action. The fire ground requires officers who can think in a compressed time frame, evaluate all sorts of information (separating the wheat from the chaff), and make the right decision. Indecision on the fire ground is a killer, and indecision with a patient on an EMS incident could create greater problems or cause the death of the patient. When officers fail to make decisions, they set into motion a snowball effect, allowing the fire or rescue ground to impose its will on us.
- *Physical friction.* Terrain obstacles, construction features, and access problems. I remember a large fire at the Virginia Beach's North End neighborhood. We had three very large structures going, a 40-knot wind off the oceanfront, and snow—yes, snow in Virginia Beach—and the fire was literally being pushed to the next street. An entire second alarm had been dumped into the path of the moving fire and was determined to make a stand, and we had deployed multiple crews with two-and-a-half-inch

hand lines and master streams to arrest the fire's progress. I moved forward only to find an officer and his crew, who I had thought were in place and flowing water, stymied by a six-foot drop into the rear of one the structures, blocking their access. Incensed, I asked why the roof ladder was not down and the line was not in place, getting the old puppy-dog look in reply. Terrain obstacle and construction features, compounded by the officer's failure to make a decision, had enabled the fire to impose its will on this crew.

- *External friction.* Terrain, construction, and weather problems. All of us have chased fire in knee walls, been unable to get to the scene in a timely manner, or had the weather affect our ability to apply tactics and get resources into position. We have to anticipate this, accounting for this type of friction in our plans. If we don't have specific resources to deal with external friction, we will lose.
- *Asymmetrical friction.* Weapons of mass destruction (WMD), Hazmat, secondary devices, and ambushes. In this day and age, we had better be ready and asking what if. If you, as an officer, find yourself surprised to be in the middle of a explosive or chemical device that just went off or if you do not anticipate the secondary device (remember Atlanta), you are setting yourself up for failure.

Self-induced friction

Fire and rescue incidents are as complicated as any battle the military fights. There is plenty of friction that can affect the event, but we also create our own friction that affects our ability to identify and act on the main problem:

- *Lack of a clearly defined goal.* We must clearly define what we expect to accomplish during the operation—especially during the first few minutes of the event, but continuing until we go home. Do we want offensive operations or defensive operations? Are they changing, and if so, why? When we do not define the goals we want accomplished to all of our troops on the fire ground, we create friction. We then lose the coordinated will that we intended to impose on the event.
- *Lack of coordination.* Lacking a clearly defined goal creates a lack of coordination among our task-level groups, who are supposed to be engaged in a very specific mission to correct a problem. When

this occurs, the resultant friction renders us unable to control an event, because we cannot apply the correct resources at the correct time and at the correct point to solve a problem.

- *Unclear or complicated plans.* Plans must be clear and understood on the fire ground or at an EMS incident. Everyone in the game needs to know the rules and the desired outcome. When our plans are too complicated, there is greater potential for failure because of misunderstandings and timing problems.
- *Complex task organizations or command relationships.* NIMS, incident command system (ICS)—whatever you call it, it is designed to take complex situations and break them into manageable parts in which all relationships are understood. Making it to complex, beyond the incident requirements, creates friction,
- *Complicated communications systems.* "What we have here is a failure to communicate!" If your manner of communicating directives and information is so technically complex or convoluted that it takes forever to send a simple message, you create friction. For example, 10 codes were one kind of friction that we had almost eliminated until the police got them back.

Because firefighting and EMS is a human enterprise, friction will always have a psychological, as well as physical, impact on the fire ground. We can never eliminate friction, so we must learn to work with it, expect it, and minimize it, to reach an acceptable outcome. We can limit self-induced friction, because it is within our control, but in emergency services, we must learn to fight effectively within a medium of friction. We cannot recreate the level of friction we experience during actual emergencies in the classroom or at the training site. Learning how to function in spite of friction comes from experience, seeing friction and its impact on emergency incidents; only then do we gain a full understanding of how complicated it is.

Uncertainty

There is another circumstance that affects the company officer's ability to identify the main problem, and that is uncertainty. To those members of the service who would stand up and scream, "Not so—I know everything on my fire ground!" I suggest that I you live in a world of cute puppies and butterflies, divorced from reality.

All actions on the fire ground take place in an atmosphere of uncertainty—that is to say, shrouded by the fog of war. Uncertainty pervades the fire ground in the form of unknowns about the exact location of the fire, the victims, the environment (current status of the roof, trusses, add on construction, knee walls, etc.), and even the operating crews. Until such time as we have GPS, three-dimensional real-time tracking of all of our crews, the only true accountability lies with the company officer who has eyes and hands on his or her crew, and not with command. Even as we try to reduce the unknowns by gathering information, we must realize we cannot eliminate them. The very nature of firefighting makes absolute certainty impossible; all actions on the fire ground will be based on incomplete, inaccurate, and even contradictory information.

The difference between an amateur gambler and a professional gambler is that the amateur plays based on possibilities, while the professional plays based upon probabilities. To address uncertainty on the fire ground, we must develop strategy and tactics based on probabilities, not possibilities. When we look at probabilities on the fire ground, we need to consider the following issues:

- *Imply a level of command judgment.* There is an old saying that "good judgment comes from experience, and much of that comes from bad judgment." As an officer, you must be able to draw on your experience and training to make command-level decisions based on what you know, what you think you know, and what your past tells you that you should be doing right about now. If you spent your entire career in staff, you can't possibly gain this kind of experience.

- *What is probable and what is not?* Probabilities for a gambler are mathematical calculations to increase the odds of winning. By contrast, possibilities are those things that we wish will happen, but typically don't. It may be possible that the truss roof that has been exposed to fire for 15 minutes will stay up, but the probability is that it won't. Your training and experience can provide you with some of the probabilities, but they cannot replace what you see, smell, taste, and hear at the incident.

- *Through the judgment of probabilities, we can estimate the fire's design and act accordingly.* Judgments based on probabilities offer us the safest and most effective method to design a strategy and apply tactics. We can then employ small unit tasks to accomplish what we need, keeping in mind the probability of success by placing particular units in certain locations and by assigning specific tasks. This allows us to combine information on what has occurred and what is likely to occur based on a given course of action.

- *Realize that those actions that fall outside the realm of possibility often have the greatest impact on the outcome of firefighting.* When three million people play the lottery and only one wins, the impact on that individual is immense. Probability says that it's a long shot, but the payoff is extraordinary. The risks that we must take in life-and-death situations are often based on possibilities, rather than probabilities. As with the lottery, the payoff is extraordinary, but here, the price for failure is much greater. Nevertheless, many heroic actions fall well outside the realm of probability. As a company officer or a fire ground commander, you must recognize the value of taking advantage of the possibilities.
- *Risk is related to gain.* Usually the more we risk, the more we gain—especially on the fire ground. Importantly, in emergency events, risk is common to both action and inaction in equal measures. Even though risk is a necessity, acceptance of risk should not be viewed as condoning the use of imprudent measures and a willingness to gambling on a single improbable event or action.
- *Understand when chance comes into play.* Chance consists of turns of events that could not reasonably be predicted or that did not seem probable. Chance does not favor either the fire or the fire service. Recognize when opportunities that we did not predict occur, and be prepared to exploit them.

So, the question that this leaves is, How do we effectively manage uncertainty, fight in an environment that is full of friction, and consistently identify the main problem? To overcome uncertainty, remember the following when responding to a fire or EMS incident:

- *Develop simple, flexible plans.* Keep it simple! Develop plans based on probabilities; remain flexible; assume that the easy way is mined; and develop a flexible mindset.
- *Plan for contingencies.* There is an old military saying that "no plan survives initial contact." Make sure that you consider contingencies and have at least two or three alternatives to accomplish what needs to be done.
- *Develop SOPs.* Standard practices that everyone understands promote safety, coordination, and communication.
- *Foster initiative among subordinates.* Every soldier is a general, and everyone needs to be able to fill a void when it

suddenly opens up. Teach your members to look for openings, changes, and opportunities, and let them take a risk. They will see things you will not, which may be needed in order to accomplish the task at hand.

As an officer, every day you build your KSAs, working toward one goal. From boardroom to mail room, from small incident to large incident, you had better develop the KSAs necessary to identify the main problem, work in an environment full of friction, work with uncertainty, and determine probabilities.

Part of risk is the ungovernable element of chance. The element of chance is a universal characteristic of war and a continuous source of friction. Chance consists of turns of events that cannot reasonably be foreseen and over which we and our enemy have no control. The uncontrollable potential for chance alone creates psychological friction. We should remember that chance favors neither belligerent exclusively. Consequently, we must view chance not only as a threat but also as an opportunity, which we must be ever ready to exploit.

—A. M. Gray

Fluidity

Fluidity occurs in fire and rescue incidents as part of the ongoing cycle of decision, result, and outcome. On the fire or rescue ground, no specific activity or action can be viewed as a stand-alone event. The incident commander or the operational forces applying the tactics cannot view their actions or the actions of others in isolation. Each action or inaction during the event affects others; ultimately, decisions and operations made earlier shape decisions and operations made later. According to the concept of fluidity, success in this environment hinges on adapting to and capitalizing on ever-changing events.

It is impossible to sustain high-energy operations on the fire ground forever. There will be times when we must thrust resources and personnel at an event and push them to the limit. The tempo of the fire and rescue operation will fluctuate—from periods of intense activity to slower periods that reflect limited activity, rehabilitation, limited information flow, replenishment of resources, and redeployment of personnel and equipment.

The secret to success is to recognize what affects the energy level of the event. Weather, darkness, temperature, and other external factors may affect your ability to sustain a specific operational tempo. Over the long haul, you must set up a competitive, fluid operation that manages the continuous flow of the event at hand.

Disorder

When a fire or rescue operation exhibits friction, uncertainty, and fluidity, the incident will naturally move toward disorder. We can never eliminate the potential for disorder, because plans may not be carried out appropriately or in a timely manner, instructions and commands may be misunderstood, or mistakes will disrupt our ability to function effectively.

Disorder is primarily a human event; thus, it does not naturally occur over time with the fire or rescue itself. If human disorder is introduced and repeated, the incident will grow, thereby increasing the level of disorder experienced by operational personnel and command staff. The impending and subsequent Twin Towers collapses are perfect examples of this.

Fire officers must be able to function in a disorderly environment. By recognizing disorder, fire officers can compensate by correcting those aspects of the operation that have introduced disorder into the event.

The human factor

Firefighting is a cardiovascular, muscular-endurance team sport that requires human beings and human initiative to sustain it. As a result, our ability to effectively respond to a significant fire or rescue incident is shaped by human nature and all the complexities, inconsistencies, and peculiarities that control the emotional and physical capabilities of human beings.

Some of the events that we respond to are true tests of courage and physical stamina. Fear, danger, exhaustion, and privation all affect the men and women who fight the fire, respond to the rescue, or work the structural collapse. People are different; what will break one person physically or psychologically will not break another. I have witnessed very strong and capable people completely lose their sense of direction and safety after weeks of working an operation, because of emotional attachment to an event. We must remain mindful of the human dimension.

It is unlikely that any command class you took examined firefighting and rescue operations in these terms. We cannot ignore the undeniable parallels to combat and war; otherwise we undermine our ability to understand the event for what it is and to draw it to a successful conclusion.

RIT—Doing It Right

Two critical events during fire and rescue operations require our members to be completely dialed in to what is occurring and to be able to perform flawlessly—or at least as flawlessly as possible given friction, uncertainty, fluidity, and disorder. These events are the search and rescue of trapped civilians and the search and rescue of trapped firefighters.

Rapid intervention team (RIT) or rapid intervention crew (RIC)—call it what you will, but our expectations are much too high. They are overstated by OSHA's two in/two out rule, which is not RIT, but rather a bureaucratic and political solution to a liability issue. On the basis of historical reviews of trapped firefighters, the two people outside are neither effective nor applicable to save someone trapped, disoriented, lost, or caught in a flashover within the structure. The fix is a placebo, with no real-world application other than making us feel good.

Every critical operation has its foundation in training that must be in place before the event. While we spend a significant amount of time teaching firefighters how to rescue civilians, many fire and rescue departments have not yet put every member of their organization through classes such as Mayday, Firefighter Down; Saving Our Own; and RIT Operations. This is akin to strapping a fighter pilot into an aircraft without teaching him how to eject or providing him with a basic survival kit. The best chance of saving lost, trapped, or disoriented firefighters comes from teaching them to recognize, accept, react, and save themselves or to alert those who can save them in a timely enough manner to yield at least some reasonable chance of success.

Why do I say some reasonable chance to success? I am a student of history, including the history of rapid intervention teams. Most recently, the Bret Tarver incident in Phoenix has enlightened us as to several key points. First, rapid intervention is not rapid. Second, two in/two out is a joke and will not save anybody. Third, firefighters wait way too long to recognize, accept, and react to personal entrapment. Finally, when this happens, it is an event filled with disorder.

In instances of personal entrapment, there is uncertainly as to the exact location, status, and position of the trapped firefighters. The event is full of friction and disorder, and members of the RIT team can become victims themselves, requiring a further assignment of resources to rescue them. Moreover, these events add to the fluidity of the event, with each decision and action having an impact on the others and compounding the event at a rapid rate.

Hypercool—How to Act When Things Look Bad

Pressure is a word that is misused in our vocabulary. When you start thinking about pressure, it's because you've already started to think about failure.

—Tommy Lasorda

Everyone has their "screamers" on the radio. You hear the screaming over the radio and rush to the watch office, thinking something terrible has happened, when Joe walks in and says, "Oh, that's just Captain Smertz. It's a Dumpster fire!" You all know which individual(s) in your organization can be depended on to be in a panic. Nevertheless, screaming into the radio reflects a lack of professionalism, a lack of control, and the inability to communicate effectively. Screaming enhances nothing and solves nothing; in fact, it makes things worse. Often it results in missed or misinterpreted communications, pushing everyone's adrenaline higher while doing nothing to solve the problem at hand. To this day, I have never found any fire or significant rescue that was resolved or made smaller by screaming or yelling.

All of you sit around the station wishing for the big one, so when you get the big one, act like you have been there before. It is normal human physiology that when you arrive on something that is really going badly and people have called you to fix it, your adrenaline will be up, your mouth will be dry, your sweat glands will be open, and you will be hypersensitive. Accept it and learn to work with it. First, recognize that things you do in the first few minutes are not going to put out the fire or solve the rescue, no matter how much you would like them to. Second, what you do in the first few moments will set the stage to achieve as quick a resolution as is possible given the circumstances.

So what is hypercool, and how do we get it? It is a state of mind, reflected in your actions, that says to the outside world, "I do this every day—it's no big deal." In reality, it may be your first time, but you better not let the people who called you know that. How would you feel if you were laying on the table before open-heart surgery and heard the surgeon say, "Finally, a chance to do what I've only read about before"? You have chosen a profession in which others call you when they are having a really bad day or when something tragic, dangerous, violent, or hostile is occurring or has occurred. It's what you do for a living, so don't act like you're in the same boat as the people who called you; it's up to you to fix it, and the first step toward success is to control your own emotions in the face of whatever is going on around you.

I have already stressed that you must be able to work—and even thrive—in an environment full of chaos, disorder, uncertainty, friction, violence, and fluidity. Accept this before arriving, and learn to manipulate it for your own good once you do arrive.

Let's say it's one of those unusual days (at least for most of us) and it's just now noon and you are backing in the rig from your fourth working fire since 0800. As the driver-operator, you back in safely, all the firefighters get off the rig, and since you want to get a bite to eat, too, you forget to fill up the water tank! Off go the tones again: structure fire corner or walk and don't walk. Everyone gets to the rig, you start it up, release the brake, and down the road you go, Federal Q wailing away. As you come down the road, you see the loom up! (When I was in Los Angeles teaching after the riots, then I was told, "Yeah, we were coming down the boulevard and had multiple loom ups!" Befuddled, I asked, "What's a loom up?" My West Coast brethren replied, "It's a column of smoke in the air. What do you call it?" Stunned, I looked at them and said, "We call it a column of smoke in the air!")

You come around the corner, see the loom up, and hit the hydrant to lay to the front of the house. Smoke is rolling out the front door and windows, and the firefighters stretch the line to the porch and start to mask up. You set the power take-off, get out, pull the tank to pump, drop the water, and look up only to notice that the light is red! Will you panic? No, the firefighters are on the front porch signaling that they are ready for water, the citizens are watching, and your Viking helmet is on, so even if you have no water, you have to do something. Well, you put your sunglasses on, turn the throttle up a little, step up on the driver's step, look over the crow's nest to the folks on the front porch signaling for water, give them the thumbs-up, and say, "I don't know what the problem is, but it's on your end!" Hypercool! Will that preclude the fact that you deserve to get your rear end kicked? No, but you couldn't run around screaming and yelling for water!

One of the most hypercool moments I remember came when I was a division chief. My office sat in the middle of a 312-square-mile city, surrounded by 22 engine companies, six trucks, two squads, four battalions, and lots of dedicated people. When the call came in for a house fire on the North End of the beach, on Atlantic Avenue, I thought, "Maybe I should step outside and look for the loom up." The North End an old, very expensive neighborhood, with an average square footage of about five thousand square feet. On this same day, we were experiencing a nor'easter, or a strong wind off the Atlantic Ocean, and it was snowing, a natural disaster in and of itself in Virginia Beach. To top it off, the North End houses are beach houses that either directly on the Atlantic Ocean or one and two blocks away.

As I sauntered out to the ramp at Fire Station 3, to spot the loom up, I looked out to the north, and what I saw was a column of smoke in the air that looked as though a plane had crashed. I walked quickly back into the office, looked across the desk at one of the other division chiefs, and said, "I think we should operate the system on this one!"

As we headed down the road, the report over the radio was, well, hypercool. Captain Rick Kellogg, responding from Fire Station 11 with Engines 11 and 14 and Ladder 11 had arrived and this was the report:

11-A dispatcher

11-A

On scene, be advised this is a working fire; we have a two-story 100 × 150 structure well involved—[pause]—dispatcher make that two, two 100 × 150 structures well involved—[pause]—dispatcher make that three, three 100 × 150 structures well involved! We are going defensive, give me a second alarm—[pause]

11-A dispatcher

11-A

Be advised it's moving to 57th street!

Not only did he have three large structures in flames, a 40-knot wind, and snow, but he now had the fire jumping across the street and starting to expand via heat and embers to structures a block away. To make matters even more interesting—and this was something we did not know until later—he had two rookies on the rig with him and was trying to hold these sled dogs by the collars, to keep them from hurting themselves, while talking on the radio, doing a size-up, and giving a report to the dispatcher in a soothing voice as if he did this every day. How cool is that?

Later he confessed, "Chief, I thought I was going to burn the whole — — block down and then another block. I was worried s——less." He did, however, project an aura of hypercool that set the tone for the entire incident. Ultimately, the fire resulted in a third alarm, with the entire second alarm moved down the street onto the exposure block, to draw a line in the sand and keep the fire from burning anything else to the ground. We lost three large houses but nothing else; the fire never went past the original structures involved, with the minor exceptions of a few sheds and some vinyl siding and roofs on the opposite block.

I hire you and pay you to do this for a living, or you volunteer to do this for your community. Don't overreact, learn what to expect, and understand what to do; accept that you won't solve a problem in a short time frame and that yelling, screaming, or coming unglued and losing focus are unprofessional. Remember to consider the perspective of the civilian(s) calling the incident in: "If it's an emergency to you, who do I call?"

Final Considerations on Being in the Soup

Tactics taught in the classroom amount to learning how to jump through certain hoops, how to advance, how to retreat, how to go offensive, and how to go defensive—essentially, how to get through things safely. These classes also teach that there are right and wrong answers. Being safe and being correct becomes a habit, and artificiality becomes a habit. As a result, commanders and frontline troops can get manipulated by the fire ground and may end up getting mauled. When you have to operate and fight on the fire ground, think only of winning—using all that you have available to you, both traditional and nontraditional. A safe death is still a sure death.

Standard solutions induce an attitude in people that there are easy ways to operate and win on the fire and rescue ground. However, this is far from the truth. Standard solutions give our members the idea that there is only one way to do things, and that mindset can lead to difficulty and defeat.

In real life, victory is accomplished by seizing the initiative through careful thinking and bold movements at the right moment. The standard solutions should be viewed as starting points for discussion and alternatives. We should be teaching people how to think, not what to think, so that they say, "Start here and evolve according to the circumstances."

Notes

1. United States Marine Corps. *Warfighting: The U.S. Marine Corps Book of Strategy*. New York: Currency Doubleday, 1994.
2. Ibid.

Chapter 14
Transitional Team Life Cycles

> *Wisdom requires the long view. . . . I am reminded of the story of the great French Marshal Lyautey, who once asked his gardener to plant a tree. The gardener objected that the tree was slow-growing and would not reach maturity for a hundred years. The Marshal replied, "In that case, there is no time to lose, plant it this afternoon." Today a world of knowledge—a world of cooperation—a just and lasting peace—may be years away. But we have no time to lose. Let us plant our tree this afternoon.*
>
> —*John F. Kennedy*

The application of team concepts to the fire service is not new. On entering the fire service, we were constantly reminded of the absolute necessity of teamwork. Classroom sessions were supported by practical evolutions that reinforced the notion that nothing on the fire ground gets accomplished without a highly trained, equipped, and motivated team.

Certainly, individuals contribute to the team effort. In rare cases, individuals perform a spectacular feat that brings instant notoriety. However, if we were given a behind-the-scenes glimpse of that feat, we would most likely discover that many members of a team made the event possible. Moreover, without the cooperation of a team and without some team members contributing in much less glamorous areas, the individual feat would never have occurred. This is true of the now-famous crane rescue in Atlanta, as well as of the window rescues of individuals trapped in downtown Manhattan. This should not be construed as taking anything away from the bravery, integrity, and honor of the individuals given such high-profile tasks. Nevertheless, it is unlikely that an individual in the fire and rescue services could pull off an extraordinary rescue or operation without the assistance of a team or multiple teams.

Having spent 29-plus years in the fire service to date and having split my career between line and staff jobs, I have had the unique opportunity to see both sides of the issue. Early in my career I also fell prey to the power and prestige of a staff job. I had the misfortune of seeing administrators and staff officers swear that the organization evolves around staff, not around operations. There are still fire service chiefs and staff officers who have been out of touch with operational issues for so long that they feel that operations is simply a portion of the fire service that must be endured.

My exposure to a wide range of fire services and contact with fire service leaders around the world has led me to the following conclusion: The most effective and efficient fire service organizations are those whose leadership understands that a staff system is in place to support operations. The importance of support operations should never be underestimated. No great army was ever successful without the support of logistics, planning, and command staff. But no battle was ever won without putting troops in the field. The bottom line is that emergency services and non-emergency services provided by operations personnel can make or break a department's reputation and political future. Furthermore, teams are necessary in order for these operations to be successful.

Because it would be impossible to compress an 8- or 16-hour seminar on teams into this space, many items of concern and interest regarding teams remain beyond the scope of this discussion. This chapter focuses on transitional team skills that can be applied by company officers, battalion chiefs, senior staff, and fire chiefs.

Constants When Evaluating the Team

Before we delve into the concept of team life cycles, we need to address some constants that apply to fire service teams.

Teams enhance the organization's capabilities through high-performance service

The purpose of any team is to enhance the organization's capabilities and to function at the high-performance end of the scale. Regardless of function, every team has been assembled with the delivery of a given service, to an internal or an external customer, in mind. Teams responsible for emergency activity—such as technical rescue, Hazmat, engine companies, and truck companies—are in place for one reason alone. That is to provide timely, effective, professional, and efficient emergency services in the specific area of function. Likewise, nonemergency teams, such as inspections, are designed to deliver a level of service in creating a fire-safe community.

No matter the intended function, a team should strive to provide high-performance service. Depending on where the team is in its life cycle, that level of service may differ. At times when the team is not functioning in the high-performance area of operation, every effort must be made to provide the resources, training, and staff to get the team there.

Small teams accomplish 90% of everything we do

Administrators and senior staff officers hate to admit how much is accomplished by small teams. They project the attitude that administration and senior staff run the department. However, almost 90% of all the customer contacts made in the fire service in an emergency or non-emergency setting are made by small teams. These teams may be engine companies, truck companies, rescue companies, inspectors, training instructors, community educators, or a mixed group of staff and line personnel put together for a specific reason.

Fire services are usually reported as the most highly respected municipal service. This reputation has been gained through contact with customers, both internal and external. The services that are most likely to be remembered by the customers—and the ones on which we are therefore graded—are delivered during an emergency. Although a department gains notoriety from a wide range of programs, it is ultimately rated by citizens and visitors on the basis of its response to and handling of emergency incidents.

Before you take action, you or the organization must ask, "Will this enhance the capabilities of the small team?" Because 90% of everything accomplished in the fire service is accomplished by small teams, this is the first question you should ask. There will be times when you have to defer to the team and allocate resources. When resources are limited, you need to identify which teams make the most contacts, so that the resources are shifted to them accordingly. Operations (suppression, EMS, and special operations) account for almost 90% of all customer contacts; thus, given scarce resources, it should be obvious where to put the resources to get the biggest bang for the buck.

For example, suppose that the vast majority of reporting, data gathering, and information management at the company level of a department is done by computer. Also suppose that there are very busy companies with lots of personnel but only one computer. Now suppose some money becomes available. Senior staff (the people who control the money in the budget accounts) have a decision to make: Do I purchase palm pilots for my senior staff chiefs (a small team, all at the low-performance end of the scale), or do I use the money to purchase computers for those busy companies operating 24/48? The correct choice should be clear; nevertheless, many fire service leaders (perhaps that is not the right word, since leadership requires much more than management) and organizations forget that the business of the business is the delivery of emergency services to citizens and visitors—period!

Teams of any kind have a predictable life cycle

All teams go through a life cycle. Like human beings, these life cycles include times of high morale and low morale, high performance and marginal performance. Recognizing which stage the team life cycle is in when you assemble or inherit a team is important, to understand what actions will advance the team toward the high-performance end of the scale.

To reach peak performance, teams require leadership and vision

Any team or group of teams that is expected to operate at the high performance end of the scale must have creative and visionary leadership. While an entire section of the Buddy to Boss series is dedicated to leadership there are some basic principles that must be applied to effectively lead a team. You have seen most of these before (see "The ABCs of Leadership," in chap. 3), but they are so important that you cannot afford to forget them.

- *Trust your subordinates.* You can't expect them to go all out for you if they think you don't believe in them.
- *Develop a vision.* People want to follow someone who knows where he or she is going.
- *Keep your cool.* The best leaders show their mettle under fire.
- *Encourage risk.* Nothing demoralizes the troops like believing that the slightest failure could jeopardize their careers.
- *Be an expert.* From boardroom to mail room, everyone should realize that you know what you are talking about.
- *Invite dissent.* Your people aren't giving you their best if they are afraid to speak up.
- *Simplify.* You need to see the big picture to set a course, communicate it, and maintain it.

Leadership can make or break a team. Even the most motivated, dedicated, and educated team members will tolerate only so much incompetence from leadership. There may come a time when the problem with a team is clearly identified as a leadership problem, and the sick portion of the team must be carved out in order for the team to survive. Make no mistake: poor leaders undermine team efficiency, effectiveness, and morale.

Being the leader of a high-performance team requires a unique balancing act. While ego is important, remember the trainer's motto: "Leave your ego at the door, or you're likely to get hurt." The failure of a team leader to practice the ABCs of leadership will lead to the downfall of the proposed leader—or the downfall of the team, if the so-called leader is not removed.

Teams are living entities

Teams are made up of individuals and are living entities. There are identifiable characteristics of high- and low-performance teams. These characteristics can be equated to the specific competency displayed by the team and thus allow the calculation of where the team is functioning on the performance scale.

Note that I said the team, not individuals, is displaying a competency. Teams have a common culture and a common vision when they reach the high-performance end of the scale. As established previously, teams have a life cycle, as does any living thing. Thus, team efficiency can be enhanced or undermined by what company officers, managers, or organizational leaders say and do. Teams can combine experience and training, and they are capable of learning, from successes and failures. Learning is the legacy that a team can pass on to the generations to come.

Since the fire service delivers service primarily through teams, it is important that company officers and senior staff understand how teams function and what teams need to be successful. Not every officer gets the unique opportunity to build a team from the ground up. Often, because of promotion or transfer, a company officer inherits a team that is already in place, and a senior staff chief inherits many teams.

There are several universal concepts to understanding: what teams need, what drives teams, and how to assess and enhance performance. Recognizing what is needed and responding effectively to that need are two different things. The ability to effectively manage and lead a team is a matter of competence.

For the purpose of this discussion, there are two types of fire service teams: *task teams* and *specialty teams*. Task teams include engine, truck, inspections, and education and are responsible for a set of tasks of either an emergency or a non-emergency nature. Specialty teams may also be task teams, but are in place to provide a very specific task, such as technical rescue, Hazmat, marine operations, WMD, or any other area of expertise that requires special personnel, training, and/or equipment.

Each of these teams has important organizational roles to play, and an adequate blend of both types is required in order for an organization such as the fire service to be successful. Organizations always have many more task teams than specialty teams. The direction of specialty teams requires KSAs from its members and leaders beyond what is expected from a task team; therefore, an extensive level of training, equipment, and resources is needed to support a specialty team.

Team Life Cycles

Every team, regardless of task or specialty, has an identifiable life cycle (fig 14–1). It is important for officers and team members to understand this life cycle, to predict the long-term direction of the organization and anticipate the measures that will be necessary to ensure survival. Life cycles affect everything a team does, as well as management and leadership's ability to develop teams effectively, with high-performance characteristics. Also, anyone inheriting an already-established team must identify what stage the team is at in its life cycle to plan effectively for the continuing development or redevelopment of the team.

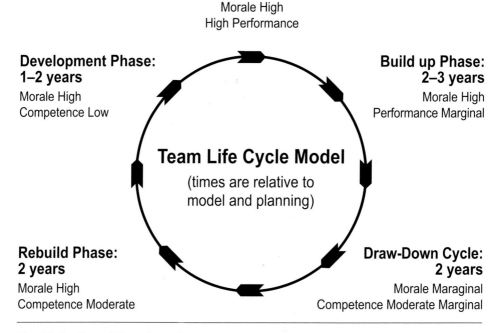

Fig. 14–1   Team life cycle model

The time frames referenced in figure 14–1 can be adjusted. Most task team time frames are compressed, since a lesser level of training and resources are needed to engage the team effectively. Specialty teams, because of the nature of their organizational mission, require extraordinary levels of support in order to reach the high-performance end of the scale.

What we discover when looking at team life cycles is that the performance capabilities are dependent on eight factors: personnel, training, equipment, resources, leadership, succession planning, morale, and organizational support. These factors are collectively referred to as the success octagon. Each factor plays a role in determining the efficiency and capabilities of a team at any point in its life cycle; undermining any of the sides of the octagon will undermine the team and its ability to deliver service. Let's define these factors, so that we have a common platform from which to discuss the team life cycle:

- *Personnel.* The human resources aspect of the team. This encompasses team members and leaders/managers. The ability of a human resource to effectively function in emergency and non-emergency environments is directly proportional to the remaining seven sides of the octagon.
- *Training.* The capability to deliver KSAs to the human resources component of the team. Training is second in importance only to emergency response. However, failure to effectively train renders the team useless on an emergency incident; thus, they are inseparable. As the military saying goes, "Train hard; fight easy."
- *Equipment.* One resource that the team requires in order to be effective. Equipment is an expensive investment, but failure to provide adequate and appropriate equipment creates an unsafe condition for team members and compromises the emergency or non-emergency service capability.
- *Resources.* All of the logistical requirements that a team needs in order to function. This includes equipment, support items, and administrative items, as well as the development of training sites and training opportunities.
- *Leadership.* The people who are selected or appointed to lead and manage the team. Leaders must be experts, plan effectively, and understand the most effective methods to obtain the most and best from their human resources. Leadership of a team is formal, but there can also be informal leadership based on the specific expertise and KSAs of team members. Above all, effective leaders require credibility. Credibility = Trust + Respect.
- *Succession planning.* The ability to continue the team life cycle, preempting anticipated life cycle changes (and therefore shortening the time frames) and developing new leadership and

members. Although the team will last forever, human resources will not. Transfer, promotion, retirement, individual career-change decisions, and changes in departmental leadership all affect team succession planning.

- *Morale.* How the team sees its own value in relation to the organization and its perception of the value that the organization places on the team. Morale is often discounted by managers as an important factor. (Note, however, that I did not say leaders, since leaders never discount morale.) In many instances, this is because the manager lacks the necessary KSAs to understand what the team does. Alternatively, the manager may simply be incompetent; incompetent people will discard what they do not understand, to make themselves more comfortable in their management roles.
- *Organizational support.* The willingness of the organization to understand and the ability to provide those aspects of the success octagon that fuel and drive a team. This means that an organization must recognize when something is lacking and correct the inadequacy, to improve the efficiency of the team.

So what do we know up to this point? First, there are very specific organizational characteristics and team characteristics that can be identified. Next, teams have a life cycle that affects its capability to deliver a given level of service. Finally, there are eight specific factors that a team requires throughout its life cycle in order to function.

Organizational leaders who fail to understand these basic concepts or who are determined, for whatever reason, to ignore them cannot expect the best out of their teams. Effectively managing and evaluating these factors enables leadership and team members to identify strengths and weaknesses and then apply the necessary fix in any area that is lacking.

Understanding Team Life Cycles—Managing the Big Picture

It is important that we manage for the long-term growth of the team. Because of their very nature, teams will not always be as efficient or effective as the leadership would like. Teams grow, change, shift, and develop over long periods. The job of a company officer, team leader, or organizational leader is to understand how to get the most from teams.

Phase I: Development

Not everyone in the team will be involved in the development phase. This represents the birth of the team concept and its mission. The hallmark of this phase is someone's vision that a team needs to be developed to provide a specific type of service. The type of service may already have been decided, as is the case with engine companies, ladder companies, and rescue companies. In other cases, a new task-level or specialty service has been identified by the organization; this may be a twist on an old service level, such as Ray Downey's development of the squad concept for FDNY, in which engine companies were modified for an enhanced level of service. The development phase is identified with the following benchmarks:

- *Mission statement.* This may be formal or informal, but the development of a mission for the team is the first step. For long-term evaluation (including SOP development), it is preferable to put the mission statement in writing.
- *Personnel selection.* Human resources staffing is based on the anticipated mission and necessary KSAs.
- *Identification of the success-octagon requirements.* Remember the eight keys factors necessary for a team. At this stage, shortfalls and needs must be identified, and a method to correct the deficiencies must be created.
- *High morale.* Assignment to the team and the opportunity to take place in the development of something new moves everyone up on notch on Maslow's hierarchy of needs. Morale can be undermined if the organization fails to meet the requirements of the success octagon.
- *Low efficiency/performance.* When all of the success-octagon factors are not in place, the team will not be as proficient or effective as it could be. Members will bring to the team a number of highly efficient and effective KSAs, but the team will not be able to harness them toward team efficiency until such time as the remaining requirements of the success octagon are met.

Phase II: Buildup

In this phase, the team rapidly absorbs the vision and mission. It is a time to build those factors that the team requires in order to meet its mission statement. Buildup is typified by the following:

- *Mission implementation.* The team begins to put into operation what was previously only on paper. This is a learning, evaluation, and adjustment period for the team, its members, and the organization. (You must crawl before you can walk.)

- *Organizational cultural change.* Teams have an impact on the way an organization does business. Those outside the team can view this as good or bad, depending on their affinity for change. During this time, leadership and team members must maintain an open and constructive dialog with those outside the team's sphere of influence.
- *Development of a team culture.* Teams begin to develop a culture of their own during buildup. This can be also be called esprit de corps, self-vision, determination, or cockiness. Logos, patches, mottos, and other team identifiers are hallmarks of this phase of growth.
- *Intense training.* Documented, verifiable, and ongoing training is indispensable during the buildup phase. While factors of the octagon are being developed and received, the team is training and learning in preparation for its mission. Training regiments are identified, along with the necessary KSAs; a method is developed to ensure continuous improvement in future years, keeping pace with advances in techniques and technology.
- *Specialty identification (weakness and strengths identified).* Team members are identified in terms of their specialties, which are groomed and supported. Depending on the type of team and its mission, these specialties can range from carpenter, to medic, to computer expert. The team exploits individual strengths while blending its human resources to compensate effectively for weaknesses.
- *Leadership established.* The leadership develops operational KSAs, as well as the ability to communicate effectively and implement the mission of the team. Effective leadership uses the identified strengths and weaknesses to benefit the team, its members, and the organization. The key to success is that the team leadership is developing expert knowledge, as well as the trust and respect of team members.
- *Morale high/performance marginal.* This is a time of almost-beehive-level activity within the team. Constant ongoing training and increasing levels of octagon support keep members busy, prosperous, and motivated. Since it is a time of great learning, team members are sponges: Expectations by the team members with regards to capability may exceed the actual capability of the

team to deliver a service (perceived vs. actual reality); while the team members are capable of providing a given level of service, they are not yet high performance, since they lack some training, experience, and octagon support.

- *Little failures/big wins.* With any new team and because leadership must encourage risk taking, there will be failures from time to time. What's important is to anticipate this and ignore the zero-defect mentality. Little failures create big wins, because each failure becomes a learning session and a point of reference in future situations. It's not a failure until you do the same thing twice; remember, "Insanity is doing the same thing over and over and expecting different results."
- *Deposits in the experience bank.* As the team builds, so does its experience level. This experience level increases in both the operations arena and the administrative/support arena. There is no substitute for experience, and experience can be supplemented with training only when there is little experience. This begins to build the institutional knowledge of the team, enhancing its ability to anticipate and react to a given environment or mission. The experience bank of the team begins to grow. Beware: The inability to pass on experience to new team members during the drawdown (phase IV) is critical!

Phase III: Fine-tuning

Every racecar needs fine-tuning. The pit crew and the driver know what the car can do. Although they have experience, training, and resources that enable them to compete, they are always looking for the edge. Moreover, they want to continue to compete; this is their mission.

During this phase of the life cycle, the team is functioning as intended. An increase in the factors associated with the success octagon will marginally enhance the team, while withdrawals or lack of support can undermine the team. At this point in its history, the team has been established and is functioning. Also, it is displaying some high-performance characteristics, rendering the team effective and efficient and allowing it to make decisions as a single entity. These characteristics include strong team identity, vigilant self-monitoring, and a high conceptual level of understanding. (These characteristics are discussed in greater detail in chap. 15.) This phase is characterized by the following benchmarks:

- *High level of training and experience.* Team members have by this time completed all the KSAs necessary in order to function in the given area of the team's mission. Since the team has been in existence for some time, on average, team members will have three to five years of experience under their belt, as well as mission-specific learning. This is a time of continued growth, advanced training, and the development of new ideas, techniques, and procedures by the team and its members. Adjustments are made in team response, training, and protocols on the basis of the operational environment and changing response needs.

- *Strong team identity.* The team members have learned to view themselves as parts of a team, rather than as individuals doing a job with other people.
- *Strong team conceptual level.* The team considers a wide range and sophistication of ideas and of factors in making decisions.
- *Strong team self-monitoring.* The team analyzes its own thinking, to monitor itself in action, determine where it may be having trouble, and make the necessary adjustments.
- *Planning for team succession.* Intelligent teams—and especially their leadership—initiate a process to identify and indoctrinate the next generation of team members and team leadership. For a variety of reasons, the team will change over the next five to eight years: promotions, transfers, career changes, retirements, injuries, and leadership changes all mean that nothing is constant except change. The impact on task teams is much less than on specialty teams, but is no less important. Because personnel who are drawn to specialty teams (special operations) are highly motivated and educated, a disproportionate number will be promoted or will seek new opportunities within the organization to expand their KSAs and to challenge themselves. Therefore, succession planning is imperative for specialty teams. A process to recruit, train, and put into operation new personnel as vested personnel leave is critical. This includes grooming new leadership at the levels of company officer and battalion officer, to ensure that there is continued leadership and management by personnel with expert skills and KSAs. The fastest method to kill a team is to have a leader who has no KSAs and does not intend to obtain them!
- *High performance/high morale.* This is a time of self-actualization for the team. In most instances, the team is working at peak efficiency with substantial training and an acceptable level of

experience. There are a number of factors that can disrupt this state—most notably, time and the transition to the drawdown phase. Changes in organizational support or any of the success-octagon factors can also affect this benchmark.

Phase IV. Drawdown

As indicated earlier, there can be a variety of reasons that the team life cycle enters the drawdown phase. For example, retirement, career changes, transfers, and promotions all contribute to this phase. For task teams, drawdown has an impact on the ability of the team to deliver service; we have all had senior members transfer, retire, or promote and found ourselves having to train the rookies in the jump seat and reorganizing our team operations. For specialty teams, drawdown has an even greater impact, since the level of training required in order to reach a minimal level of performance is extensive. In both cases, training and experience leave the team; efficiency and effectiveness drop; morale may drop; and the vested personnel who remain become concerned about the continued ability to function at the previously established and accepted level.

The drawdown phase is a naturally occurring stage of the team life cycle. It is unavoidable and should be expected. Its impact can be reduced or minimized by effective succession planning during the fine-tuning phase.

The drawdown phase is defined by the following benchmarks:

- *Personnel/leadership turnover.* Depending on the situation, these can be losses of small or large numbers. What you would like to avoid is large levels of training and experience walking out the door all at once, and this can be done only by proper planning and vision. New faces will always appear on the team, and these new faces require training, equipment, team indoctrination, team building, specialty identification, and mentoring. People are not spark plugs; removing and replacing them all at once does not make the team run better.
- *Marginal morale/moderate performance.* Morale may fall because vested team personnel see the change and may be uncomfortable with it. Losing respected team members whom you have depended on in the past increases a team's level of operational anxiety. Until the older team members have to work in emergency and training conditions with the new members, there will be a certain level of anxiety and of expectation. As knowledge and experience leave the team, the ability of the team to perform effectively decreases.

Also, there may be a slight culture change, as personalities and the Karma of the team change at the company level.
- *Necessary increases in the success-octagon factors.* Because new personnel require training, equipment, and all of the other factors associated with this cycle, actions must be taken that increase the flow of the eight important support factors (the octagon) discussed previously.
- *Learning new leadership.* If the changeover occurs in the leadership arena, team members will need to learn the new leader's style. Also, expectations will need to be established, for both team members and leadership.

Phase V. Buildup

The buildup phase is an exciting time in a team. As new members become indoctrinated into the mission, begin to receive their training, and participate in the mission, their confidence grows, as does the confidence of the vested team members. An entirely new life cycle begins to emerge.

Remember the first time the department sent you out of town for training—how excited you were and how psyched you were when you got back? Look for that same gleam in the eyes of new recruits as they return from their first assignments, and reflect back on what it was like for you 10 years ago!

The team will be reenergized by harnessing the energy and motivation of the new personnel toward the team, its concepts, and its mission. Long-stagnant programs may find new energy once someone new is willing to take them on; new ideas about operational methods and new specialties are added to the team capabilities. This phase has the following benchmarks.

- *New ideas and methods.* As new personnel come on board, so do new methods of accomplishing old tasks. Each team member brings a new set of ideas, as well as the ability to learn the team culture and accepted methods of accomplishing emergency or non-emergency tasks.
- *Increased-tempo training.* Once the new personnel are assigned, the necessity for training them becomes inescapable. Getting new recruits the basic KSAs to do their jobs is second only to emergency response. Because of the general need for training within the department, the tempo or training increases for the vested team members as well. Since vested team members must mentor the new personnel, to pass on

their knowledge and experience, more team drills are held; by teaching, the vested team members are in fact learning a second time.

- *Increased competition.* This phenomenon occurs whenever new personnel enter a team. In My experience, new recruits typically express the following sentiment, especially when assigned to a specialty team: "Okay, old-timer, I want to know everything you know. I am going to be a sponge; I am going to listen, learn, and practice; and then I am going to be better than you ever were!"
- *Moderate-to-high morale/moderate performance.* Morale tends to be high among vested personnel as they participate in training and begin to accept the new team members. New team members' morale is high because they are gaining new and exciting skills and being exposed to new challenges. Still, until the new recruits complete the necessary training and experience and it is incorporated into the team, the team will not operate at peak efficiency.
- *Managing the octagon.* As the team rebuilds, it is imperative to reevaluate the flow and function of the success octagon. It is possible that some areas are lacking and need to be increased, and that others have been missed or taken for granted.

Team life cycles are like the tide, inevitable and predictable. They can be extended or compressed by proper or poor planning, respectively, and are also compressed by failing to educate oneself on team dynamics or failing to take actions at the appropriate time and place. These principles must be understood by leadership and team members for organizations to be successful.

Fire service company officers, team members, and organizational leaders must effectively use teams to accomplish organizational goals, provide emergency services, and get the very best from their organizations. Regardless of your position in the organization, learn to recognize those factors that will create an environment where you, the team, and the organization can be successful.

It is possible for a fire service organization to survive without the effective use of teams. Those organizations that choose to walk that path will be neither innovative nor successful, and will simply muddle through customer service, exhibiting management skills rather than leadership. Officers who fail to effectively manage and lead a team will find themselves with a poor reputation and in an untenable position with their personnel.

Definition, recognition, and application of the fundamentals of team power are crucial to organizations and their members. Furthermore, the effective training and deployment of teams has a direct effect on the level and efficiency of customer service be provided to citizens and visitors.

Start by teaching the fundamentals. A player's got to know the basics of the game and how to play his position. Next, keep him in line. That's discipline. The men have to play as a team, not a bunch of individuals. . . . Then you've got to care for one another. You've got to love each other. . . . Most people call it team spirit.

—Vince Lombardi

Chapter 15
Team Decision Training

> *In any moment of decision the best thing you can do is the right thing, the next best thing is the wrong thing, and the worst thing you can do is nothing.*
>
> —*Theodore Roosevelt*

The Concept of Team Decision Making

I don't believe anyone would argue against the value of teams, teamwork, and team development in the fire service. Although many organizations speak about teams and teamwork, there are still some that undervalue the contribution of small teams that allows the large organization to flourish and function. Teams come in all sizes and shapes—from operational teams, which provide response services, and support services teams, which evaluate and provide equipment specifications or building specifications, to administrative teams, which are responsible for the implementation and evaluation of strategic plans. These are just a few examples; there are numerous other services that teams provide to organizations.

In the previous chapter, we looked at the team life cycle, which applies not only to the micro (small team) but also to the macro (entire organization) level when evaluating and predicting where the organization and its members are at a given point in time. It was also demonstrated that teams and organizations go though a cycle that is predictable, that can be manipulated to some extent, and that can be evaluated on the basis of certain characteristics at a given point in the team or organizational history.

What is missing from almost all organizations is the understanding of what is really a team. Consequently, the training to teach everyone—from instructors and company officers to chief executives—how to evaluate their teams is lacking. As Douglas McGregor says, "Most teams aren't teams at all but merely collections of individual relationships with the boss, each individual vying with the others for power, prestige and position."

If you were to ask members at all levels of the fire service, "Can you evaluate a team?" the overwhelming answer would be *yes*. But, if you were to ask for three characteristics of a team indicating that it is moving from entry level to high-performance level, you would get a blank stare. Very few fire service members of any rank have learned how to observe and classify team behavior patterns and actions to discern where they are in their development as a team, not as individuals.

In April 1993, Gary A. Klein, Caroline E. Zsambok, and Marvin L. Thordsen published an article on team decision training concepts in the *Military Review*.[1] Their research was specifically geared toward developing a model that could be used to teach teams how to make decisions and, even more important, to teach those responsible for the team how to evaluate the progress and development of the team by identifying specific traits, characteristics, behaviors, and attributes linked to progression or regression as a team. The implications of these links are significant and applicable to fire service teams.

True teams—focused on accomplishing what is good for the team and the collective organization, rather than what is good for the individuals on the team—are rare indeed. Each of you can probably attest to the frustration of sitting in a meeting or planning session and watching members of a so-called team jockey for position, ownership of ideas, and recognition. Many times this occurs because the team does not know what is required of it and does not know how to adapt to unexpected events. Compounding this problem, the bosses often have no idea how to evaluate the team as such, instead evaluating individual actions and positions.

To be effective, a team must be able to make decisions as a team. The power that this creates is a resource for the entire organization, because the many are always smarter than the few. Collective wisdom can shape direction, business, and other aspects of life as well. This is demonstrated in the following example, paraphrased from James Surowiecki's excellent book, *The Wisdom of Crowds*.[2]

Francis Galton's Curiosity

In 1906, at the age of eighty-five, British scientist Francis Galton headed from his home, in Plymouth, to a country fair. His destination was the West of England Fat Stock and Poultry Exhibition. As a scientist and statistician, he was driven by the measurement of mental and physical qualities and breeding.

While at the fair, he came across a weight-judging competition. A large fat ox had been chosen and put on display, and the crowd was gathering to place wagers on the weight of the ox. Eight hundred people took their chances and placed bets, including butchers and farmers, who were presumably experts. Along with these "experts" were ordinary people.

As a scientist, Francis Galton wanted to prove that the "average voter" was capable of very little. Galton felt that most average people

> were incapable of rendering a very good guess in this or anything else in which they were not experts or had not been educated.
>
> When the contest was over and the prizes had been awarded, Galton borrowed the tickets from the organizer and ran a series of statistical tests on them. He took 787 of the 800 (the others were not legible) and ranked them from highest to lowest. He then added all of the crowd's estimates together and calculated the mean, or the collective guess based on the wisdom of the crowd. If the crowd were a single person, it would have been the weight that was guessed.
>
> The crowd collectively guessed that after the ox was slaughtered and dressed, it would weigh 1,197 lbs. In fact, after being slaughtered and dressed, the ox weighed 1,198 lbs! In other words the crowd's judgment was essentially perfect!

Galton discovered that under the right circumstances, groups are remarkably intelligent—often smarter than the smartest people among them.[3] The concept that collective decisions are usually better than singular decisions has since been proven experimentally by research.

With the myriad issues we face in the fire service today—reduced budgets, staffing cuts, increases in service calls, and the demand for new and nontraditional services—there is very little room for inefficiency. Teams that can be evaluated and that can make decisions on very complex issues get very good results for the organization they serve.

Where Teams Reside

Teams reside at different locations on a scale measuring functionality and expertise. This location depends on a variety of factors, including membership, length of time as a team, internal support, and task assignments. Some teams will never be successful, while others may be highly successful. The time frame required to move from one point to another varies by team, but the process to get there does not.

From a linear perspective, there are two extremes. At one end is the highly functioning team that is functionally greater than the sum of its parts; this team accomplishes things that could never be done by any one individual. At the other end is the dysfunctional team that is essentially wasting the individual member's time and effort; this team accomplishes less than even the most unprepared individual might accomplish by working alone.

The purpose of team decision training is threefold: first, to create a method that allows teams to progress from the dysfunctional or developmental side to the highly functional side of the model; second, to teach team members how to evaluate their decision making as a team; third, to teach supervisors how to seek and recognize behaviors and characteristics of the team that indicates they are moving. Supervisory recognition is critical: if the organization is going to effectively harness the power of the team, it must provide a method to make it highly functional; furthermore, bosses have to determine whether the team is moving at an appropriate pace. As a side benefit, this will provide the training needed for individuals who may be assigned to new teams at a later date.

What Works and What Doesn't

Let's first examine this concept from the instructional side of the house. In other words, what does it take to provide decision training to teams, to instructors who are training teams, and to the supervisors of the teams? Also, what do organizations need to know about teams and how to effectively evaluate and apply them?

Exercises alone do not provide the necessary team decision training, for reasons that should become apparent as we discuss them. Exercises are usually designed to place people or teams in a given situation and evaluate ability to reach a preestablished outcome. In theory, exercises provide practice for the participants: everyone gets to participate in one form or the other. However, because they provide practice without instant feedback, exercises alone are a very poor method to develop a team that can think and make decisions.

We'll work back to the importance of instant feedback by way of an example. Take a typical exercise in which participants are told to run from point X to point Y. In the end, we get all the participants together and say "you did this right" or "you did that wrong." Stop for a moment, though, and think about all the decisions and tasks that took place during the exercise that pushed the team into either a right or wrong solution. In some instances, we have trained the wrong behaviors—for example, when the team did something wrong (for lack of a better word). When people practice in a certain way, they will repeat that the next time—and with enough repetition, practice makes permanent. When we are training the right behaviors, that's good, but when we train the wrong behaviors or allow the wrong behaviors to be practiced, that's not so good.

Anything we undertake in practice has a set of processes that have to be put together in order for the event to take place. Consider how many steps it takes to put up an extension ladder, or to stretch a hose line up two or three flights of steps, or to search a room. These simple, basic tasks involve many steps that all must be taken in order to get it right. Similarly, there are many steps involved in specifying a new piece of apparatus or developing the blueprints for a new fire station.

Along the way, there are some critical processes that require instant and continual feedback. If we do not identify and evaluate these critical processes quickly, how will the team begin to understand and train itself on that particular process? If we let the process run and at the end say "this is right" and "that is wrong," then we have a system that looks strictly at performance but does not consider or teach how the decisions were made. When we allow this to happen, we teach teams what to think, rather than how to think. This leads in turn to another pitfall, setting them up to develop poor habits that, while suitable to the artificial world of exercise, may be highly dysfunctional in a real-world operational setting.

Continual feedback by trained observers is critical for the following reasons. First, you want the team to own and improve any process that they use to complete their task. Once a team understands the process, it can ask "What can I do to improve the process?" The only way to accomplish that is by understanding what was done correctly and by determining what needs improvement immediately. We do not want teams to learn bad habits and bad processes. Next, you want each member of the team to understand the habits and behaviors that are necessary for a team to own a process. Learning how teams make decisions, rather than how individuals make decisions, is important; furthermore, team members should understand how individual decisions can be incorporated and benefit the team.

Exercises require instructors, and herein lies the problem: Instructors—and most bosses—do not understand the requirements of team training and often cannot provide feedback on the processes that a team undertakes to get from point X to point Y. They may claim to know about teams, but when asked to evaluate and train teams, they run exercises and evaluate only at the end.

People responsible for teams need to prepare in advance, defining the key processes that a team needs to apply and establishing objectives that can be monitored and evaluated. Otherwise, they will be unable to provide the necessary feedback. Most bosses and instructors are not even aware that these criteria are unfulfilled by their programs.

In general, all training has a very similar goal, regardless what you set out to accomplish. First, the requirements of the training, what we call the KSAs, are identified. Second, the medium for the experience (lecture, Powerpoint, exercise, tabletop etc.) is chosen. Third—and here is where most programs fall flat—the target behaviors of the team must be *observed*. Finally, based on that observation, the adequacy of the performance is judged and feedback is provided. We are usually very good at training individuals on specific skills, but we have do more to mold and develop teams.

When training is ongoing, instructors and bosses must establish, monitor, and observe team decision-making skills as well. We are essentially developing another set of KSAs, dealing entirely with team decisions, rather than individual decisions. Once we establish what these are, we can monitor for them and ultimately provide feedback to the team.

You may say, "We provide training for teams, we train them in the process of developing specifications, and we use the completed staff work model to ensure that the work is accomplished correctly." Sure, but those drive individual performance and individual skills; in no way do they teach your team to make decisions or provide measurable objectives for use in evaluating the team's performance. The reality is that you have to be trained and educated on what that means.

Team decision training is not new. There is a general consensus among training professionals about the theory behind the team decision training process. Furthermore, the behaviors that they have identified are substantiated by other research projects.

Fiscal expenditure in training is always an issue. The framework to provide instructors, bosses, and ultimately teams with decision training is already in place and is very inexpensive to add on to your current training structure. For example, if you run programs that include exercises that employ trained observers, you are three-quarters of the way there. If you provide training to your support services or administrative teams on how the decision-making flowchart works for projects and have committee oversight, you have all the pieces in place to bring team decision training on board. Exercises or training programs already in place that train teams to accomplish things together can be extended to train them to adopt better processes to achieve those same ends.

The one big mistake you can make is to streamline the process of team decision training and the process of training your observers (instructors, bosses, and team members). Understanding the processes, observing them, and providing ongoing feedback that starts as quickly as possible prepares the ground for training. Whenever we conduct an exercise or training program, the feedback needs to begin quickly and continue throughout the process. Instructors must be trained and allowed to observe teams either in workshop format or in actual exercises.

Developing the Model

A bevy of information exists on suggested behaviors and how to model them. Because of its good fit with the fire service, I have adopted the research model of Klein, Zsambok, and Thordsen.[4]

It would be nearly impossible for even a trained observer to track all of the different behaviors exhibited along the proposed linear model of performance. It is generally accepted that decisions made by teams are based on cooperation, leadership, coordination, shared mental models, technical understanding, organizational process, and other such concepts. These are broad, overreaching behaviors and sufficiently vague that observers would have a difficult time determining whether a specific observable behavior—for example, one member failing to inform another team member of an important event—is a case of inadequate communication, coordination, information management, or failure of a shared mental model, instead of anticipation or any other aspect of team decision making. To be successful in teaching both observers and teams, we must work from an unambiguous framework.

A theme throughout this book is that teams are living and breathing entities. They are intelligent; they will try to understand events that occur around them and to them; they will use experience and training to draw inferences and to solve problems and make decisions about a wide range of issues.

The model we have chosen has three primary components: team identity, team conceptual level, and team self-monitoring.[5] Each of these characteristics has a corresponding subset of observable behaviors. Identification of these characteristics is crucial to team decision training, as the framework in which to interpret subsequent observations.

Team identity refers to the way that individuals in effective teams view themselves as part of a team. This is important since some members may view themselves as simply individuals doing a job that happens to involve other people. It has been my experience that members of your team with good skills who can identify with the team and understand how to function as part of the team are more valuable than star players who see themselves as individuals working with other people. If the collective wisdom and decision making of the team is more powerful, than that of the smartest individual, as research has indeed shown, then learning to see oneself as belonging to a team is much more important to getting excellent decisions.

Team conceptual level refers to the way the team thinks through and about the decisions it makes. This involves the level and range of ideas that the team brings to the table and the number of factors that it applies in making a decision. The more sophisticated this process becomes, the more functional the team becomes.

Team self-monitoring refers to the team's ability to analyze its own thinking, to check its activity and decisions, and to make decisions regarding adjustments. If the team is having trouble, does it recognize this and make the necessary changes or adjustments to the process?

I began reading up on team decision training when I was the battalion chief of special operations. My observation at the time was that while our teams were composed of very talented and dedicated people, we lacked the true sense of team that was necessary to perform at the high end of the scale. I desperately wanted to do something that would drive home the point that teams must exhibit the characteristics and behaviors we are discussing here. Each quarter all of the teams held a drill, and wanting to incorporate this into team decision training, I contacted the Tidewater Psychiatric Institute, which had a very good initiative course for troubled youths and corporate team building. This consists of a series of events that involve evaluating and completing specific tasks; it is impossible to get through these tasks successfully in the time allotted without the entire team participating, thinking, reacting, adjusting, making wise choices, and evaluating those choices.

Bringing together the technical rescue team and the Hazmat team for the exercise, I began to detect bad vibes. The consensus was that although I was wasting their time because they were already a team, they would humor the chief and participate in these games. This exercise served two purposes: first, it gave me an opportunity to use my newfound team decision behaviors and observe them in action; second, it gave the teams the opportunity to practice applying these characteristics even though they did not realize what they were doing. In the end, I received such overwhelmingly positive feedback from the team members about what a great team-building exercise it had been and how much they had learned about the requirements of team decision making that we used the facility multiple times after that, until they closed the place down.

How do these general characteristics exhibit themselves in entry-level teams? In teams at the low end of the linear scale—in particular, those just getting formed or just getting trained—most members are out for themselves. This sets up many aspects of interaction, including competition for ideas, competition for recognition, and competition for positions in decision making. If we could channel those options into a group that was more concerned about the thoughtfulness of the team's approach to a given task, we would eliminate the competition. Whenever we have competition among team members, they are unable to strengthen the identity of the team and the overall success of the team, because they are too busy working as individuals.

Some teams will never form an identity becuase of the individuals on the team, a lack of leadership, or other factors. Hearing expressions like "Just tell me what you want me to do" from team members clearly indicates that the team is not making decisions and that they still do not identify with the team, but rather a task.

Other teams will never form the necessary conceptual level to develop the proper range of ideas and thoughts for problem solving. In some cases, they will oversimplify the issue, and in other cases, they will make it ridiculously complex. Part of this issue is not knowing the respective personality profiles among the people who compose your teams. If you have all D- or C-players, it's unlikely for there to be enough diversity of thought—it's just not possible, since thinking alike is their style.

Still other teams will be unable to monitor their work and develop time lines. Either they are not working together, or they are unable to judge whether the strategy they have chosen is even working.

An effort, thus, must be expended to develop a model for all observers—especially team members—to evaluate their actions as a team. To be effective, teams must apply the characteristics described in this section to all of their endeavors. Rather than thinking about working as a team, they should be thinking about who they are as a team. This is critical so that the team can judge its progress and evaluate the efficacy of each strategy it undertakes.

Benchmarking Behaviors

The goals of benchmarking are to make teams effective as quickly as possible and to allow bosses and instructors to evaluate progress. Some teams are tight only during the very short time provided to complete a given task, so getting them to remain a team is a pivotal component of their success.

At least two issues must be understood at this point, especially for teams and team members who will remain together only for a set length of time: first, the team members must understand team decision making before they join the team; second, they must have received that understanding through a training program and experience earlier in their career. You must have a frontline officer development program that captures your officers early for education and an education program that continues this throughout all ranks. Furthermore—and no less important—people must get the varied experience necessary to assist them in learning how to make team decisions. Finally, you must use every exercise as an opportunity to teach team decision-making skills.

It is also critical that observers (team members, officers, instructors, and bosses) be able to identify specific behaviors. Observers must also provide feedback on effective and ineffective team behaviors and allow new strategies to be applied. Klein, Zsambok, and Thordsen have stressed that the behaviors we seek cannot be too difficult to observe or else they will be missed; they cannot be too general, or they will not be identified; and they cannot be too specific or they will become trivial.[6]

To process this information, we are going to break the observational model into characteristics and behaviors and also provide a visual model to tie it together (fig. 15–1).

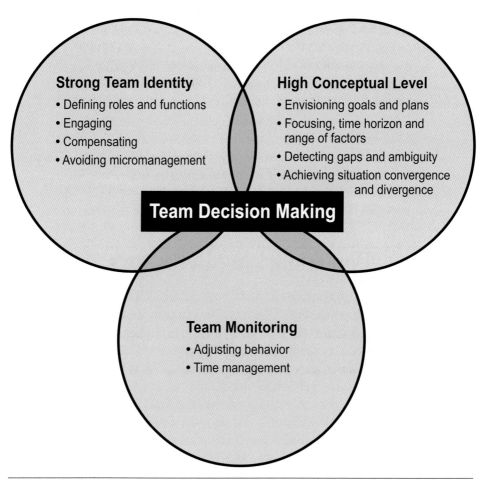

Fig. 15–1 Team decision training. *After* Klein, Zsambok, and Thordsen[7]

Team identity
The characteristic of team identity has four behaviors that are observable:[8]
- *Defining roles and functions.* Does the team ensure that everyone on the team knows is or her job and functions? Does the team take the time to cross-train or educate other members to provide supporting roles for a different job when a void is created? You have likely seen teams whose members, in the attempted completion of a given task, became confused as to exactly what they are supposed to be doing. Likewise, you have probably seen teams that, when someone does not show up or fails to accomplish a given task, have no one else who understands how to pick up the task and do it.
- *Engaging.* This behavior involves a focused approach and the attention of team members. Are team members paying attention to the task they are assigned and its associated functions? I have seen team members who, during a trench-rescue training evolution, have begun to focus on someone else's task, on jets flying overhead, and other stuff, rather than what they have been assigned. In support or administrative teams, members may simply tune out during briefings or discussions.
- *Compensating.* This aspect of behavior is very obvious, and all of us have seen it. It poses the question of whether anyone steps in to help if a team member is overwhelmed or is having difficulty completing a task. Have you ever seen a firefighter on a fire ground moving a heavy piece of equipment by him- or herself and no one seems to notice and assist? Have you seen support or administrative teams with members who are not completing reports or research on time and no one approaches them and asks, "What can I do to help?"
- *Avoiding micromanagement.* Does the boss or team leader stay at his or her job during a crisis? You do not need the boss to take over for subordinates and do their jobs; you need the boss to do what the boss is supposed to do, and that is to direct and get the task or operation back on track. A boss leaving his job during a crisis breaks down the team's effectiveness, since it creates a void in leadership when it is most needed.

Team conceptual level
There are four behaviors for team conceptual level:[9]
- *Envisioning goals and plans.* Does the team assist its members in understanding what the team in trying to accomplish? In

the context of tactical operations, this is referred to as the commander's intent. Educating your team or department on the strategic plan entails training and explanation. In developing its strategic plan, VBF&R did a terrible job of educating the members. Our approach was to mail out the plan and say "Read this," which did not work. Teams have to understand the goals they are striving toward and the plan to get there. Goals and plans include a broad range of actions, from simple tactical solutions on the fire ground to the development of policy for the organization.

- *Focusing.* This aspect of behavior deals with the ability of the team to perceive the appropriate features of the task and the necessary time frame in which it needs to be accomplished. One aspect of this is time horizon focusing, which examines the balance between a team's ability to look at short-range or immediate events and its understanding of the need to look at long-range consequences and implications; when teams focus only on one or another, they are not conceptualizing the team requirements. Focusing also deals with behaviors that indicate a team is examining a multitude of options and perspectives instead of just one. On the training ground, this might mean that the team has a plan A, a plan B, and and a plan C, in case one or more do not work. Are they listening to different ideas from different team members to gain different perspectives?
- *Detecting gaps and ambiguity.* This amounts to the team's recognizing whether it has the complete picture, all of the information to make a decision. Is information missing, and if so, do they notice? Are the information, facts, and data incongruent? If the team recognizes that it does not have all the information or that not everything adds up, they should do something to correct the situation, or at the very least, they should acknowledge that they are missing something.
- *Seeking divergence and convergence.* This entails watching the team members and observing their ability to seek many different opinions, taking them and converging on a common assessment of the issue at hand. Seeking different opinions also drives the team to coordinate all team members' understanding of the situation or assessment. This may involve the use of visuals, such as maps, tactical drawings, and flowcharts, to get everyone on the same page.

Team self-monitoring

There are two behaviors related to team self-monitoring:[10]

- *Adjusting.* This behavior spans everything the team does. Can a team make the necessary changes to its procedures, time frames, processes, members, or other specific aspects that affect the outcome? Can the team adjust to situations, internal and external influences, changing events, or other influences that may impact, impede, slow, or affect the outcome of their task?
- *Managing time.* Is the team able to track its progress? This is a very easy behavior to observe. Does the team set a timetable or schedule? Does the team follow the schedule? Also, does the team recognize when they are not going to be able to adhere to the schedule originally set and adjust accordingly? Do they establish benchmarks that provide references for progress and time frames? When teams do not do this, you will see the infamous college "final flurry," where they try to do everything at the last minute to get it finished; we all know that quality is compromised by this approach. In both administrative and operational settings, does the team recognize when it is time to cut off discussions and get on with business? Does the team set up parallel efforts to conserve time and get the task done?

Putting It All Together

The application of teams and the power they provide to the fire service is undeniable. While teams and teamwork have gotten lots of play over the years, we have previously failed to explain to our instructors, bosses, and team members that decision making by teams is a collective endeavor and almost always results in better decisions that those made by a single individual.

The dynamics of the team are expressed in the chief characteristics and are then identified in the observable behaviors we have outlined in this chapter. It would be a shame to waste time, money, and effort on a team that did not accomplish what was expected of it. That can be avoided, for the most part, by developing, teaching, and incorporating team decision training into anything that involves teams.

Team decision training is exactly what it sounds like: teaching people how teams make decisions and how they should be thinking, not what they should be thinking, as part of a team. Since our officers will be responsible for teams and will be members of teams throughout their careers, team decision training should be provided early in their careers.

Finally, it is just as important that we train our observers, including our instructors and bosses. They must be trained on exactly what to observe when evaluating a team, as well as on how to provide feedback on those aspects of behavior and thus improve the team's effectiveness and readiness.

Notes

1. Klein, Gary A., Caroline E. Zsambok, and Marvin L. Thordsen. "Team Decision Training: Five Myths and a Model." *Military Review*, April 1993.
2. Surowiecki, James. *The Wisdom of Crowds*. New York: Doubleday, 2004
3. Ibid.
4. Klein, Gary A., Caroline E. Zsambok, and Marvin L. Thordsen. "Team Decision Training: Five Myths and a Model." *Military Review*, April 1993.
5. Ibid.
6. Ibid.
7. Ibid.
8. Ibid.
9. Ibid.
10. Ibid.

Chapter 16
Planning and Implementation

> *When planning for a year, plant corn. When planning for a decade, plant trees. When planning for life, train and educate people.*
>
> —*Chinese proverb*

Planning or Decision Making—the Consequences of Doing Neither

On a piece of brown paper, more than likely a lunch bag, were written these directions:

> *Y'all go west young man on 18. Turn right on Nettle. Go Yankee (north) on Nettleton to the stop sign (you must stop). Turn left onto Granger road to the traffic light. If light is red, you must stop (you stupid hillbilly). If it is green, turn left onto Weymouth Road. Take Weymouth Road to Union Street. Turn right onto Union, and go to the second driveway (large parking lot).*

This was provided by a chief from the Medina, Ohio, fire department when I failed to plan my proper route to the Buddy to Boss class I was scheduled to teach. Because these words provided me with the necessary information I needed to get where I had to be, they enabled me to accomplish the main thing.

Planning takes a variety of shapes in your organization. Most often, we hear about strategic planning, which we try to relate to what we do at the company or battalion level. Organizations ordinarily fail to teach their officers how to plan effectively at a given level of the organization; planning at the company level is an entirely different animal than planning for the entire organization, but it is no less important in getting the business of the business accomplished.

This chapter is short and sweet. What I want you to take away is the importance of planning, as well as an awareness of the pitfalls that people at all levels of the organization walk into when planning and implementing a plan. In my experience, most people make planning much more complicated and drawn out than it has to be. For that reason, I have been slightly turned off by the process; moreover, the human behavior of paralysis by analysis, which is discussed later, is a cause of considerable frustration. My aim is for you to be able to avoid these issues and behaviors when you find yourself in a position to influence the creation or the outcome of a plan.

We need to examine planning first from the strategic perspective, which involves short-, mid-, and long-range planning and capital improvement. Most fire departments are calendar based, accomplishing strategic planning on an annual basis: the initiative for strategic planning takes place during budget negotiations for the next year, and there is an annual review of the previous year's plan by the business-unit heads. While there are indeed fiscal constraints on strategic goals, but more and more studies are showing that anchoring your strategic planning to an annual, calendar-driven event and one that is focused on specific business units is not always the most effective method. Calendar-based strategic planning limits decision making, does not address issues that occur in a timely manner (as they occur during the calendar), and often results in no decision making at all because some of the stuff is just too hard. Entire texts have been written on strategic planning, so I will cover the highlights here.

Strategic planning will affect you, as an officer, and the team you supervise. At the company or battalion level, you must have some link to the strategic plan, so that you know what to do. The question this poses, as inane as it may sound, is, "How do you know what you are supposed to do when you come to work?" If we are ever going to reduce the prevalence of the separate-shift mentality in the fire department, we must find a common link that provides oversight for what we should all be doing, what is critically important to accomplish, and what the main thing(s) are.

You and your team are the tool by which much of the organization's works gets translated into something tangible. Organizational charts and written words are useless without educated and competent people in places of leadership who can carry them out. This was made clear during the Hurricane Katrina deployment, and there are examples enough within fire department and other government agencies that you could write a manual of case studies.

The departmental plan affects your planning cycle, since you have priorities that you have to incorporate into your planning. Furthermore, even though by their very nature, certain things would get done regardless of whether someone put them in the big plan, you are responsible for their implementation. These affect your timely execution of the plan, because you have a set amount of hours, weeks, and months in a year to accomplish everything that needs to be done.

The difference between having an impact on and influencing the strategic plan has a lot to do with your geographical location in the organization. For instance, you would spend much more time working on strategic issues in a staff position than you would in the operations section.

Planning to operate at the company level, although not rocket science, can be frustrating and time consuming. External forces drain the resources you control; staffing changes daily with sick time, annual leave, and backfills; last-minute requests and issues surface; that pesky structure fire happens just when you had the plan working perfectly, throwing a kink into the works—there is a litany of events that can foil your good intentions. It is very important that you follow through on your plan (make a decision), and if it's not working, change your plan.

Failure to plan at any level can result in very negative results—in particular, frustration at every level. First, the people who work for you will become frustrated; most workers want to have some structure and want to know what is expected of them on a daily basis, so that they can go do it. Next, the people you work for will be frustrated; they expect, sometimes demand, that things get done on time and professionally. Finally, you will get frustrated, because failure to plan reflects badly on your ability to lead and get the done job.

When your people are upset and your bosses are upset, that creates problems for you. It should just not be that difficult. Thus, it's your duty to plan effectively. By accepting your promotion to officer, you have accepted the responsibility of leading your members toward goals and objectives designed to achieve an outcome. Because there are so many things that you have to do, planning and prioritization are critical for your success, the success of your team, and ultimately, the success of the department.

Planning at the Company Level

The really nice thing about not planning is that failure always comes as a complete surprise and is not preceded by long periods of worrying and frustration.

—Source unknown

A couple of rules apply to planning at the company level. First and foremost, be flexible. Calls, last-minute requests from your boss or other companies, customer requests, and changes in someone else's plans will all affect your plans. You and your crew are a resource that will be applied by the organization, sometimes only at the last minute. If this becomes an ongoing issue, you need to meet with your boss to discuss how poor planning on someone else's part is affecting your ability to plan for your work.

Second, plan to get the main things done first; prioritize what is important. Rest assured, there will always be too much to do in any given day. If, for example, you leave physical training to the end every day and don't allot time for it, the message you are sending to your members is that it is not a priority. (Remember that response, training, and physical training are three things critical to our success.) Furthermore, you need to ensure that you accomplish in a timely manner what the organization expects of you.

Finally, understand that planning means decision-making. Making decisions that lead to implementation of an action is more important than simply having the action on paper.

Before we examine what we can plan, let's talk about the morning planning meeting. Although it is called a planning meeting, it has other purposes as well.

The first purpose of the planning meeting is to evaluate your team and see if they are all ready to play the game today. Every morning, you should get the shift together around the kitchen table and lay out the plan of the day. This is such a valuable tool that we should be teaching it to company officers from day one. This allows you to see each of your members, face to face, first thing in the morning. (Remember keeping in touch?) You need to check that each of them is healthy enough to be at work. It may sound stupid, but we routinely hire people who hate to take sick leave and come to work operating on three of their six cylinders. Also, some younger members engage liberally in social activities and don't return home until two hours before work. On any given day, it's unlikely that you will have your regular crew together; relief, sick leave, vacation, trades, and so forth will put another player or two on the field on almost every shift.

This meeting provides an opportunity to discuss the plans for the day and also to establish some future plans. In the short range, you can outline that day's plan of action—what you want to get done and when. You can make sure the proper uniforms are being worn; you can assign drills and training; you can make sure everybody knows what the main effort for the day will be; and you can ensure that the tools and supplies are available to accomplish what is outlined in the plan.

Planning and Implementation

Keeping lines of communication open is imperative. Review memos, e-mail, orders, directives, and general information that have come down since last shift. We don't want anyone claiming "I didn't know that" several weeks later. Use communications to field questions and get answers for your members, to control rumors, and to address look-over-the-fence issues. Attention to communications also enhances the abilities to make assignments for upcoming shifts and to plan for future events.

Incorporate a morning meeting into your routine. Keep it brief, to 15 or 20 minutes, and follow an agenda, even if it's in your head, to make sure you cover everything. Eventually, this will become second nature.

I suppose this list of things could and does apply to just about any work group, small business, of even branches in large corporations. You have only the change the title of the planning sections we are outlining here and apply them to your work group. The following list of things that we can plan for at the company or battalion level can be adapted to any work group, small business, or even branches of large corporations:

- *Response.* How in the world do you plan for response? You don't know the time or the place, but you do know the frequency. You know the average daily level of call volume on your shift in your station. Based on that information, you can predict what can reasonably be accomplished on a given workday. If you get more done, that's great; if you get less done, then you have to make up for it later. Be realistic about what you can and cannot get done.
- *Training.* Training takes many forms at the company level. It can be a simple in-house drill assigned to one of your members, but if you expect quality, you had better give a heads-up several shifts in advance (plan), so that your members can put together the drill and deliver quality. Other forms that training can take are individual training as requested (as discussed previously, in chap. 6), in-service or company training, a two-company drill, and battalion proficiency testing. Training is so critical to what we do and manifests itself in so many different ways that understanding and fulfilling the requirements is invaluable.
- *Physical training.* You have to keep the working end of the fire department finely tuned and well oiled. If you have a motivated shift, you may be able to do physical training on an individual basis; otherwise, you may want to do it as a group. Regardless, don't put it off until the end of the day, when everyone is beat.

- *Maintenance.* You have a wide range of important housekeeping and apparatus chores—daily, weekly, and monthly apparatus issues and station maintenance—to get done. Furthermore, moving the apparatus to a remote location for pump testing or major maintenance takes time and energy, as does getting the reserve apparatus in station and in service.
- *Administrative time.* This includes paperwork, logs, reporting mechanisms, and any duty that takes time and scheduling to accomplish.
- *Human resources work.* This is everything where you must apply to the human resources under your supervision, including evaluations, counseling, transfers, relief, and physicals.
- *Traditional programs.* The department has outlined many traditional programs in its goals and objectives, including hydrant maintenance, inspections, and pre-incident planning.
- *Meetings.* This includes meetings with your boss, shift meetings to discuss issues that have cropped up over time, and civic meetings that your company needs to attend, as well as preparing the station for a meeting that someone is setting up.
- *Fill-ins or move-ups.* This involves the movement of your unit to cover a previously planned event for another company. Doing so affects your ability to attend to the needs around the station, so planning something that you can accomplish while you are away is a good idea.
- *Downtime.* This might be meals or study time for you or someone from your crew who is going to school or competing in a promotional process. Remember that your members need personal and family time throughout the shift.
- *Planning time.* As an officer and leader, you must take the time to put your feet up and think. Evaluate where you have been, where you want to go, and how to get there; also evaluate whether your current plan is working. Don't put the plan on an annual or calendar basis; instead, make it a continual process to review, update, and change when needed. Identify threats to your success early and be flexible.

There are many more issues that require planning than can be listed here. Planning should be focused, not haphazard; otherwise, the result will be a final flurry of trying to get things done that could have been planned for all along. Don't put off things that you know you can get done just because they are not fun, are mundane, or don't excite you or the shift; unfortunately, such dues-paying work in the fire department accounts to about 75% of what we do.

Learn to plan early in your career and be flexible. Even though you can expect some setbacks, keep planning. Learn how to prioritize, which is critical to getting things done, and improve your team, letting them know you are out in front of issues. These measures keep the kittens in the box.

Planning to Pay Your Dues

The fire and rescue services offer many opportunities besides operations. A variety of support services, logistics, safety, and staff jobs are necessary to keep the business running. A smart officer will ask team members what they want to do in the department now and in the future and help them find a path to get there. Some people just want to be firefighters, and that's great—we need excellent firefighters! Other people want to bounce around and get a taste of everything the job offers, and still others want to be officers and move up the ranks to become fire chiefs. These are admirable endeavors, but they all take planning to realize.

As an officer, there are key issues that you need to start planning for as a rookie firefighter. First, as ridiculous as it may sound, start planning for your retirement. Even when it hurts, put something away in that deferred compensation program (401(k) or something similar), because 25 years goes real fast. Start early and the fund will build quickly in the later years.

Next, plan for your education, both formal and résumé structuring. Today's fire service requires, at minimum, a bachelor's degree in order to advance; since most organizations have a system that pays for at least part of your education, get that formal education out of the way early. Résumé builders make you marketable after your career is over—or even in the middle of your career, if you want to move on to something bigger and better. Seek out every class, every specialty, and every new issue you can find, and get competent in it. Unless you plan to spend 30 years in the fire service then move to a desert island, this is of utmost importance; I have seen plenty of folks who retired and did nothing and were dead within two years.

I am amazed at how many people think they have a vested right, by royal decree, that guarantees them shift work for their entire career. In reality, the organization will look to the ranks to move people into specific positions that need to be filled. New members need to be told, as rookies, that while the organization will do its very best to meet their needs, they must also meet the needs of the organization, and that often means different work locations. People should not feel slighted or be shocked when you tell them you need them at a day job.

As an officer, you should expect, plan for, and embrace spending some time in a staff (day job), rather than spending your entire career in the field. It will provide you insight into how departments work, open additional doors for you, increase your influence, and fundamentally make you a better officer.

Paralysis by Analysis

The biggest fault in strategic planning at the global end of the scale is that there is too much in the plan that people do not want to accomplish or will not accomplish. This may be because of personal reasons (they never wanted to achieve the goal in the first place) or because of changing fiscal, political or environmental issues. Some people are uncomfortable with change or are so driven to evaluate data that they cannot make a decision based on reasonable information within a reasonable time frame. A great leader removes obstacles to change and drives the plan to implementation by allowing members to engage, implement, evaluate, and change. Individuals or processes that stand in the way of implementation need to be removed, bypassed, or educated. When you created the plan, you knew who the people were, what the processes were, and that some form of education would be needed—so don't allow these to be used as excuses.

If we put something in the plan, we had better be willing to implement it in a timely manner, within the time frame subscribed to by the plan. If we are unwilling, we had better acknowledge we cannot do it and eliminate it from the plan. If that happens, put it in the category of "if things change, let's do this," and have someone trusted, who does not have an agenda, keep an eye out for changes in the fiscal, political, and environmental arenas that will allow you to jump on these items quickly.

What we want to avoid at all costs is the paralysis by analysis that creeps into planning and implementation. This permeates certain planning models and organizations as a result of someone's claiming to need more information, more completed staff work, more data, and so forth, before we can take a risk and implement the plan. There is truth to the saying that "a good plan implemented quickly is better than an excellent plan never implemented." All of the information will never be in, so we need to evaluate what we have and make a move—in other words, either implement it at that point or kill it.

Functional Planning Tools

Successful generals make plans to fit circumstances, but do not try to create circumstances to fit plans.

—George S. Patton

There are a variety of tools to help you plan. This may seem basic—and it probably is—but it bears repeating. Some tools are based on personal preference, and others are organizational tools.

Most organizations should have a planning and training calendar. This calendar should outline—at the least frequent, on a quarterly basis—what the organization is attempting to accomplish and what it has scheduled. You need access to this calendar, because any conflicts that arise will occur at this level. If your organization does not have such a calendar, advocate for one; it's difficult for you to plan when you don't know what you have already been committed to.

Organizational strategic plans provide an overview of expectations, goals, and objectives, as a foundation from which you can create direction. This plan should give you and the other officers some idea of what is expected organizationally, so that you can incorporate this into your training and operational plan to ensure that you get done what the organization needs to accomplish.

Personal planning calendars can take many forms. These range from daily simple planners and pocket calendars to Franklin Covey calendars and electronic calendar/planners on pocket computers and e-mail–based systems. Electronic planning calendars provided by the organization can assist, but if no one updates them, they become more of a burden and a source of misinformation than a help. Moreover, you would need some form of backup, since you never know when the gremlins or the electronic world are going to crawl into the system and cause crashes, eat data, or otherwise wreak havoc on our lives. Station wall calendars serve as an excellent quick reference for planning daily, weekly, and monthly events. Leave, appointments, and anything else that affects the company or division can be put on these. An erasable monthly or yearly calendar is another option, but again, it would have to be kept up to date.

Meetings with your boss, your shift, and other companies are also a great planning tool. Depending on your needs, these should be held once a month or once a quarter. Through these, you can coordinate multiple-company drills, for example, and incorporate your boss's needs into your planning process.

Finally, as strange as it may sound, professional journals can be used as a tool to help plan future events. Changes in tactics, strategy, and consensus standards (and everything else that affects your job) usually pop up in a professional journal before hitting the mainstream. Keeping up on the industry by reading journals allows you to anticipate what is coming down the pipe; moreover, you could undertake some research and get out in front of the training or information curve.

These tools exist to help. Find out what works for you, and apply them as you see fit.

Chapter 17
Managing Change

> *Whosoever desires constant success must change his conduct with the times.*
> —*Niccolo Machiavelli*

Change—the Perilous Journey

Recognizing, evaluating, enacting, and sustaining change has a lot to do with decision making. Decision making and change are so closely tied together that you cannot be successful at one if you are not successful at the other.

Change requires a wide range of decision making, using novel and foreign tools and techniques. It may require gut decision making and emotional decision making. Change will certainly require us to make the correct decision about how we go about evaluating, implementing, and communicating change to our politicians, organization, customers, and members. If you make bad decisions about change management, you will fail to sustain the change.

This chapter predates this book and was written in the wake of frustration with the process of change in my own organization. When I brought it back out, I was surprised at how closely some of the problems I had written about were still issues in 2006.

Later in this chapter, we talk about change in terms of what we need to be prepared to do. To lay the ground for this discussion, however, we need to recognize the potential pitfalls when the change process is not done properly, timely, or honestly.

Change is a scary thing for most of us, and carried out incorrectly, it can fatigue both an organization and the individuals within the organization. Living with and managing change, however, is a reality of organizational life. People, processes, tools, tactics, and strategy will change if you stay in the job

---

> *Change is like a marathon, the front of the pack is off and running and the back hasn't even moved.*
> —*David Gravin*

long enough. The challenge for all officers—especially company officers—is to understand that all organizations will need to change and, therefore, that we must embrace new ideas, new tools, and new methods.

Officers will likely be confronted with organizational change many times during their tenure. Such change may be brought about by the appointment of a new fire chief, political change, fiscal changes, or union activity, among other causes. When this change occurs, chief officers are often tasked with being the messengers of change.

If change is driven for the wrong reason(s)—is not sincere, is a knee-jerk reaction, has not been properly evaluated and options weighted, or has an advocate or administration that will not complete the necessary follow-up—many problems will keep change from being sustained. Change of this nature presents many pitfalls to the company officer. First, the message that accompanies the change may not be one in which the officer and the stakeholders really believe. This may be because the stakeholders were not involved in the process, or it may be because of a fiscal or political change that is outside your circle of influence. Nevertheless, you will be expected to deliver the message in a positive manner and support the changes through your actions and the actions of your members. Second, organizational change may necessitate a change in the way you walk the walk, when you talk the talk, thereby putting your credibility on the line and jeopardizing the trust you've earned.

This is sometimes done with the knowledge that the organization—and in some instances, senior chiefs—may not effectively implement the changes promised, thereby offering up the company officers and battalion chiefs as sacrificial lambs. This leaves company officers and midlevel chiefs at the battalion level in the precarious position of balancing the message delivered against the possibility that the change may not be implemented effectively or sustained over the long run.

This first section may seem depressing. However, having witnessed this, I would like to make the point that change has parameters that must be optimized in order for it to be effective. When carried out incorrectly, not only does the change fail to get implemented, but a lot of collateral damage is done to officers, members, morale, trust, and willingness to embrace future change. Change is driven by human beings and thus must be managed by human beings. If it is not undertaken correctly and understood by the stakeholders, change can be a very bad experience.

Change occurs for a variety of reasons. Importantly, any change that we are involved in or that we have some level of influence over should be executed in such a manner that the change will be sustained over the long haul. Therefore, not only do the officers need to get behind the change in word and deed, but the actions of our officers and the organization as a whole must be focused and provide the necessary foundation to enable change to be embraced and, in turn, driven by the largest part of the organization.

The Reality of Change

Entire volumes have been written on change. Still, many people, including senior officers and senior officials in municipal government, do not understand how change can be achieved in an effective manner so as to result ultimately in lasting change. Although you can drive change from the top, cramming it down people's throats, in the long run, that approach will fail. True leaders must have an understanding of how real change is driven.

To begin with, we need a short history lesson. If you reflect on all the great changes and discoveries that have driven dramatic change over the past two thousand years, two commonalities are evident: first, they all broke with the status quo, the way things had always been done; second, lasting change was always driven from the bottom up, never from the top down.

Leaders must understand their role in change. Real leaders do not implement much change; change is best accomplished by our personnel when they think it up and embrace it as their own. Even long-range strategic planning for an organization must have stakeholder input and buy-in early if it is to succeed. This is fundamental, since the change must be implemented by the very people we often leave out of the discussion.

Henry Mintzberg of McGill University has suggested that leaders should actually be like queen bees. The queen makes babies (educates and raises new leaders) and exudes a chemical that keeps everything together (ensuring accountability, trust, and walking the walk), while the other bees (organizational members) busy themselves in going out to sense the environment, finding sustenance for the hive, and making the changes necessary to keep the hive alive in the face of the evolving environment. Remember, from our discussion of vision in chapter 3, that the troops in operations are those who would have to buy in and implement change; change will fail without the support of the troops.

Mintzberg has also indicated that the belief, by officers, that change comes from the top is a fallacy driven by ego. Most organizations survive because of small efforts at change originating from the middle or bottom of the company (the rank and file), only belatedly recognized as successful by senior management. Even a strategic plan developed for a large organization and encompassing long-range planning requires the support and activity of the workers in order to succeed. This is a lesson for senior leaders that should also be taught to our most junior leaders, to prepare them for future leadership positions.

So how does organizational change affect our development of future and current leaders? It boils down to two specific issues: first, trust in the organization by individual members, work groups, and teams; second, learning by environmental absorption. Future leaders will trust when trust is earned, and they will learn about change and how it is implemented and managed by being involved. On the one hand, change done in a positive and effective manner is an excellent teaching tool and model for new leaders. On the other hand, change done poorly, micromanaged from the top down, instead of driven from the bottom, will fail, creating a poor model, from whose mistakes future officers cannot learn.

Human learning is accomplished in a variety of ways, but experience is one surefire method. When people's experience with change is negative, they learn to distrust change, even if the motives and commitment of the people driving it are sincere and truthful. Moreover, they learn wrong, practicing methods for implementing change that do not work.

Change is implemented and communicated as part of the vision, which usually comes from the fire chief and senior staff. At the company level, vision is making meaning daily, and all department officers have a responsibility in implementing the vision at their level. It then becomes the fire chief's direct responsibility to ensure, at all levels of the organization, that promises made regarding change are implemented in a timely manner. He or she cannot afford, once the commitment is made, to allow staff to stifle change because of personal agendas or individual likes and dislikes.

It is also the sole responsibility of the fire chief, as communicated by senior staff, to ensure accountability for actions and walking the walk by his senior staff, midlevel managers, and company officers. Failing to do so provides the wrong example to members of the organization striving for leadership roles; thus, as role models, the fire chief and senior staff chiefs are the exact opposite of what we want future leaders to emulate. Trust is a valuable currency that cannot be spent without due regard for the price and quality of the product.

The Model of Placebo Change

Here is a common story about organizational change fatigue. Some event drives the recognition for change; it may be global change, such as a strategic plan, or it may be simply a change in some portion of the way we do business. When this occurs, the leadership of the organization holds a big kickoff for the change. This is often followed by seminars, leadership expectations, inspirational talks, and the provision, in some form, of a tool kit. This tool kit comprises a manual, outlines, slide show, and other tools to be used by the messengers (officers, committee members, or team) to communicate and implement the change. The problem is that after all the hoopla, no follow-up ever takes place for many of the changes. The organizational change, like the tool kit, ends up on the shelf, collecting dust.

Future leaders learn about change by being involved in it. Future leaders who have had a very bad experience with change—and with the promise of change from the leadership implementing it—will be soured to change. Furthermore, since most leaders spend the vast majority of their careers in one department, the impact on the leadership psyche is significant; consequently, a snowball effect results. Changes, from the company level all the way to the organization level, are learned events; thus, experience determines how future leaders view change. How change is implemented and the method used shape the skills of future leaders in change planning and management.

Not everyone will be happy with change. No single person is going to be happy with every step. In part, this is because there are people who are worried only about short-term change and people who are only worried about long-term change. This conflict between the short- and long-term views is advantageous, since it represents the convergence and divergence of ideas. The leader's job is to effectively manage this conflict and ensure that it is not of a personal nature and that it is focused on pushing the organization ahead. True leaders recognize the need for short-term immediate success without losing site of the long-term change.

What Makes Change Successful?

In light of the discussion we have devoted to the restraints that inhibit change, it's even more important to examine how to be successful at all levels of the organization. Before we go any further, let's examine one of the largest changes in U.S. government history that has affected each of us—the creation of the Department of Homeland Security (DHS). This tremendous change speaks volumes about how change can and should be managed within the organization, all the way down to the company level. It's also provides a very real lesson: just because we change, there is no guarantee that we are going to be better off or more effective than we were before the change, unless we create the leadership to decisively move and the management structure to support it.

In 2001, Tom Ridge one of the most vocal proponents of reorganizing the government's antiterrorism efforts into a cabinet-level position, used an organizational chart to illustrate how homeland security would work. Anyone who saw the chart was terrified by its complexity: a mass of lines, boxes, and layers of agencies—the quintessential portrait of an untamed bureaucracy. Sound similar to flowcharts you have seen in your organization? It has become clear by now, in the wake of the Hurricane Katrina response and other lapses in DHS initiatives, that charts mean little.

All the charts in the world won't make something work; it takes bold, educated leaders with the necessary KSAs to drive the organization. You cannot place people in positions in any organization who are unable to lead aggressively and get the job done, regardless of their connections or résumé. Title alone does not make you functional or competent. That is evident on the basis of the actions of FEMA's Michael Brown; Ray Nagin, the mayor of New Orleans; and Kathleen Babineaux Blanco, the governor of Louisiana. Increasingly, more information points to the likelihood that even Michael Chertoff, the Secretary of Homeland Security, was unaware of what his organization was and wasn't capable of doing. On the other end of the spectrum, the United States Coast Guard and the FEMA US&R teams were neck deep in the response early on and were highly successful in their operations. However, two highly functional parts of a larger organization are not enough to make the entire event successful; it takes the entire organization and external cooperation.

Organizational charts tell us very little. Take the chart of a successful firm and compare it to one of an unsuccessful firm—for example, a fire department you admire and one you do not (perhaps even your own)—and the one thing that stands out is that they look the same. Charts can never capture what drives performance and creates excellence: the caliber of your people, the insight and vision of leaders, and the energy and commitment of the workforce. Each of these are put at great risk when a leader reorganizes and implements some form of change. What really ends up making or breaking the organization is who ends up in each leadership box on the flowchart.

In reorganizing the way we deal with terrorism, the Bush administration is essentially advocating for a giant merger. We all know that mergers and change slow down productivity, which is the main reason why so many changes fail. Next, lines of control create uncertainty. People driving projects wonder if their initiatives are still sponsored and considered important to the organization; consequently, frontline supervisors receive anxious questions that they cannot answer. The elimination of FEMA as a cabinet-level position and its merger into the DHS infrastructure allowed managers (not leaders) in DHS to marginalize the organization and undermine its efforts during domestic response. Even more telling is that a horse breeder was put in the executive seat and expected to have the KSAs to perform. Coupled with senior managers and midlevel managers (not leaders) who are more interested in their own survival than effectively running the organization for its main purpose, the overall bungling of the Hurricane Katrina response or other operations should not come as a surprise to anyone.

Now we are witnessing a reorganization of the reorganization. Although it was a noble idea, the original reorganization did not work because it lacked effective leadership. Personnel were moved into key positions without the necessary KSAs to understand and implement the necessary changes and to conduct the business of the business.

You only have to pick up the paper to read the disparaging news about mergers. Tyco and Enron, among many others, come to mind. Nevertheless, the news is not all negative. Companies such as Citigroup and Pfizer demonstrate that integrating formerly separate organizations can work.

Fire service leaders who must implement change understand that reorganization and restructuring do not lead to success unless followed up quickly with other organizational moves. Change requires focused leadership, educated messengers, and stakeholders who buy in; ultimately, the leadership must aim to for the change to be successful in reality, not just on paper. No matter what the particular change, you have to make sure that the business of the business gets done as the number-one priority; the main thing remains just that, the main thing.

Returning to the issue of homeland defense, the question is, How does this educate us as fire service leaders? All changes require certain rapid organizational moves in their wake.

Let's examine the aspects of successful change and examine the example of DHS. At the time of this writing, there have been five years of DHS reorganization, from 2001 to 2006. Keep in mind these specific changes when you begin to think about change in your own organization.

Lessons from 2001

Find and focus on early wins to build momentum. For DHS to be successful, it needed to build momentum and confidence for both the internal and the external customers. Leadership, including division and midlevel managers, was chosen based on having the background to understand the jobs, the KSAs. This included replacing people, even with tenure, who failed to perform in the past.

The possible early and rapid wins included outlining a plan for universal smallpox vaccinations, establishing uniform security protocols at airport check-ins, and developing common databases for the Federal Bureau of Investigation (FBI) and the U.S. Citizenship and Immigration Services. These were not the biggest problems, but by solving any of them early, Mr. Ridge could create an environment where departments and people would have to work together despite previous boundaries. The three specific objectives mentioned here offered a bonus, in the form of a deposit in the emotional bank account of the members affected by those changes—that is to say, they were all goals that would reduce public anxiety.

Former mayor Rudy Giuliani achieved early wins in New York City when he cracked down on petty crime and vagrancy and when he replaced all the file cabinets and furniture in the Social Services Department. These early wins created momentum, convincing others that long-term change could actually happen.

The lesson is this: When implementing change, don't try to solve the biggest problems first. Go for some early and easy wins of smaller problems. If you can solve the small, nagging issues, people will stick with you for the long-term change.

Lessons from 2006

What's the report card on these objectives from 2001? There is no universal system of smallpox vaccination yet. There are universal security protocols at airport check-in, but they have focused on certain aspects of security at the expense of others—and in some instances have been ridiculous. Why can't a lighter be carried on—am I Richard Reed, the shoe bomber, or am I going to set the plane on fire? Why can't I carry my Spyderco knife anymore, but the lady next to me can carry 12-inch knitting needles?

Here is another good one, I travel with biological and chemical stimulants all the time, and only once in five years has anyone from the Transportation Security Administration (TSA) even found them. They were focused so intently on explosives that they pushed these vials—two of them on 5 × 9 cards respectively marked "Biological agent" and "Chemical agent, 2,000 respiratory doses or 200 skin doses"—out of the way while I watched.

Databases, while better than they once were, are still not common among agencies assigned to DHS and are not linked with supporting federal agencies that may have the same interest. Outside DHS, the FBI has purchased a computer networking system that does not work, and the Central Intelligence Agency (CIA) posted a picture of a most-wanted terrorist for two years before anyone realized it was the wrong man!

Jack Welch, former chairman of GE, has called this "change fast":

> *Little victories are essential for creating momentum and providing sufficient credibility to pat the hard-working people on the back and defuse the critics. The initial emphasis is on quick wins without sacrificing the long-term vision. In less successful situations, either you don't get the win or they are not fast enough to be credible. Holding 62 meetings is not a win.*[1]

Shortly before I retired, our acting deputy chief, who had only been in the acting position for about two weeks, was able to obtain a quick win, by getting authorization for the battalion officers to meet with the captains off duty and have captains who are paid overtime. This was something we had asked for years, but previous deputies had refused even to consider it. The quick win went a long way toward convincing people that change was possible. The real problem from a human standpoint is that it may have come five years too late to have the kind of impact it could have had in the first few months.

Give frontline teams more autonomy. Consolidating expertise away from the action (in staff) can leave an entire organization isolated, with potentially disastrous results. This should sound familiar, since it is tied directly to the concept that small teams run the fire service; that frontline members, who are operating the system, are closest to the customer; and that change comes from below, not from above.

An example of this is the Burlington Northern Railroad and the effect of deregulation. Once deregulated, Burlington Northern's operations were appalling. Burlington Northern assigned union workers to teams and gave the teams the authority to get the trains running and to make the changes they believed needed to be made. These teams were able to get the trains running on time and productively, because frontline railroaders—like frontline firefighters—see problems close up. As a result, Burlington Northern's turnaround was highly successful.

Although people would call this empowerment, that is a misnomer. The organization had merely removed obstacles to success—such as bosses who had always done it their way, lack of information, and lack of self-confidence. When you are able to remove obstacles, whatever they are, out of the way, things change—and ultimately, great things happen.

Focus on the core functions. In government and business, it makes sense to concentrate on those things that need immediate attention, rather than to attack everything at once. When Gordon Bethune took over Continental Airlines, it went from near bankruptcy to the number-one airline in the nation. Bethune focused on the most basic processes, such as scheduling, on-time performance, and getting meals to passengers. A back-to-basics approach forces performance to improve everywhere, because so many back-office functions are directly related and linked to those that really matter. Remember that it is the staff's job to support operations and the organization, not the other way around.

When the rules of the game change, we must educate everyone as to the new rules and then expect them to play by those rules. We must, however, ensure that everyone knows what the main thing is. When we lay out too many priorities (such that we end up doing nothing well) and when we lose sight of the core functions (such that we are no longer able to do our jobs), change will fail. You can only juggle so many plates at a time without dropping any of them, so choose the most important plates to keep them all in the air!

The function of the leadership group (senior staff) in this situation is to create the vision and develop the strategies that support it. They must invoke urgency and stay focused on a clear course. When we develop budgets and plans with no vision or strategy, the entire endeavor becomes superficial; everyone can see this, so the leadership group loses credibility.

Tap leaders down the line, and then get them moving for you. Mr. Chertoff needed a network of middle managers loyal to him. However, in addition to being loyal, they must be competent. Trust, respect, credibility, and loyalty are earned traits. The people who make up this network are people who, even though they don't normally have access to his office, know exactly what is going on, or what the scoop is. Good leaders know they cannot move an organization or its bureaucracy without a vanguard of agents of change who are accountable to them.

Choosing a great team is the key to this success. You have to have people who can work together as a real team to drive change. Many times the chief executive of the organization picks the wrong people, and you end up doing committee work instead of driving change. True leaders lead for something more than themselves.

Make a decision. A good decision executed quickly beats a brilliant decision executed slowly. You have to clear the decision-making bottlenecks by ensuring a recommend-agree-input-decide-implement-evaluate-tweak system, allowing decisions to be made according to the plans that have been laid down already. If you don't want to change, don't put it down as part of your plan; then you won't have to make a decision on it.

Change as an Educational Leadership Tool

People, leaders included, learn from experience. Leaders can learn to effectively manage and embrace change, but the opportunity to learn has to be driven by good experience. Getting involved in change that leadership does not know how to effectively manage can teach the wrong things and instill in people and leaders a sour taste for change.

If teaching our future leaders is about creating an environment where they can learn and practice the core values we require, then implementing any change should be a method of teaching. This leaves the question, Why do future leaders "learn wrong" when it comes to organizational change? David Garvin and Robert Cizik have addressed this in their book *Learning in Action*. Garvin and Cizik outline three stages of change:[2]

- *Stage 1.* The organization articulates the challenges that are motivating it to change. It designs a response and establishes goals.
- *Stage 2.* The actual change takes place. It's in the details of this stage that the proverbial devil lurks. This stage is one of execution and adjustment to hard, practical realities.
- *Stage 3.* The organization reviews what it has won and lost. This stage is about acceptance of limitations and adjustments to the new realities of the postchange work.

During the first stage, the organization must evince widespread dissatisfaction with the status quo. Someone (fire chief and/or senior staff) must develop a vision for the future and a plan (strategic plan, accreditation document, etc.) to get there. During the second stage, there must be a real willingness to take on the resisters—the most dangerous of these give you the "kiss of yes" (yes-men and -women). The third and final stage, consolidation, is a time to measure rewards and losses, and the organization must be ready to make changes to the plan, undertaking an honest assessment of what is and is not working (listening and accepting feedback). The failure to adjust on the basis of an honest assessment becomes the most dangerous and likely pitfall when new leaders are exposed to poor decisions and personnel whose egos do not allow them to walk the walk.

We must educate not only our newest leaders but also our leaders who have been around for a while as to exactly what learning wrong means. We have to show them the up- and downsides of change and teach—yes, teach—people how to communicate, cooperate, understand the fear that change creates, and be able to act as agents of change in both word and deed.

The two main reasons that change efforts create organizational learning gaps and have an impact on our future leaders are as follows:

- *Poor design.* This includes the failure to address the underlying processes used to get work done: operational response issues, departmental programs, resource allocation, relying on IT to provide a magical bullet (collecting data with no way to use it or assuming that computer programs [staffing programs, monthly reports, etc.] will enhance operational effectiveness), and not insisting on the necessary behavioral changes (requiring rank and file to change, but ignoring behavioral changes required in chief officers). To be successful, change of any kind must work; in other words, when you design something, you had better make sure it does what you want it to do, and if it doesn't, you don't need to be afraid to acknowledge that it did not work and try something else.

- *Poor communications.* Garvin and Cizik compare organizational change to the start of a marathon: change will be occurring rapidly in some units (those at the front of the race, our staff positions), whereas it won't have gotten under way in other units (operational units, companies). Leaders responsible for communicating change have to be prepared to give the same speech at least six times, or it won't be heard. Leaders must be able to explain an initiative, thoroughly with clarity of intent, letting the troops hear the arguments for and against options that were rejected. Finally, leaders must be able to address employee's fears. Firefighters and future leaders want to know why you think they can make it through change, and they want to know how you are going to help them through the change. This requires credibility, trust, and respect at all levels of the organization, but especially for chief officers, who are often the messengers of change.

A further cause of creating change fatigue is known as *senior staff block*. It is identified by the inability of senior chiefs (or anyone is a position of influence or power) to implement or embrace a change that is already outlined in a strategic document or an accreditation plan, even when the opportunity presents itself. You must strike fast when presented with the opportunity to implement portions of your plan.

For example, suppose that your organization finished a strategic plan, as well as the long process of accreditation. Moreover, you have completed a staffing and deployment document in support of those efforts. These three documents outline the long-range plan and the method to accomplish it. Now suppose that one of the operational benchmarks has brought to light the need to implement a heavy squad program to enhance response, RIT, and a wide range of other activities. Assuming that the organization has already ordered a new squad and has the apparatus and staffing to place two squads in service prior to the new squad's arrival, one would think this would be an opportunity to check off the block, accomplish the benchmark, and move on, while evaluating the effect of implementing this measure. It could be done immediately, after which would begin an evaluation period to adjust and improve. But assuming instead that senior staff members stifle the change by demanding additional staff work, justification for the squad concept, operational impact studies, and other blocks, we have a classic syndrome of fear of change, fear of the unknown, and fear that the change which will require us to rethink our operational deployment model and change the way we do business; perhaps it might even require that we sacrifice one of our sacred cows, such as a backfill engine, to get the squad in service.

Thus, rather than embracing the change as outlined in the plan, we have delayed or blocked the implementation by hiding behind staff work and organizational concerns. The question you should ask in this situation is, Weren't those items considered when the plan was made? You can bet they were and that the implementation of this or any idea outlined in the plan is the navigation point toward which we should be steering. Think about the message that this sends and the impact that it has on midlevel officers and upcoming officers, who have a stake in making this program work, and on the members of the organization who clearly know what the plan states. Morale, trust, credibility, respect—all of these valuable traits are cast out when change is blocked.

Finally, to best educate our future leaders on how to evaluate, implement, and communicate change, we must beware of the "lipstick on the bulldog" approach, which Moss Kanter of the Harvard School of Business has suggested is typical of many change efforts. In this approach, the leader sees something that's ugly, such as a process or a product that requires improvement, and wrestles with the change. Finding it difficult to get the thing to behave properly, the leader decides to make it look superficially better and move on. The result of this cosmetic approach is that while the bulldog's appearance has improved, it's also become really angry!

In the case of fire services, this could be a leader who asks why members do not want to join the fire prevention office. It may be a leadership issue, a perks issue, or any number of issues. Instead of identifying the root problem and implementing a meaningful change, leadership finds the answer too difficult and puts lipstick on the bulldog!

Winning the Stakeholders— Key Decisions for Success

Change affects everyone in the organization. It may affect our work schedule, where we work, what we do, and the process by which we get something done. All of these are sacred to us, and when they are changed, we want a good explanation as to how it will make things better.

Good leaders must make some decisions regarding how to win stakeholders early in the change game. Failing to make these decisions and then implementing changes has a wide range of associated side effects—including frustration, backpedaling, undermining your leadership, and miscommunications.

From a leadership perspective, the first thing that you need to possess is the instinct or the prescience (pre-instinct) to anticipate potential change. The trends become visible through reading professional journals, attending conferences, and other activities. This is one of the main reasons that education and events outside your jurisdiction are so important. Industry trends need to be evaluated: some are flash-in-the-pan ideas; others, such as air management, RIT, two in/two out, CBRN-compliant SCBA, haven't just cropped up overnight. You must look to the horizon.

Another source of this anticipatory action is right in your own backyard. You can clearly see ahead to political, fiscal, budget process, or realignment changes within your governmental structure. Each time the big organization changes, the organization we work for (in most instances, a department) must adapt to the larger change. Getting out in front of this and communicating it to everyone is critical, to prime the pump about what to expect.

The key to implementing and sustaining change is getting the members of the organization, those who are actually going to have to implement the change, on board from the very beginning. No one is so ignorant as to suppose that everyone in the organization will be happy with the change. Indeed, a very small percentage will actively try to undermine the change. Organizational reality tells us that 20% of your people will do 80% of the work (the Pareto principle); nevertheless the commitment of the stakeholders is required early and throughout the event.

From the embryonic stages of planning or discussing the change, members from all levels of the organization need to be involved. Regardless of rank, all members should have a say and their opinions should be recorded too. A good leader will understand team decision training and be benchmarking, looking for characteristics indicating that the team is moving toward the correct end of the linear model described previously (see chap. 15). If this involves something as complicated as developing a departmental strategic plan, make sure labor and management are involved from the very beginning; if it is something simpler, such as a change in process or procedure, use the same approach. Create a cooperative environment in which there are no secrets, so that everyone can see the positives and negatives associated with the choices that must be made.

Once you have addressed the development and nurturing of an inclusive decision-making system, you must find a method to communicate openly, frequently, and honestly with the people outside the work groups about what is taking place. With great trepidation, those on the outside will be trying to figure out what "they" are talking about. To create an inclusive "we" feeling, you will have to disseminate information quickly, even if it's not perfect, and you will have to come back and adjust it later. This needs to be done in both written and oral forms. Officers must be circulating among the members, telling them exactly what went on, before rumors can fill in the void of truth.

As a result of communicating frequently and openly and trying to get information out quickly, at times it will not be as accurate as it should be. This may be because of work in progress, or because of political or fiscal changes during the process, because we simply did not have all the facts at the time we relayed information to the members, or because we made a mistake. There are dozens of reasons why information becomes obsolete or changes; thus, our members need to know not to get frustrated—that this occurs during the planning stages of implementing or developing change and that as time moves onward, more accurate information will be disseminated. People need to understand that they are not being deceived—that they have what we have.

Change is a learning event for all of us, so it's critical that we begin the education process early and in small doses. As we decide to make a change to a method of doing business, policy, or process or as we anticipate the change coming, we need to educate our members about what to expect, giving them a time frame. Further, once the change has occurred and education has taken place, we must expect everyone in the organization to abide by the change and play by the new rules.

Finally, we must listen to our frontline troops, who as Colin Powell has said, are correct until proven otherwise. Most change must be implemented by human beings. Those human beings are going to be the first ones to be able to tell you if the change is or is not working, if it needs to be changed again, deleted, or reevaluated. We need to listen to the people who are actually in the field of play, in either operations or staff. It's not about rank; it's about who is engaged in the action. There is an even more important message here, which is that we need to listen very closely to those who perform any kind of service to tell us what needs to change.

Let's use a very simplistic example to demonstrate listening to or failing to listen to suggestions regarding change. Suppose that your organization uses two radio channels to manage emergency events—the dispatch channel, or channel 1, and a tactical channel for fire ground operations—and that you have the following policy:

> *Units responding to an incident will provide a roll call on the tactical channel and provide a size-up and report on channel 1 to the dispatcher and incoming units. Units will then switch to the tactical channel to conduct fire ground operations.*

In light of this policy, the operations members—engine, truck, squad, and battalion chiefs—put forth a suggestion:

> *This is not working. We need to change to use the tactical channel for all communications between units and leave channel 1 for command to dispatch. The first-arriving units will give the size-up on the tactical channel, and the battalion chief will relay on channel 1 to communications.*

Their reasons are to eliminate the need for company officers to have radios on two different channels and to create a safer and more efficient method of communicating between units. However, the deputy chief of operations (a staff job), with the support of the fire chief, refuses to allow the change because he does not believe it's an improvement.

The question should be, "Who's using the radios on the fire ground every day, the deputy or the engines, trucks, squads, and battalion chiefs?" The answer is obvious, as is the solution. listen to the frontline personnel, who are using the process; make the change, creating a quick win; and move on. It is a matter of either reevaluating—or, as in this example, refusing to allow change.

Since honesty is the best policy, don't be afraid to advertise, and celebrate your failures and successes. When you implement a change that does not work, hold a ceremony in which you make the failed change walk the plank. In other words, admit your mistake, get the change reversed, and move on. Likewise, when something really works, put a spotlight on the success and give credit to the developers and messengers of that change. Use these tools to bring about bottom-up change, to show people that change is possible, and to sustain change over the long haul.

Organizational Big Picture Mistakes

Best is the enemy of Good.
—Israeli Special Forces saying

A big mistake that organizations make, especially in strategic planning, is over-commitment. Be realistic about what you can and cannot do. Many organizations try to impress politicians—for example, by listing hundreds of items in a large document but failing to implement even half of them. However, this approach drives stakeholders and street-level bureaucrats (members) away from the change, because they see words on paper but no action.

There will always be restraints that keep you from doing what you want to do. These might be fiscal, political, geographical, resource, or staffing limitations. They might be known up front, or they might appear suddenly, while attempting to implement change or during the planning or implementation phase of a project.

So how does one manage this effectively and get the stakeholders involved? First, being realistic means focusing change on the core aspects of the business—the main thing. I prefer to look at priorities in a horizontal sense, outlining goals in the order in which they will be done given a limited time frame. By contrast, politicians take a vertical, rather than horizontal, view, wanting everything to be done with equal fervor and commitment. Quite frankly, that would make it impossible to be good at anything, and a true leader would, behind closed doors, tell those who espouse that view just how ridiculous it would be to work like that. You have to decide what you want to be great at; you cannot attempt to be great at everything, or you will end up being good at nothing.

This is not meant to suggest that we should not anticipate what we would like to do, forecasting over 10 or 20 years—of course we should. There are two points that we need to understand to play this game. First, the restraint that is keeping us from making the change today could disappear tomorrow, presenting us with an opportunity to implement what previously we were unable to implement. This could occur for any number of reasons—from getting a grant, to a change in law or in the political or budget process, to a local vote. If we realize this can happen, then we can prepare our experts and stakeholders to strike. We must be ready to take advantage of opportunities and we must monitor the horizon, looking for chances to bring about those tabled ideas.

Second, we must never pass up an opportunity. This game is fleeting: a unique opportunity arises only once; when it happens, pull out all the stops, call in favors, do your politicking, and be willing to compromise so that you get what you want at least in part.

The following is an example of how one organization failed at this game. Our marine program consists of a fireboat and several Zodiacs. The members assigned to the program are highly trained and dedicated. They are also well aware of the risks to Virginia Beach which is surrounded by water (including the Chesapeake Bay, the Atlantic Ocean, multiple inland bays, and the intercoastal waterway) and has a large anchorage for one of the world's largest ports and massive marinas with million-dollar crafts. Marine programs are expensive to run, as anyone who has ever owned a private pleasure craft (a hole in the water where money goes) can attest. However, when you are a fire and rescue service surrounded by water, the business of the business dictates that you cover all the possibilities.

A grant for our marine program came along, and it was quick: the turnaround time was less than 60 days. The prize was a $1.5 million fire and rescue boat. Since the cost of operating the boat was never been calculated into the budget, there is some trepidation. Still, an opportunity that never existed before and will more than likely never exist again presented itself.

What is required in such a circumstance is for the fire chief his staff, and the experts to mobilize and for you to seek compromise, call in favors, and use your influence. The organization should do everything in its power to get the grant and purchase the boat.

To quell the budget uproar over operational costs, you might tell the naysayers, "Let me get the boat. If I have to dry-dock it for a year or two until we can work out the operational funding in the budget, then so be it." In other words, compromise, but get the boat. This kind of opportunity requires a refocusing of resources and staffing to pursue and capture the opportunity; when a leader fails to take advantage of any unique opportunity, it speaks directly to the leader's inability to perceive, anticipate, understand and act decisively toward implementing change.

Change is not simple, it is a highly complex game of anticipation, estimation, evaluation, human endeavor, and risk that you must understand and educate leaders about. Develop an aggressive philosophy, change fast, and be able to accept the success and the failure that comes with change.

Change requires leaders who are intelligent and can relinquish their personal agendas to create a better organization. They must be smart enough to see when personal resistance to change by members of their senior staff is driven by fear or by personal gain or loss. To be successful and sustain change, leaders must employ the energy and expertise of their members to drive the change from the bottom up. Without the desire, drive, education, and ability to sustain change by understanding what one must do, words on paper are like a mirage in the desert: you can walk toward it and talk about it, but you will never reach it.

Notes

1. Kotter, John P. "Managing Change: The Power of Leadership." *Harvard Business Review*, January 2002.
2. Garvin, David, and Robert Cizik. *Learning in Action.*

Chapter 18
The Business of the Business

I have no ambition in this world but one, and that is to be a fireman. The position may, in the eyes of some, appear to be a lowly one; but we who know the work which the fireman has to do believe that his is a noble calling. Our proudest moment is to save lives. Under the impulse of such thoughts, the nobility of the occupation thrills us and stimulates us to deeds of daring, even of supreme sacrifice.

—*Edward Croker*

The Essence of Being a Firefighter

It can be surmised that firefighting and EMS, regardless of education, will always be a blue-collar job. The art and science of fighting fire, working collapses, Hazmat, technical rescue, intubating a patient, or cutting apart a vehicle is nothing but manual labor with a brain attached to it—albeit a really smart brain if you are going to survive. In the end, you are sweaty, cold, hot, wet, dirty, bloody, and sore. The closer you are to the operational end of the business, the more true this theory is. When and if you decide to promote past company officer to reach the executive level, how can you effectively understand and lead a blue-collar organization if you have never been blue collar?

You have to earn the right, at least in your peers' eyes (but maybe not in the government's eyes), to be called a firefighter. You have to earn, by your actions and the application of your brain, your spot as an operator.

In the first several years on the job, whether volunteer or career, you will be tattooed; this is not ink, but it is just as permanent. This tattoo will reflect your abilities to do the blue-collar work of stretching lines, taking heat, dragging victims, lifting hydraulic tools, passing the really hard endotracheal tubes and intravenous lines (IVs), hanging in there with the team when they need you, and pushing yourself past the point of exhaustion and then giving a little bit more. These are traits and skills that could never be measured by an exam created by a human resources or government agency. The tattoo will be carried with you as a tool that fellow firefighters and officers will use to evaluate you. Your tattoo is your reputation, both good and bad, and will precede you wherever you go in the department. Once you get this tattoo, it's as difficult to get rid of as real ink, nor is it the last tattoo you will get.

Each time you promote, you will be tattooed, by labor and management, with your capabilities as an officer. You must balance your actions, keeping in mind both the interests of your members and your responsibility to the organization. Remember that as a supervisor, you now play a different role.

If you become the type of officer who defends everything your members do, even when it is in violation of policy, is negligent, or represents a poor decision, and if you fail to solve issues at your level, even when it means discipline, you will get a reputation. When this occurs, you become very ineffective at bargaining with people higher in the chain of command on behalf of your members.

Similarly, if your actions with your members are never anything but clear-cut, rule driven, and unfriendly, you will earn a reputation with those members, who will tell others. If your actions on emergency incidents are unsure, confused, tactically unsound, and display the same characteristics of a civilian whose rear end is on fire (especially yelling), that will ink you as well.

When I was a division chief, we had duty rotations. Whenever there was a second alarm or greater, the duty division chief would respond and assume command. It was inevitable that many more divisions and staff battalions would respond. In most cases, this assistance was a blessing from a command perspective; they were competent, and we could assign those branches or divisions and let it run. We did, however, have one division chief from staff who all of the other divisions would compete desperately to clear him first, before he arrived on the fire ground. Why? Because we all knew he was dangerous—and downright spastic—on the fire ground. It got so bad, especially when strong leadership, solid strategy, and direction were required in order to get the job done, that a department-wide discussion was held about who had cleared this individual the night before and how it really improved operations. It was not a secret: the command staff knew it, the midlevel chiefs knew it, and the company officers and firefighters knew it. As sad as it sounds, there are people like that in organizations who have never had to learn to be firefighters first and then command officers.

It is unfortunate that there are people in organizations all over the United States who have never earned the right to be called firefighters but nevertheless wear the badge. I would estimate the number is small, but it's still a shame to share this honor with people who would be better off being accountants or running a hardware store.

If you come in from the outside, for example, to a chief's job, then there is no way for the members of the department to recognize your tattoo from your earlier days. How do you earn your tattoo as a chief executive? Follow the guidelines in this book; show people you understand the blue-collar work of being a firefighter; remove obstacles to success; and surround yourself with competent people. Only then will the rank and file know that you have been there.

When you are surrounded by people who know their business, technically and professionally, your actions and decisions quickly dispel any pretensions of who you might be or who you try to make people think you are, exposing you for who you really are, good or bad. The right to be called Firefighter, Lieutenant, Captain, or Chief is earned through nothing less than your KSAs. In our profession, it does not take very long to realize who the wannabes are and who has earned the title!

The 5% Rule

The modern fire service requires a lot from its members. There is so much to do to keep a modern service running smoothly. The mundane daily tasks, which we do well and professionally, are necessary evils.

What it really all boils down to, however, is the 5% of the time. What's the 5%, you ask? It's when what might happen actually happens. That's what you are paid for, what you have been trained for, and when you are expected to shine. Your very existence in the station is predicated on the inevitability that something will go wrong somewhere. This is the business of the business, what we live for, and it should never be taken for granted.

Everything else you do, from painting hydrants to doing inspections and washing trucks, is dues paying. This gives you the honor and right to be there when the 5% happens. When things are going poorly for someone, when their world is crashing down around them, even for a moment, it is our job, our calling to be there, fix it and fix them to the very best of our ability.

There are two rules, and breaking either is unforgivable. The first cardinal rule is to always take care of the customer. When the Smith family has a fire or medical crisis, it is our job and our job alone to intervene and do whatever is humanly possible—and occasionally superhumanly possible—to engage and restore a semblance of normalcy to their lives or to save their lives. When it is necessary to risk our lives and the lives of our crew, it is our duty to make the commitment, above all else, to save a life if it can possibly be saved. No matter how hot, cold, wet, or tired we are, we have a responsibility, entrusted to no one else on earth, to serve our fellow human beings.

The second cardinal rule is to perform your job at the highest level of technical competency. If you cannot do this, if you lack physical or moral courage or if you are technically, mentally, or physically incapable of making a 120% commitment, then you need to find a different living. There is absolutely no excuse for failure arising from inaction or lack of skill or courage. The only failure that is acceptable occurs when we have done everything humanly possible and committed all our KSAs toward the desired outcome and a higher power decides it is not to be—and even then, we should be trying to intervene and negotiate the desired outcome! Short of that, there is no reason for us to even contemplate failing.

Where we work, there is no time-out, there is no next time, there is no second chance. When the 5% calls, you had better have your game face on. No one should have to endure poor strategical and tactical decision making, indecision, lack of physical or moral courage, or inability to do the assigned job at an incident. If you are unable to function in the 5%, you should remove yourself from our company and ponder what kind of person you would have been instead had you been capable at the moment of truth.

9/11 Retrospective

In July 2005, I was asked to write a small remembrance for a 9/11 memorial service. It captures the importance of the 5%. Please indulge me while I share this with you.

> *It's difficult to adequately paint the canvas of emotions, thoughts, and perspectives with regards to how I feel or felt on 9/11. Like all Americans, I was shocked that we had become so complacent that it was even possible for someone to plan and implement such an operation without being caught. I suppose I was rather naive when it came to what I expected from our security and intelligence apparatus. Unlike many other Americans, however, it was clear to me within only a few moments of the second aircraft striking the World Trade Center that we were at war! For me to comprehend what that would mean in terms of my job and what it would mean to our children, families, and most certainly, our military was impossible until many days later.*
>
> *I could write volumes on what I saw, how I felt, what I expected, and what reality turned into for me and hundreds of other US&R task force members, so rather than pour the soul of emotions and thoughts onto the plate, I choose to open only pages of the book for a short glimpse of individual chapters.*
>
> *I lost many close friends that day, but in retrospect, it was nothing compared to the loss absorbed by families, sisters, brothers, mothers, fathers, uncles, aunts, and the thousands who lost their blood. My pain comes only in short bursts when I think back; those who lost family will endure the pain of that loss for their entire lives. My pain was expressed in the funerals I attended, in the survivors I met, in the families I talked to. I am fortunate, however, in that I can walk away with my memories, while families who have lost a*

son, daughter, mother, or father do not have that luxury. I cannot begin to fathom their loss or their pain.

There were so many acts of valor and selfless heroism that day that they will never all be captured or understood. I believe, in talking with survivors, that many of them, and many that did not survive, surmised that they would not see their families again, and yet continued in their tasks until the very end, never believing there was any other option than to continue. When you consider a decision of that magnitude, you begin to fully comprehend the depth of commitment, love, courage, faith, and valor that members of the fire service can call on in a moment. Will we all have the physical and moral courage to reach such conclusions in having found ourselves in a similar position, or will we shirk our duty and find ourselves among the timid souls who know neither the taste of victory of the physical commitment of expending oneself in a worthy cause.

Operationally, I was impressed with the professionalism and selflessness of the US&R task force members and command staff. My first few nights at the towers, I would have bet anything that many of our members were not going home, as time after time during briefings, I continued to say "Folks, this building has swallowed thousands of people. It is still hungry and will think nothing of swallowing dozens more without even a blink." What kept them alive was their understanding of the environment in which they were working, their extreme professionalism, their regard for the safety of their crews and the countless hours they had spent preparing and training for just such an operation. When you think about the operations that they undertook and the results, it is truly a testimonial to what trained, equipped, and committed teams are capable of accomplishing.

Like all events of this kind, it was contaminated by phonies, con artists, and parasites. Make no mistake: there are people who go about their lives living off the misery of others. These events open up a canvas on which some people would paint themselves as they are not, claim to do what they have not, claim to be something they are not, and feed off the blood of others! To these individuals, I say, "There is special corner of hell reserved just for you. Be prepared, for your day will come, and it approaches quickly and like a thief in the night!"

We must never forget what happened to us that day, as Americans, as families, and as individuals. We must be vigilant in the face of terrorism; as Sun Tzu said, "Kill one, terrorize a thousand." We must continually ask what if and prepare for it. Freedoms demands that we never underestimate, never take for granted, and never yield our beliefs and values to evil. Those who would find themselves timid, who complain about yielding certain civil liberties for our mutual protection, who say it will never happen again who complain of the methods we use to prosecute and *marginalize those that would threaten our freedoms, our families, and our children—go forth, because I seek not your council. With tail between your legs, carry yourself from our sight to where men and women of courage are not panged by your presence. Leave us so that we may remember our heroes, savor our freedoms, and place before us something more important than ourselves! When we have protected and assured freedom for you, return to enjoy what others have given you with their sacrifice and live among the men and women of the community who have risked so much for you, never daring to look those who risk much in the eye, as it may threaten you into submission.*

Why You Are Dangerous

Sometimes in our business goodbye really means goodbye.

—*Ray Downey*

Every day each of you go to work thinking it's going to be another day like the last, that you will end up doing the same old mundane traditional tasks that you did last shift. Perhaps if you are lucky, you think, we will get a structure fire and we can be guaranteed that we will be running EMS incidents most of the shift. When you get this feeling, you had better slap yourself in the face and wake up. This complacency is dangerous, even deadly, for you, for your team, and for others around you. Soldiers can tell you that when they are in the combat operational theater for several months and start to think it's another routine patrol, someone ends up getting killed. When you start to feel comfortable that everything is going to be as it always has been, you are dangerous.

In 2005, 106 of your brothers and sisters left their families, their pets, and their homes in the morning thinking just that, and they never came home at the end of the shift. Some of them died doing some very stupid things; some of them died because other people did very stupid things; some died because their heart and lungs were not adequate to do the job; and some died giving everything they could to save their fellow human beings. You must remember what you do for a living and that your entire world could be turned upside down in the blink of an eye. No one goes to work thinking "Today is the day I am going to be in the fight of my life—for my life or for the lives of others."

Every day you go to work, ask "What if today is the day?" You had better be an expert at transition drills, or those skills and events that require you to change your mode of thought instantly and without thinking, based on instinct (autopilot). In law enforcement or the military, transition drills involve switching from a long weapon (rifle) that has failed to operate to a handgun without even thinking about it. It is drilled and redrilled into people's heads until it becomes so natural that most people will not even realize what they are doing. This is extremely important because, when someone is shooting at you and your primary weapon fails, you need to stay in the fight and defend yourself without hesitation. Hesitate or try to fix the malfunction at this critical time and you will end up dead.

Transition drills for the fire and rescue service involve three events. This is exemplified by entrapment, or lost firefighters—the Mayday. If you are the one trapped or lost, you had better be able to recognize and accept it and react without thinking, because there is nothing more deadly to you short of a fall off a roof or getting caught in a flashover. You must know exactly what you should be doing to yield the best chance of survival, because your initial actions will set the stage for life or death. You should have a series of steps that you institute automatically, without thinking.

Training is, of course, essential to transition drills for the fire and rescue service. I continue to be amazed at the number of fire departments that have not put their entire department through a class on Mayday/firefighter down or saving our own. Regardless what you call this class, it teaches skills that must be learned before something goes wrong, because there are very specific methods to perform certain tasks. For example, you don't want to bail out of the 15th floor by sliding a charged hose line over the balcony to escape to several floors below if you have not been trained. Likewise, if you have not been taught the Denver drill, or how to get someone out of a basement, or how to set up a pseudo anchor and slide a self-escape rope, you will probably not be successful. Moreover, these skills are applied so very seldom that continual, ongoing training is vital in order to maintain your abilities.

In case you ever find yourself lost, trapped, or disoriented, it is imperative that you be smart enough, trained enough, and quick enough to recognize the trouble—not to deny it, letting your ego override reality—and make the decision to declare the Mayday. Then, put your brain to hypercool, push your emergency button, arm and alert your PAL device, and give your ASLIP. This last is a report regarding air status, the situation you or your team are in, your location (so the not-so-rapid RIT or others can find you), your intent (what you will do to save yourself), and a personal accountability report (PAR), which indicates who is with you and how many you number. This must occur immediately, without thinking, and in a very deliberate and cool-headed manner.

Finally, when things get so bad that you have to deploy RITs, you are fighting against time and dealing with a dicey situation at best. RITs require special training and equipment. Furthermore, they require people who will not be afraid to do what others might find very discomforting.

Many departments and individuals either fail to address or take for granted high-risk/low-frequency events, assuming that they will never happen. When you get complacent about what you do for a living and lose sight of the critical nature of these events, you become dangerous.

Leading or Doing: Failing Either Will Alter the Moment of Truth

When as leaders, we find ourselves responsible for the business of the business, especially during the 5%, it is crucial that we perform. As discussed in chapter 13 ("Battlefield Firefighting"), leaders must be prepared to deal with aspects of the fire ground other those that were taught. If you have not educated yourself well beyond the basic premise of the organization, if you do not anticipate what could happen every day you go to work, then you will be unpleasantly surprised one day. In preparation for Operation Overlord, the invasion of Europe during the Second World War, Dwight D. Eisenhower wrote some very prophetic words regarding the expectations of a leader in times that require all their KSAs:

> *Another Point: there is a vast difference between a definite plan of battle or campaign and the hoped-for eventual results of the operation. In committing troops to battle there are certain minimum objectives to be attained; else the operation is a failure. Beyond this lies the area of reasonable expectation; while still further beyond this lies the area of hope—all that might happen if fortune persistently smiles upon us.*

When faced with a crisis, leadership must employ deliberate speed. If you act too quickly and are not adequately prepared, you will certainly make mistakes; thus, if you have not trained for the events that occur infrequently, you will never be effective at leading during the crisis. Facing a crisis means understanding what needs to be done while acting calmly and moving deliberately at the appropriate speed. Sometimes fast is slow, and this will allow you to take the appropriate measure at the appropriate time. Doing the first thing that comes to mind or doing what you did at the last event may lead to difficulty.

Under extreme conditions, leaders must think in terms of ordered flexibility. Ordered flexibility guides the selection of operational tactics. To remain flexible, you must understand the strategical and tactical situation you are engaged in; you must understand the rhythm of the event (remember fluidity); and you must have a plan, as well as a backup plan to address changes in rhythm.

The efficacy of your tactics is contingent on the ability and willingness of your members to perform in critical situations. There is a direct correlation to your ability to lead successfully. If your members are not trained and are incapable of transitional-type events and the organization has not provided the critical elements of the business of the business, you will fail. Leaders in times of crisis must trust and be trusted by their subordinates and peers. Only with trust can people be counted on to exert maximum effort during the crisis.

When leadership or action fails, we fail to achieve the outcome we seek, and the outcome is what we exist for. Competent leadership is critical to the business of the business. Remember the nature of what you have chosen to do for a living or as a volunteer, so that you don't become dangerous.

Chapter 19
When Leadership Fails

> *The fish always rots from the head down.*
> —Source Unknown

Why Leadership Fails

Since this entire book is about leadership and the management tools required by leaders in order to survive in organizations, I could say that if you don't follow the principles of this book, you fail as a leader. That would be unfair and quite frankly wrong. As human beings, we not only fail from time to time, but we have to be taught, in the classroom and by exposure to others, how to act and what to do. You cannot fault people for not acting a certain way when they have acted another way throughout their entire lives.

Two questions are addressed in this chapter. Why does leadership fail so often? When it does fail, what can I do to survive, protect my team, and thrive?

The reality of the fire service is that you will work for both good bosses and bad bosses. In the latter case, the question is, How did this happen, and what can I do while still maintaining my leadership principles? It's tough when you find yourself in a position of working for or around poor leadership. The perils of making the wrong choices out of emotion are pretty great and put you at risk of violating your personal commitments, integrity, commitment, and honor.

At some point during your career, you will find yourself struggling for air in a leadership vacuum. Loss of leadership can occur at the company, the battalion, and the division levels and all the way up to the chief of department. Unfortunate as this situation may be, it is important for us to understand how to survive, maintain morale, and continue accomplishing the business of the business under such horrific circumstances.

Leadership failures occur for any number of reasons, many of which are predictable, at the organizational and individual levels. The personality traits associated with individuals who fail when placed in leadership positions can be identified in advance. Failure can be the result of self-interest of leaders who think only of themselves and their advancement through the organization. All leadership failures can be traced back to topics already discussed in this book, such as failure to create vision, failure to earn the trust and respect of personnel and peers, failure to take care of your people.

In this chapter, we will study several of these causes of failure in detail and examine a case study. Some of these stories have happy endings, while others have very sad endings. The fact remains that many individuals must survive under officers who will never be able to lead effectively and who, in some cases, can barely manage.

There are many specific reasons why leadership fails, but they can be distilled into three main issues:

- Individual failure
- Organizational failure to educate
- Poor application of the small team and of their energy toward the organization

Organizations fail to develop leaders for three reasons. First is the inability of senior leaderships to understand teams and the critical importance of operations. In many fire departments, operations are viewed as a necessary evil that fire chiefs and senior staff must deal with and that eats up much of their budget. We must always remember that the primary mission is to provide fire and emergency services and the associated support services in cooperation with fulfilling our mission.

The second reason why organizations fail to develop leaders is that many officers reach a level in an organization that they simply have no business occupying. In the past, this has been termed the Peter principle, where individuals reach their level of incompetence. This includes everyone from company officers to fire chiefs. Ill-considered promotional processes and political appointments are among the causes of the Peter principle. It may also occur because other officers failed to do their duty in properly evaluating people when they had the opportunity, and as a result, people who should have had poor performance reviews got passing grades. Finally, it may the case be that someone does an excellent job of self-promotion on paper, despite not having the necessary skills to lead.

The third and final reason is that organizations are failing to create future leaders. It is imperative that this cycle be broken, to ensure success. We invest 28 weeks in training a firefighter but often provide no training for future officers. This failure to educate is a great disservice to our heritage and to the future of the fire service.

It's important to expand on these concepts to understand what we can do to short-circuit this cycle, which is destroying our ability to lead the fire services of today.

Those of you reading this book may be in one of three situations. The first situation is that you are working for great and competent leadership that is already practicing much of what we have outlined in this book. The second situation is that you are attempting to survive while working for someone who is a total or partial failure as a leader; under these circumstances, effectively leading those around you becomes even more needful for your survival, the team's survival, and your ability to deliver service. The third situation is what I term *failure from below*, where you, as the supervisor or senior officer, discover that you have a leadership failure below you; the question is, What do you do, and how do you do it?

I wish I could tell you that everything works out fine in these situations, but unfortunately that is simply not true. Perhaps if the new generation of leaders understand how these events happen, they can lessen the impact that leadership failure has on future organizations. However, the current state of affairs is that fire chiefs who are poor leaders keep their jobs and make the big bucks owing to politics. Senior staff chiefs who have been in place for decades cannot be removed despite their dinosaur mentality, and midlevel and company officers are protected by the same system that promoted them despite their inability to do their jobs.

As discussed previously, there are things that are out of our control. Organizational parameters established for hiring, firing, and promotion can set the stage for leadership failure in advance. In our litigious society, local, state, and federal government agencies are hesitant to raise the bar for fear of legal reprisal. The French diplomat Jean Giraudoux once said, "No poet ever interpreted nature as freely as a lawyer interprets the truth." The reality is that the United States has more than half the world's population of lawyers. Many of them have only a nominal practice and lots of time on their hands and will eagerly file a frivolous lawsuit in hopes that the nuisance value will force a settlement.

But take heart because eventually these people will retire, or there hand will be forced when they get caught doing something really stupid. There is light at the end of the tunnel only if we begin to develop and educate a new generation of leaders who are unhappy and discontent with the ways of the past and willing to do their duty to change the face of the fire service. Although this will be a long endeavor requiring patience, skill, and savvy, it can be done.

Setting the Stage for Organizational and Political Failures

Create your organization so it can be run by an idiot, because someday it will.

—*Source unknown*

Government organizations often create an environment where personnel with poor leadership skills rise to their level of incompetence, in accordance with the Peter principle. Someone may function very well in a specific area but may fail miserably when given a promotion or even when moved laterally into another position at the same level. Functional incompetence based on the location in an organization is a real problem. Fire departments have an unrealistic expectation that uniformed employees should be able to fulfill any job in the organization. We move officers from place to place continually, and some end up where they are unable to function and lead effectively. In simplistic terms, we set these people up for failure, and as a result, all of the individuals in subordinate roles suffer as the individual fails time and again, dragging everyone down with him.

While organizational culture, rules, and politics play a role in allowing and even sanctioning a failure of leadership, it is often in combination, synergistically, with inadequately trained or inadequately prepared individuals. The following identifiable and predictable cultural aspects of organizations create leadership failure:

- Being minimally qualified (the issue of trust and respect)
- Failing to provide leadership when needed
- Setting promotional quotas
- Equating the uniform with competence
- Promoting noncombatants
- Hiring the best résumé

These organizational cultures and processes have an impact on life and effectiveness at the company, battalion, and departmental levels, as evidenced in the case study. The case study portrays exactly what happened, with all the original players, although the names have been changed—not so much to protect the guilty as to protect me.

Some of my observations may be politically incorrect, but I am unapologetic. There are a few givens in the fire service that have not changed over the years. First, because this job is a muscular-endurance and cardiovascular sport, you must be in shape enough—with upper-body strength, leg strength, and cardiovascular endurance—to support your team and function effectively. Second, firefighting and EMS is a team sport. Unfortunately, our teams cannot always be at full strength in terms of staffing and deployment, and a single weak link can affect a team during emergency and nonemergency operations. If you bury one weak player amid six strong players, you might be able to succeed, but if you attempt to bury one weak player amid only two or three, during the heat of battle you will be in deep trouble. Third, you must have the trust and respect of your peers and subordinates and have a reputation as an able firefighter.

I don't believe in lowering our physical-fitness standards, written-exam standards, or criminal-background requirements to ensure that the organization represents the community. Instead of lowering the entrance bar, we should be raising it, regardless whether that would eliminate individuals who give us the balance that government seeks. For years, the fire service has been hammered as a "lily-white, male-dominated organization," and statistically, this is true. Furthermore, in some instances, qualified candidates were kept from joining because of race, creed, color, or gender. Don't get me wrong: I do not sanction this. No one should be held back because of race, creed, color, or gender; every American should be given the opportunity to succeed in our society. However, compromising our standards corrects one problem by replacing it with another. Everyone should have to earn what they seek, not be handed an appointment on the basis of standards designed to balance an organization, rather than attract qualified candidates.

Several questions beg to be answered by those who accuse the fire service of not being diverse enough to represent the community it serves. Does congress have the racial and gender representation that makes it representative of the United States? Likewise, how about state government and local government, local industry, the public works department, or any other organization?

James O. Page once said, "Minimum qualifications become maximum expectations." The fire service and most government civil service agencies are notorious for establishing minimum entry criteria, minimum performance expectations, and minimum productivity levels and allowing them to become maximum expectations. Some of the individuals that we hire have minimum physical capability, minimum operational capability in an emergency situation, and function at a minimum level in and around the station. They do only what needs to be done to meet the minimum qualifications, and they may be incapable of doing any better. Moreover, supervisors are at a loss to force improvement from these "get-by" players, because they recognize that the organization is not behind them, owing to legal constraints or quotas, for example.

These minimum players may want to promote, and some have the book smarts to do so, even though they may be minimum players when it comes to KSAs and physical capabilities; unfortunately, many of our promotional exams have very little that evaluates an individuals' real-world firefighting performance and instead evaluate how much they can regurgitate facts about basic management.

Our promotional process mirrors the current state of education in this country. Like schooling, promotional exams are based on objective measurable standards—namely, test scores. For some people, education becomes a surrogate for achievement, and in the fire service, books smarts are substituted for functional capabilities. Don't think for one moment that I do not value formal education. I have bachelor's and master's degrees, and I believe that formal education should be a requirement for promotion, especially to the higher ranks. However, education without practical knowledge, physical capability, and good old common sense sets one up for failure. There must be a balance, and the balance must be measurable.

By and large, our promotional process does not identify who will be successful as a leader and who will not. To be a good leader, you must be a good firefighter to be a good leader. There is no guarantee that just because you are a good firefighter, you will be a good leader; conversely, however, if you are not a good firefighter, you will most assuredly fail in leading a company through emergency and nonemergency events.

Case study: How trust and respect affect ability

Firefighter Jane was one of those employees with a good heart, and she was hired under our minimum-qualifications standards. As it turned out, she was not really physically capable of pulling her own weight and supporting a team, but she was able to do the minimum required and thus received evaluations that allowed her to keep her job.

Firefighter Jane gained a reputation as being less than aggressive on the fire ground and functional as an EMT. She had difficulty making decisions under pressure, and her peers had to chip in an additional 50% effort to cover for her. The firefighters knew it, the captain knew it, the battalion chief knew it, and as in most departments, everyone else knew it. Nevertheless, since our requirement was to do the minimum stipulated, she remained gainfully employed.

Now, Firefighter Jane was good at studying and had a good memory. She repeated the right buzzwords, quoted policies, and did quite well in the promotional process; while she never scored at the top of the list, she was able to make the assessment centers and score toward the bottom middle. As the department got older and people retired, increasingly more promotions were made, and one day Firefighter Jane got the call: she had been promoted.

Captain Jane was assigned to a firehouse, and for the first few months, the members of the shift, including another female firefighter, tried desperately to work with her. Her performance on the fire ground was less than exciting, but it met the minimums. Also, her battalion officer was out on injury leave, so she had no supervision (another organizational failure of leadership).

The complaints started, first informally and then formally. Transfers for the entire company were submitted, and company members indicated that if they weren't transfered, they'd get out by way of EAP. The company was in a shambles; transfers were made, leaving places open for new people to enter the company. Can we expect the outcome to be any better, though, with an infusion of new personnel?

The root cause of this leadership failure was the promotion of a candidate who had failed to earn the trust and respect of her peers during her tenure as a firefighter and was physically and mentally the weakest link on the team. Her reputation preceded her, and she reinforced that stigma when she arrived at the company, through her actions and inactions. The problem is that there is no way to quantify the parameters of trust and respect, and the legal system would skin you alive for using arbitrary and capricious standards.

As secondary cause was organizational failure at the senior staff level. This falls under the classification of failure to provide leadership when needed. Because the battalion chief was on an extended injury leave, the new captain had no supervision, except an acting captain functioning as a battalion officer. After more than 14 months without anyone in this position, the organization finally did move in an officer from another battalion; however, this amounts to closing the barn door after the horses have left. The organization should have anticipated the problem—being aware of the past history of the individual and knowing the supervisor would be out for an extended period of time—yet failed to take action. You cannot leave weak players in a leadership role to grow weaker when you know what the problem is.

Does this story have a happy ending? Probably not. The organizational reality. First, Captain Jane has completed her probationary period without an evaluation from a superior officer (and I am guessing with minimal approval of the acting battalion officer). As a result, it will be nearly impossible to correct the leadership and technical issues with this officer.

Is there light at the end of the tunnel? Potentially, things might change for the better. The new battalion chief could set expectations and hold the captain to them, although that would require close and constant supervision. Captain Jane might be directed to return to school. She could be mentored, drilled, and tested, and her performance could be documented on the emergency incident, as it certainly should be. In fact, the organization will have to do all of these to prove that it has attempted to help the employee, if they ever have the need to hold her accountable and responsible. Only after all of these actions are taken can the new supervisor mark "does not meet minimum requirements" on the evaluation. With minimum qualifications being what they are, though, this officer could very likely be with the department until she retires—and might even make battalion chief one day!

Failure to Educate

Human beings are a product of their upbringing, their environment, and the culture they are exposed to, including peer pressure. Most organizations invest considerable time, energy, and money to educate new recruits in the technical aspects of the firefighter/paramedic job. New recruits are like fertile soil on a farm, begging for nutrients and water, from which will be reaped wonderful crops. Most fire recruits are ecstatic that they were able to successfully compete for the few firefighter jobs offered by the organization and feel lucky to have been given the opportunity to become a member.

Since, as we indicated earlier, our firefighters are our future officers, what better time is there than now, in recruit school, to start educating them and allowing them to see how they fit into the vision. Most organizations fail to make the department vision and its core values explicit to our newest personnel. More important, however, we fail to provide concrete examples of what they can do to align and do their part to make this entire thing we call vision work. Likewise, we fail to show them how our core values are applied to our work each and every day. This is not limited to the applications for emergency operations, but extends to the daily routine and the often-mundane tasks that every firefighter does as a member of a team. Finally, we need to teach some of them how to live and work in our family.

Most organizations will spend 16, 18, or 28-plus weeks educating and training recruit firefighters. Why is it, then, that those same organizations fail to provide the necessary education to their new company officers when they are promoted? Most new company officers have just finished a very competitive process to transition from buddy to boss, and overnight or over a weekend at best, they must change gears from riding backward to riding forward.

Many organizations fail to provide even the most basic officer training to new officers. This includes simple items such as evaluations, discipline, leadership principles, management tools (paperwork, computer use, and departmental flow), violence in the workplace, managing diversity and sexual harassment, basic tactical applications for company officers, and the list goes on. These educational issues provide the basis for effective leadership and management at the company level.

Failure to train creates an environment where our new company officers are set up for failure. As a result, they are unable to assist in implementing the vision, because they are not sure how to handle the most basic situations, much less the most difficult situations of all—those dealing with our largest resource, people. Failure to train also creates officers who have not been provided with the organization's expectations of a new officer; as a result, they may be guided by the culture of the company they inherit or by their battalion officer. Nature abhors a vacuum, and a vision/expectations vacuum will be filled with something that perhaps is not what the chief wants. We cannot have our new officers going over to the "dark side" early in their career, or they will carry that with them through their promotion to chief officers.

Departments must commit time, fiscal resources and energy toward the development and provision of officer-candidate schools and continuing education for officers. Otherwise, the leadership vacuum will continue to grow and manifest its sway over our future officers.

Promotion of noncombatants

Officer candidates must understand that to hone their leadership and supporting management skills, they will have to do some staff time. Staff jobs offer a different set of opportunities and a different workload than most operational jobs. Just as in the military, a well-rounded officer should plan on diversifying his or her career.

Another educational failure that organizations are guilty of is taking their newest leaders and placing them in noncombatant roles (staff jobs) immediately after promotion. This is usually done because no one else wants the job or because it's an easier organizational move than making a veteran chief or captain who has never done any staff time go to staff. It may also be driven by past failure to educate the veteran officer; because the senior staff doubts the veteran's ability to do a staff job, the young officer is assigned instead.

My observations on the fire and rescue ground and in the operations arena lead me to suggest that an officer's first assignment should always be in operations. This includes initial promotions to company officer, as well as battalion officers. The essence of the fire service will always be emergency operations, or what we know as fire and rescue operations. When departments are faced with cutbacks, no city manager ever says, "Go ahead and shut down the fire stations. Just make sure you keep training and prevention operating."

Learning to command an operational battalion to function effectively and safely on the fire ground should be the priority. Every officer will be in operations at some time in his or her career; thus, it is crucial that they be mentored in the necessary skills to make decisions under extreme circumstances during emergency events. These skills cannot be learned behind a desk, and the responsibilities of company officers and battalion chiefs are entirely different than what was required in their former positions. The business of the business will always be responding to and controlling emergency incidents in a safe and prudent manner; only training and experience can provide this skill.

In any department, about 90% of the staff are assigned to operations. This means that operations is also the most appropriate place to learn how to deal with the difficult issues that fire service leadership face, including human resources issues, evaluations, diversity, violence in the workplace, and human social interactions. If you really want new officers to learn to lead people and practice the supporting management skills, they must be placed in a situation where the potential for emergency and nonemergency interaction at a humanistic level is highest.

When organizations place new officers in staff positions, they deprive these officers of the most important learning aspects of the job. These include commanding emergency events and being part of the largest and most diverse human resources pool in the department. When you learn to make the kinds of decisions required on the fire ground, your decision-making skills and range of options greatly improve. Training in decision making and understanding the small team concept pay real dividends when officers moves to staff positions, because their ability to make decisions has been greatly enhanced.

There will always be time for staff jobs, but everyone must learn how to fare in combat first. Since the vast majority of staff operations, such as training, logistics, prevention, and administration, are designed to support the operations section, it makes sense to send new officers to operations first. This gives the new officers a greater appreciation for the impact that they will have in a staff job.

No officer candidate going from captain to battalion officer should even be considered for promotion unless he or she has spent time in operations. For example, a newly promoted captain might be sent to a staff job and stays there until promoted to battalion chief, reporting to the field to command a battalion. This situation should never occur—ever! Here is a battalion chief who has never spent a day in the front a fire truck in command of a company all of a sudden in command of an entire battalion. Likewise, a new battalion officer promoted from an operations captain position should not be sent to staff until such time as he or she has gained operational experience in the field.

These organizational issues described here represent only a few of the problems that create a leadership vacuum. Failure to educate and promoting noncombatants are identifiable problems; organizational leadership must recognize these faults and correct them. These problems are born of the manner in which organizational leadership educates and grooms officers, directly or indirectly. If corrected, the development of leaders, instead of managers, will be favored.

Individual Failure

A leader at any level of the organization can at times fail to step up to the plate. For example, in the case study, the division chief failed to fill the vacuum left by the battalion officer. I was one of the division chiefs at the time, and my failure contributed to the issue. A firefighter/paramedic assigned to that company said so: "Chief, you guys failed us!" Those words cut me like a knife: I realized that, as a chief in the organization, I had failed to step up and voice my concerns; even though it was not my battalion or my shift, I still had the responsibility to identify the problem and recommend a solution. As division chiefs, we had all failed the firefighters assigned to the shift. We knew of the preexisting problem with the newly promoted officer and that it would be magnified by the absence of a battalion officer; we could have easily anticipated what would happen, yet we all failed to act. Mea culpa.

Individual choices result from a wide range of factors, including education, exposure to others, and culture. Human beings have the opportunity to make choices, and when we choose not to lead, we are responsible for a failure in leadership. All of these failures are like a set of dominos that are set up in a circle. A failure of one establishes the failure of another, in a chain of linked events. Failure of the organization to educate contributes directly to individual failure.

Fire service company officers walk a fine line. In many ways, they remain members of labor, working foremen and -women. They supervise many of the tasks that they once performed as firefighters, so they understand the requirements of the job. In times of short staffing, they assist with those tasks, as any good leader would do.

Make no mistake, though: the job of the company officer is to plan, supervise, and direct, not to work on the same tasks as the firefighters. In departments that are short staffed (three-person engines or trucks), it may be an operational necessity for the company officers to perform tasks that the firefighters should be performing on the fire ground; however, this detracts from what the company officer should be doing on an emergency incident, and that is plan, supervise, and direct specific tasks. Removing a company officer from this intended position is dangerous for his crew and for the organization, but that's the subject for another book.

The ability of a company officer to not manage but lead is dependent on a several factors. We are all creatures of our environment, and as we promote through the ranks, we bring with us a foundation of management practices, traditions, and mentoring and operational experiences. With these KSAs, we are able to draw inferences and make decisions about issues we face.

Furthermore, we are molded by our educational experiences—formal and informal, technical and managerial. While it is important for company officers to receive formal education in their quest to become chiefs, it is just as important that they obtain technical training related to their jobs and the jobs of their subordinates. Effective leadership is dependent on one's credibility with subordinates and peers. There is no faster method to torpedo your credibility than by failing to obtain and maintain the KSAs required of your job and those required for you to understand the jobs of your subordinates.

Interestingly, the failure to create leaders with management skills can be attributed to both organizations and individuals. On the one hand, individuals at the top of any organization must recognize company officers and battalion officers as a resource to be developed, mentored, and used as conduit to receive and filter cultural perspectives from the field. On the other hand, organizations must create an environment and culture that provides the necessary training, mentoring, and growth to develop tomorrow's leaders.

The molding of company officers (future chief officers) should begin when they enter the department as rookies and continue when they receive their first promotions. This tie binds all officers together. The purpose of this development is to foster the ability to lead effectively, rather than simply manage. Those who manage create a career path dedicated entirely to themselves; this is called the *I syndrome*.

The I syndrome can be found at every level of officer, up to the chief of department. The I syndrome is exemplified by individuals who advance their careers at the expense of others. The I syndrome is also evident in officers who forget to lead by remembering what it was like to follow and in officers who fail to take care of their stakeholders, the firefighters.

One other trait typical of the I mentality is the refusal to question or to invite dissent. Officers engage in this mentality because rocking the boat, being outside the mainstream, could jeopardize their ability to progress rapidly through the ranks. Further, people who respectfully question what is going on risk getting tagged as non–team players by chief officers who feel that if the employee is not saying *yes*, he or she is not doing his job. Always do what is right, rather than what is popular.

Individual choices can create a situation where we fail as individual leaders. As previously discussed, this includes failing to follow the ABCs of leadership, losing trust and respect, and neglecting to maintain KSAs. The following are points to keep in mind when striving for leadership at the individual level.

Walking the walk and talking the talk

As a chief, I consistently put on my turnout gear and breathing apparatus at working structure fires. I know that my job is not to have the knob anymore or take the roof, and unless assigned as the operations officer, my gear might not be an immediate necessity. So why mess up my perfectly clean white shirt with my gear? Or, expecting that chiefs would come to battle prepared to do battle, you might ask, Isn't that the way it is supposed to be?

This simple act shows personnel that you are ready and willing to do what they do. They see you as leading from the front. Also, it's difficult to correct someone for a PPE violation when you stand around in Nomex pants and a cotton shirt during working incidents. People learn by seeing and doing, so if they see their officer or chief doing something outside policy or slacking, then over time, this becomes the culture of the battalion or organization, which is difficult to change.

The following are basic rules for leading by example:

- *Establish expectations.* Let your personnel know what you expect of them clearly and precisely, and let them tell you what they expect from you. Establish this relationship early, and adjust as necessary.
- *Understand that morale is important.* Some senior chiefs blast morale as just firefighters whining. Nothing could be further from the truth, though. Remember that morale filters down from the top. It reflects the attitudes of people at every level—especially the attitudes of supervisors toward their own jobs and toward the people they supervise. As a company officer, you must relay morale problems to your chief. When dealing with people, you must sometimes make decisions that are technically irrational.

You can be totally rational with a machine, but with people, logic has to take a backseat to understanding. Spirit and a belief that your personnel make a difference are what create morale. Never underestimate the power of a group of people who feel that they can do anything and who are supported in accomplishing it.

- *Expect accountability and responsibility.* Confederate general Robert E. Lee once said, "In all things do your duty; you can never do more and you should never do less." So what does that mean? Well, first you have to follow rule number one, which is to make sure that expectations are clear. Next you must expect leaders to lead and provide the necessary management tools. As a company officer, you will be expected to address human resources issues, such as diversity, sexual harassment, violence in the workplace, and evaluations, in a timely and professional manner. Also, operational issues and the delivery of customer service should never be second rate. Last, emergency operations should always work efficiently and professionally—no excuses. None of this "we will lay no line before it's time," or "I didn't know there was a hydrant there," or "we forgot the suction unit at the truck." Mistakes will be made, and they should be used as learning points, with one exception: mistakes regarding basic skills and KSAs should never be tolerated on an emergency operation!
- *Be visible.* Your personnel want to see you, if only to talk. Don't be afraid to assist with tasks around the station when staffing is short. Keep in touch: Make sure you know the status of your personnel, their families, and their health. Cultivate an air of detached involvement; this means getting close without getting cozy. Field Marshal Erwin Rommel said, "The commander must try above all else to establish personal and comradely contact with his men, but without giving away an inch of his authority." Your personnel will look to you to deal with operational issues that interfere with their jobs. They will also look to you for guidance on human resources problems. Leader must keep a finger on the pulse of the morale of the troops and be mindful of their personal welfare. Never lose the common touch: lead by remembering what it was like to follow.
- *Keep your word: Follow up.* A man or woman is only as good as his or her word. Whether you are the fire chief or a new recruit, if you tell someone you are going to do something, you had better do it. We may promise things only to find out later we cannot

accomplish them, perhaps because of fiscal issues, politics, or individual mistakes. When this happens, follow up by going back to the individuals or the group and explaining what, why, where, and how. Otherwise, your credibility will quickly be unraveled, and word will get around that your word is not worth spit. Liar and black hole are terms that are often used to describe chiefs who don't keep their word or follow up.

These humanistic aspects of leadership are the areas where I see failings in many organizations. The ABCs of leadership provide an excellent base from which to start correcting our failures. Our ability as leaders is then dependent on being a teacher and student, in seeking out the leadership within our ranks and mentoring it, so that the culture of leadership becomes tradition.

Failure to Exploit Small Teams

The value of small teams to the fire service has been emphasized in previous chapters. Members of the fire service who are uneducated about small teams—team life cycles, team decision training, and other aspects of team engagement—can apply themselves effectively at the individual and organizational levels. Again, you can see the domino effect: the organization's failure to educate has an effect on individual decisions and leadership.

To operate effectively, the fire service must employ small teams in emergency and non-emergency roles. Small teams account for over 90% of the entire customer contacts associated with fire and rescue services and are the primary reason for the high ratings that fire services enjoy when the citizens of a community are polled. Furthermore, small teams accomplish 90% of all the work done in the fire service. This might be an engine, truck, or rescue company, or it might be a small team of instructors or inspectors. The effectiveness, safety, and efficiency of these teams are based on several factors, as described earlier (see chap. 14).

Leadership failure comes when we fail to exploit and support teams. To make sure you remember their value, ask yourself this question: "Does what I am about to do support the small team concept?"

How to Survive Failing Leadership

Want to create conflict in a bunch of officers really quickly? Ask this question: "Why should anyone be led by you?" I did this one day at the end of my career and was stunned at the 20-minute tirades I got that never answered my question, but just placed blame. As I have stressed in this book, there is good reason for someone to be scared of that question, because leaders must have followers in order to lead.

A subtle issue with failing leadership is that it may not be your boss who is failing at all. It might not even be your boss's boss. Leadership failure can be well above you on the chain of command, and you will have limited ability to correct it. When we have failure at the highest levels of the organization, everyone suffers. The corrective strategies to apply differ depending on your place in the chain of command.

Not everyone can be a leader. Some people don't have the authenticity or knowledge necessary. Some people simply don't want to shoulder the responsibility that comes with being a leader. Some people prefer to devote more time to outside activities and their private lives. Let's face it: there is more to life than work.

Another misconception is that everyone who gets to the top is a leader. Nothing could be further from the truth. Real leaders are found all over the organization, and if you reach the top because of political savvy, that does not make you a leader.

Great leaders are also not always great mentors or coaches. An entire consulting industry has sprung from this concept and would have you believe differently, but that would be wrong. The problem with this philosophy is that there is no proof that someone who can inspire people is necessarily good at teaching technical skills. It is possible for great leaders to be great coaches, but they are the exception rather than the rule.

Failing leadership can have many faces. In general, we assume that the boss is the culprit, because he or she just doesn't get it. However, it is inappropriate to blame everything on a boss. Leadership requires relationship: when for whatever reason, you cannot form a relationship with your boss or your members, someone is going to be unhappy, and leadership is going to fail at some level. Failing leadership can be due to a lack of direction or the inability to take action or make a decision. It also can be due to conflict between two or more people who just do not like each other as human beings or have totally different philosophies on how things need to be accomplished. It can be slow breakdown of trust, based on action or inaction over time. Although we could list dozens more reasons why leadership fails, the fact remains that it happens.

Failing leaders can turn themselves around. It takes a rude awakening for this to happen. It may be driven by external factors, all-out rebellion (a no-confidence vote or constant conflict with their staff), or an ultimatum (as in, "Either fix yourself or you are gone").

If we eliminate the people who won't change, who don't care, or who have decided that they are not going to do anything to begin with (those who, no matter what you do or how you educate them, are just going to have to leave the organization before it gets better), we are left with two forms of leadership failure at the human level. First is the inability to develop a relationship, respect, trust, and communications between boss and subordinate. This is a lose-lose situation; in the end, the boss will win, but only by losing a potentially great employee, capable of great things for the organization and possessing the energy to make the boss successful. You really can't call them bad people for their failure to develop a relationship. You could, however, say that they failed at leadership if they made poor decisions, fail to implement change, and are inept at nurturing trust, respect, credibility, and other key leadership characteristics.

Second, those who have been promoted or appointed beyond their competency level cannot function effectively at that level. This happens much more frequently than you might imagine. These may be managers who have been around for a long time, rising through an old system of promotion, or they may have gotten the job or appointment through politics.

Even leaders who worked their way up through the political side of the house can be effective in leading organizations, as long as they recognize their limitations. If they are smart, they will surround themselves with great people, set a vision, and let it run, intervening only to remove obstacles to success. However, if they are not smart, they will view themselves as great officers or leaders, and it will be readily apparent to everyone around that they are attempting to pass themselves off as something they are not. These leaders may be highly ego driven and may be abusive to their staff, and will continually make assertions like "I grew up in a system that did this and that. Don't tell me how it's done." In the long run, they will not surround themselves with people who can thrive in spite of and compensate for their leadership failings.

People who do not have the KSAs to do their job exhibit certainly identifiable behaviors. These people will continually fall back into the previous role, trying to do someone else's job at the next level down, and they will micromanage. Because they have not been educated in a variety of ways and lack the skills to realize they are incompetent, it's not entirely their fault. Their actions, however, do have an impact on all of us.

There are a variety of strategies you can employ whenever you find yourself a victim of failing leadership. I cannot guarantee that these will work; it depends on the boss, the environment, and you. Your boss must be willing to listen and change when real change—not just change on paper—is needed. You must learn to adapt to different bosses and their likes and dislikes (including their pet peeves) and their past (upbringing and organizational education). In extreme cases, for real change to occur, people need to leave the organization. These following survival tips have been prioritized, so that you can approach the problem by asking, What can I do first, and then what if that does not work?

Leading up

The stereotypical boss/subordinate relationship features an out-of-control or psychotic boss and a helpless underling. While this might be true in some instances, it's usually an exaggeration.

Leading up is usually best applied as a strategy when it is your direct supervisor who is failing. It is a technique that can be applied when you are in senior staff and work as part of the department leadership/management team.

Leading up involves developing a pattern of interaction, a relationship, between you and your boss that will deliver the best possible results for the organization. It may well be an uncomfortable relationship, fraught with issues, but as long as you work there, you are going to have to develop a connection on some level. This requires you to learn about your boss in context: his or her strengths and weaknesses, the pressure they may feel from above and from their peers, and their organizational and personal objectives.

Leading up should never be manipulative or used for political or personal gain. Make no mistake: some people, especially yes-men and -women, use leading up for this very purpose. The reality of creating a relationship with your boss is that it enables all of you to move on to the business at hand of running the organization and facing the challenges to the organization.

The reality of this engagement is what Liz Simpson terms the *power differential*.[1] To put it bluntly, your boss can fire you, but you can't fire him or her. This differential will always affect the dynamics of the relationship and will result in frustration when your actions are constrained by your boss's decisions. How you handle this depends on your own ability to deal with authority, your level of respect for your boss, his or her credibility in making decisions based on past events, and myriad other dynamics.

The two extremes are arguing or picking fights with your boss over decisions and swallowing your anger, behaving in a very compliant manner. While neither of these actions is productive, you need to be aware of where in this spectrum your attributes and predispositions fall. On the one hand, your boss can be hurt when he or she fails to receive the cooperation, dependability, and honesty of his or her officers. On the other hand, sometimes you make your bed and have to sleep in it too.

One of the most important concepts is becoming aware of your boss's expectations. Many bosses fail to lay out their expectations adequately. When this occurs, the boss/subordinate relationship will not click like it should. Sometimes this entails giving credit to your boss for accomplishments. It also means discovering what your boss's personal goals are within the organization.

Finally, pay attention and learn your boss's behavior. Find out what pushes his or her buttons, and try to avoid that. There is always a learning curve when you try to examine someone's behavior patterns.

The following are suggestions from Liz Simpson on accommodating a boss's work style:[2]

- *Does he or she prefer more a formal or organized approach?* When you have a meeting with him or her, take an agenda.
- *Does he or she become inpatient or inattentive?* In this case, keep digressions, background detail, and informal chitchat to a minimum.
- *How does he or she process information?* If he or she likes to study it alone, then provide information in written form. If he or she likes to ask questions, then provide it in person.
- *What's his or her decision-making style?* Is he or she a high-involvement manager (a micromanager)? If so, touch base frequently, on an ad hoc basis. If he or she prefers to delegate, then report about significant problems or important changes on a regular basis, while handling the details yourself. Is he or she difficult to figure out? For example, when you provide your report, is it returned because it purportedly does not contain everything he or she needs? If so, he or she is probably uncomfortable with the current status of the project and doesn't feel up to date. With this kind of boss, make sure you document these meetings and follow up with memos.
- *How does he or she handle conflict?* If he or she thrives on it, be ready for lively, spontaneous exchanges. If he or she tends to minimize conflict, respect that preference without falling into the trap of relaying only good news; he or she needs to be told of both the successes and the failures, but this is best done in private.

I had a boss at one time who was a true micromanager. Face time was more important than accomplishments; he would suppress any change he personally did not like, and a wide chasm separated our points of view with regard to how to meet the operational objectives. Eventually, I became so frustrated that I isolated myself from him and made contact only when absolutely necessary. We never did fix our relationship issue, even though I tried all of these techniques.

Sometimes this strategy works, and at other times, it doesn't. Nevertheless, you have to try leading up before taking the next step. If you and your boss can figure each other out and if at the same time you can deliver the goods for your boss, you will gain greater respect and trust from your boss and potentially be allowed to buy in on projects that are important to you.

Bunker mentality

When you have tried to understand each other but the division between you is too wide, then adopting a bunker mentality is the next step. A bunker mentality shields you and your members from out-of-control or lunatic bosses.

Bunker mentality, like any other solution, has drawbacks and is more effective at the company and battalion level than at the senior staff level. When I isolated myself from my micromanager boss, I was using bunker mentality, but it did not work very well. It is usually successful when the failure is several layers above you, as long as you have limited contact at that level. While you and your members are affected by the actions or inactions of individuals at that level, none of them is your direct supervisor.

When one decides on bunker mentality, there are pitfalls that must be avoided. First, you must not neglect your duty to develop new firefighters and upcoming officers. You have an influence on them and must provide the right example through your actions, by functioning as a leader.

Next, you must continue to plan for and meet organizational goals; they are why you exist. Don't confuse anger or frustration with your boss with anger and frustration with the organization; you must strive to do what is right to support your members. Bunker mentality does not give you the right to undermine success or to talk bad about what needs to be done; simply go out there and do it. If you have real problems, talk to your boss behind closed doors.

Finally, never let the customer know your frustration. We are here to fix other people's problems, not to confuse them or place them in a conundrum by exposing ours to them.

Bunker mentality assumes that if we keep our heads down, do the right things, influence the things we can influence, and lead up when possible, eventually the reason we are in bunker mentality will either leave or change. Sometimes that takes a very long time, and it may not happen in your career. Bunker mentality allows us, as leaders, to continue to demonstrate by our actions that we can come to work every day and lead, protecting ourselves and our members to some extent from failing leadership actions.

Political or managerial action

Another method of dealing with failing leadership is to address it from a political or managerial posture. This approach is risky: it can pay great dividends if you're successful, but it can also have dire consequences if you find yourself on the losing end. There is an old saying, "When you choose to kill the king, you had better kill him and not just wound him." This is not to be taken literally, but whenever you take political action, you need to understand the consequences of failure.

Killing the king may not mean getting rid of him. Instead, it may mean getting things accomplished by political mandate. There would be no recourse or retribution (legally) from the boss, since his or her bosses mandate the change.

For example, no-confidence votes can be applied in collective bargaining systems. This is a double-edged blade, however, since the no-confidence vote may be directed at someone who is truly not leading and deserves to go or may be misdirected at a true leader who the union can't stand (because of his or her progressive stance, desire to change, and intolerance of the way things have always been). The failure to lead may be on the labor side, as is the case when they maintain things needlessly, such as positions, pay rates, and policies. In either case, one side or the other has some failing leadership, but there may be no justice in the action taken!

Political action groups and voting blocks work very well. Virginia is a right-to-work state, so there is no collective bargaining. As we say, "We have the right and you work!" When such issues as fire department paramedics, four-person staffing, and ladder company officers were brought forth, most senior members of the organization failed to act on them, and the city council supported them. Consequently, the local branch of the International Association of Fire Fighter (IAFF), along with the teachers union and police union, formed a voting bloc and elected 8 of the next 11 council members.

This provided the political clout to push for and get paramedics for the fire department. It provided the impetus for ladder company officers, and despite the opposition of the city manager and the budget director, the position was passed. I remember sitting in a meeting the week before the vote, and one of our chiefs, the budget chief, said, "There is no way they are going to get ladder captains. It's just not going to happen." I looked at the chief in disbelief, since it was clear the union had the votes on council to get them. This chief was either playing to the fire chief, to show solidarity, or was simply out of touch. (I wouldn't think someone in that position could be so out of touch. And as it turned out, two years later, that very same chief was promoted to division chief for less than four hours and was made a deputy chief the next day.)

Political strategy and maneuvering can be used to counter failing leadership and actions. It may remove the problem, or it may be directed at inaction.

Moving on

There are two forms of moving on, one that is accessible to most people in the organization and another that is more drastic. The first strategy of moving on is to simply transfer to another station, shift, battalion, or division, away from your boss or the chain of command that is failing you. I don't have to describe transfer in detail here, since it is a fairly common practice that has been applied for years.

The other strategy of moving on is to leave the organization. Sometimes changing jobs or departments or retiring is the option you explore when leadership fails. It's a strategy that is not available to everyone and is difficult to undertake. It's nearly impossible to apply this strategy when you have a financial stake that will be lost by leaving. Unless you have put in all your time or have very little time vested, it's difficult to rationalize. If you have 20 years on the job with a 25-year retirement, you need to stay for your pension.

It takes a lot of courage to release the familiar and seemingly secure, to embrace the new. But there is no real security in what is no longer meaningful. There is more security in the adventurous and exciting, for in movement there is life, and in change there is power.

—Alan Cohen

At the end of my career, I chose retirement as my last resort, after applying all the other strategies. In many ways, I was blessed, having worked for phenomenal leadership most of my career. For the most part, my officers and my peers were good to work with and tried hard to make the organization successful. I was also fortunate to have my time and pension in place, so that I could exercise this option. Following are a few lessons I have learned about the choices that you make during your leadership journey.

When Leadership Fails

The right to lead is given by followers; it's never successfully demanded. No paper or strategic plan written or spoken can create an environment that inspires people to follow. Only through personal action does leadership come to life; only then does the written word support your actions. No paper can hold the power possessed by a human being who has been given the power of leadership by his or her followers.

Sometimes not getting what you want is a wonderful stroke of luck. I have discovered that frustration increases with rank, when you find yourself trying to follow poor leadership. While the company officer level is the most influential position for internal influence, you often are not affected by the many day-to-day decisions that are made in running the business. As you move from nonexempt to exempt, organizations tend to abuse you, figuring that they own you both on and off duty. Also, with all the overtime available today, especially when you are a firefighter-paramedic, you make just about as much or more than a chief officer anyway. My point is that you should consider why you are promoting past company officer and understand what management requires of you.

There are leaders, and then there are managers. Many people confuse management and leadership. Leadership is about doing the right thing in practice. Many people hold themselves to be leaders when they are really managers. If you want to know who is a real leader, there is a simple test: ask the followers. The bottom line to choose what you are going to be and not pretend to be something else!

Life isn't fair! Hard work, dedication, and competence don't always ensure success. When you are on the outs with people with the power, they can pretty much do what they want regardless of the outcome. Cross someone who not only has power but also has limited credibility, an empty glass of character, and an economical way of seeing the truth, and you will be in real trouble. If you choose not to compromise your own principles and become something you are not, it becomes a win-lose situation, with the individual holding the power in the win position—unless they are somehow exposed as who they really are. Just be sure that when you make the choice to stand on principle and character, you fully understand the consequences.

When, as a leader, do you know it is time to head home? It's undoubtedly not exactly the same for everyone, but if you are a true leader, you know when it's time. Sometimes it comes as quite a surprise, and at other times, you see the signs long before you make the decision. I have come to the conclusion that there are many reasons to head for home: sometimes it's personal; sometimes it's environmental; sometimes it's political; and sometimes, well, it's just time.

When following is not an option, it is time to evaluate your options. To lead effectively, you have to be a good follower. A leader, based on actions, direction, character, credibility, ethics, and commitment, earns followership. When in the course of leading, especially at the senior levels of an organization, you find yourself in a position where, for whatever reasons, you can no longer follow the man or woman who is at the helm, it is without a doubt time to go. When you find that you would have to sacrifice your personal integrity, credibility, and character to support something that you don't believe in, it is incumbent on you to get out of the way. Sometimes this means moving on to another fire service job; sometimes it means retirement; and sometimes it means finding somewhere to tuck yourself away until such time as you can emerge from the bunker.

Having fun is critical to your health and success, so when the fun ends, it is time to consider heading home. Working in the fire service should be fun. That's not to say that there is not a high level of stress (especially at the command level and on the fire ground), hard work, and pressure to create great things, but it should still be fun. When the fun goes away, it's time to start thinking about doing something different and challenging. People make the workplace fun, and when leadership fails, it becomes anything but fun.

Machiavelli once said, "You can tell the intelligence of a man from by whom he seeks council and from whom he surrounds himself with." There is another old saying: "If you roll in it, you get it on you." As times change, the staff you worked well with changes, and subsequently you can no longer abide by the general consensus of what, how, where, and when. When leadership begins to make choices with regards to what type of organizational structure they want, they send a clear message about where they want the organization to go. As the team changes, it will affect your ability to function effectively, and when the overall character of the team changes, you have a decision to make. Can you change to something else, or do you need to get out of it before you get it on you?

Goals and desires change over time. Sometimes we find that we don't want what we craved when we first started—that the brass ring we reached for was not the brass ring we wanted at all. As times changes, so do people's wants and needs. It may be that the job you are in does not give you the opportunity to reach your full potential, so you head somewhere else. Alternatively, it may be that you find that you had a good run, but now you just want to do something different. Don't be afraid to follow your desire, even as it changes over time.

Going home or changing houses is an individual matter. I can tell you that when you go out on great terms, at the top of your game, it can be very unsettling. When you head home because you just don't feel like you belong anymore, it can be a depressing experience and make you angry. In the end, heading home is okay; it's a natural event of the team, organizational, and individual life cycle that plays itself out millions of times a day, across millions of organizations.

Be thankful for the people you met along the way who made the job, your life, and your desires worthwhile. Coworkers, family, friends, and, yes, even enemies if you have them contribute to your experience as a leader and to the path you ultimately take. As I head home, I keep in mind the firefighters and officers who taught me so much, who offered me a piece of themselves, who guided me, taught me, and pushed me back on the right track when I fell off. I am thankful to the men and women I worked with and for and commanded on the fire ground for enriching my life with laughter, professionalism, and friendship.

What Next?

There is no such thing as the business of leadership or the leadership racket. Leadership is an art that the fire service is rapidly losing touch with. The fact remains that we can turn around the trend of creating manager-politicians and restore the leadership tradition. Whereas management skills can be learned in a classroom, organizations and the people in them must embrace the necessary changes in approach, culture, and personal strategy that are so important in creating and employing leadership. Henry Gilmer once wrote about leadership, "Look over your shoulder now and then to be sure someone's following you." This is good advice for the fire service as we approach a turning point in our ability to develop tomorrow's leaders.

Notes

1. Simpson, Liz. "Why Managing up Matters." *Harvard Business Review*, August 2002.
2. Ibid.

Conclusion
Changing the Culture One Person at a Time

> *A great deal of talent is lost to the world every day for want of a little courage. Every day sends to their graves obscure men whose timidity prevented them from making the first effort.*
>
> —Sydney Smith

In ancient times, Plato wrote a story with the premise that people lived their entire lives inside a cave—indeed, not only in a cave but chained with their backs to the opening. There was a large fire burning behind them, the purpose of which was to reflect on the wall in front of the people a shadow of anything that passed in front of the cave, giving them a glimpse of what the outside world really was.

All information from outside the cave was filtered through the mouth of the cave and the fire. It was up to each individual to interpret what they saw reflected on the wall of the cave.

Plato asks the following question. What would happen if we unchained one of these people and allowed them to exit the cave? What would happen once the individual cleared his or her eyes from years of living in the cave and was able to see the outside world—what the reality of the world outside the cave was?

Plato goes on to suggest that if this individual reentered the cave and told the people who remained chained up that what they had been seeing as reality all these years was wrong, they would most certainly try to kill him!

Similarly, when you read this or another book about leadership, attend a conference or seminar, get advanced training, or network with people and then come back to your organization reenergized and ready to makes some changes, you have spent some time outside of the cave. And when you return, there will be people who will be very unreceptive to the message you bring them. Just as in the story of the cave, they will want to kill you!

Organizations are made up of people, traditions, and methods that have taken years to put in place. These structures are often closely guarded by people who view them as totems of their power. You cannot change an entire organization overnight; even with exceptional leadership, it might take years or decades, depending on the size of the organization and the strength and charisma of the leader. Leaders are surrounded by external politics, hidden agendas, fiscal constraints, and even manipulative and perhaps evil people; all of these factors must be taken into consideration by your strategy.

When you return to the cave, don't be frustrated that you can't change the organization overnight. Your work group, your company, the youngest members of your team, your division, and the people and processes that are within your circle of influence are ripe for teaching and change; this is because you can make a personal decision to change yourself, and then, by your action, you can begin to change those around you. The higher up the chain of command you go, the more capability you will have to change or to demand change from those around you who are resistant to change.

Leadership is a personal choice. It's tough work to show up every day and lead by your actions; it's tiring and can be frustrating and even risky. Nevertheless, it's up to you now to make a choice. You can dismiss all of this advice, or you can say, "Today is the day I make a choice and a change and become an everyday leader."

No one can tell you what do and how to act, but the fire service is screaming for leaders to fill a huge vacuum. So lead, follow, or get out of the way!

Epilogue

Life is short. Don't waste time worrying about what people think of you. Hold on to the ones that care. In the end they will be the only ones there.

—Source unknown

If you have reached this point, either you skipped to the end or you read the entire book. Regardless of the endeavor, thanks for getting this far. I learned something about leadership and accountability these past few weeks when my editor e-mailed me to find out why I was so far behind on my submissions. Clearly, he was going to hold me accountable and responsible for my performance and commitment. Subsequently, I locked myself in the upstairs office and wrote for three weeks, finishing the book before I went off to places unknown in my new job. He practiced good leadership principles, and I got my act together and finished the book. It's hard to put a lifetime worth of thoughts and experiences down in a short time. I can assure you that it is much easier to talk about them than to write about them.

The principles outlined in this book are universal in application to leadership. If you are the chief executive officer of a Fortune 500 company, or a company officer in a one-horse town, or a young military officer, they are equally applicable. What is different—and in fact critical—is that if you work in the fire and rescue services, the military, law enforcement, or a security agency, your responsibilities to leadership take on an entirely new meaning. Not too many CEOs can lose their lives or the lives of their employees by making a bad decision, being technically incompetent, and/or failing to identify the main problem. They might lose lots of money, swindle stockholders, and end up in prison, but you won't kill people and have catastrophic consequences. To them, leadership might be a way to make money, but to us, it's a way of life—literally.

I have crafted my philosophy on leadership from many of the excellent philosophies in other people's writings. I have been a student of leadership my entire life and have learned from real human beings, as well as from written material by people I never met. One sure thing about leadership is that the fundamental characteristics, behaviors, and actions of true leaders have been around for untold thousands of years. Those who espoused them were ultimately successful in one way of the other; those who did not have slipped into obscurity.

Fire and rescue services resemble the military in that they absolutely require leadership on a daily basis. The people you work with and the people you work for depend on your decisions for their lives and their property, and they expect you to return the sons and daughters, nieces and nephews, wives and husbands, entrusted to you in one piece. Despite our best efforts, that does not always happen; even when it does not, though we should be able to hold our heads high and say "I did everything humanly possible." More so than in any other profession on earth, leadership is critical in fire and rescue services.

My quiet thinking time is often spent on my back porch at night, with my iPod in my ears, a good cigar, and a vodka Collins or two, glancing at the sunset. I reflect on what I have done, what I should have done, my successes, my failures, and how I might do it all differently given the chance. I have rearranged my personal and professional life so many times from that place. I cannot even begin to tell you how many of humankind's ills, as well as my own problems, I have solved there—only to forget to write them down so that I can capitalize on them for the good of humankind. I also think about good friends I have lost over the years; about opportunities and decisions I would have changed knowing what I know now; and about whether any of this will have a real impact on the next generation of leaders.

At the beginning of this book, I thanked many people who have given me more, taught me more, and tolerated me more than I ever expected or deserved. To list all the people who made a significant impact on me—some of whom most likely do not even realize it—would take volumes. Here, for their impact on my leadership style and personal direction, I would like to thank from the bottom of my heart Al LaFountain, Chief John Sinclair, Chief Billy Goldfeder, Captain Ray Gayck, Deputy Chief Steve Cover, Captain R. B. and Teresa Ellis, Deputy Chief Jimmy Carter, all the operators on the Norfolk FBI SWAT team, my FEMA IST, VATF-2, VBF&R Hazmat team, and VBF&R tech team.

My son, a second lieutenant in the U.S. Army, is headed to Iraq in May 2007. I pray to God at least twice a day that he will be returned to me safe, alive, in one piece, and better for the experience. I hope that, as a new leader heading into harm's way, he finds the strength and wisdom to listen to the noncommissioned officers who have been there and done that; that he keeps safety and survivability foremost in his mind for the sake of the men and women under his command; that he is surrounded by a chain of command composed of great leaders, with the KSAs to make the right decisions; that he disregards the ego inherited from his father and does what is correct rather than what his emotions tell him; and finally, that he does his duty professionally, competently, and aggressively and return home safe and sound, as a better man who will have seen, done, and experienced things that the rest of us can only imagine. I would expect the same from any leader—this just happens to be my son.

I would love to correct all the personal and professional missteps in my life, but that's part of being human. At a Christmas party, a good friend of mine who was recently promoted to deputy chief of VBF&R told a member of my family, "Chase is an very intelligent man, but he sure does piss people off sometimes." I suppose he is right, some people I am sorry I pissed off; others I wish I had pissed off more. I do know this: leadership in all aspects of life is sorely lacking, yet it is critical to our existence and our success. It's not only about work but also about raising kids, managing your life, and trying to be a good person, attached to God, Country, Family, and Friends. It's hard to wake up every morning and strive to lead. As I retire and move on to other things, I am reminded that being a follower is not such a bad thing, because leadership takes its toll on you, even as it is worth every bit of yourself that you give. In the end, you only leave behind what you leave behind, and this book, along with whatever influence I had on supervisors, subordinates, organizational change, and my friends, is what I leave behind.

Appendix A
Sargent's Critical Commandments of Leadership

Thou shalt always understand the basics of organizational need and direction.

Thou shalt always keep humanity in your leadership.

Thou shalt always do your duty.

Thou shalt always follow the Universal Three.

Thou shalt always strive to maintain technical competence and educate yourself.

Thou shalt not be surprised when the kittens try to get out of the box.

Thou shalt understand that not all policy makes sense, is wise, or is even good for the organization, but it is still policy.

Thou shalt evaluate people based on performance and capability, not friendship or ease of application.

Thou shalt understand that prejudice is a human construct and that diversity is an organizational reality.

Thou shalt immediately address anger and violence in the workplace.

Thou shalt be accountable and responsible for your actions and the actions of your team(s).

Thou shalt never, ever, forget the business of the business.

About the Author

Chase Sargent is a decorated fire officer who spent more than 26 years with the City of Virginia Beach Fire and Rescue Department as a Firefighter/Paramedic, Lieutenant, Captain, Battalion Chief, and recently retired as a Division Chief in command of B-shift. He obtained his BS degree in Forestry and Wildlife, with a major in Fisheries Biology from Virginia Tech and his MPA from Golden Gate University. He completed the Executive Fire Officer program and is an NFA instructor. Chase is a member of VATF-2 and the Operations Chief for FEMA's Urban Search and Rescue White Incident Support Team. Chase was the night operations officer at both the Oklahoma City Bombing and the World Trade Center Collapse. He is the president of Spec Rescue International, a training and consulting company in Virginia Beach. Chase has been the Chief Tactical Paramedic for the Norfolk division of the FBI since 1995 and is both COTOMS and OEMS qualified. Currently, he works as an independent contractor to the US Government (DOS/DSS) and volunteers as a fire fighter in Virginia Beach and as a paramedic for the Courthouse Volunteer Rescue Squad.

The author can be contacted at cnsargent@aol.com for seminars or information.

Index

A

abilities, 352–354
absent without leave (AWOL), 113–114, 170
Abshoft, Captain, 86
accountability, 360
   communications and, 239–240
   definition of, 249
   firehouse/fire ground and, 231–257
   lack of, 245–248
   leadership traits for, 255–256
   look-over-the fence mentality and, 243–244
   maintaining, 251–253
   mistakes and, 253–255
   as normal operating procedure, 247–248
   success and, 248–251
   talk the talk, walk the walk and, 238–239, 359–361
   trust and, 234–243
   virus and, 244
action(s), 104
   conversion of words into, 16, 18
   end state of, 107
   leader's everyday, 5, 13
   Plan of, 20
   regular/repeated, 191–192
adams, Wild Bill, 254–255
adolescents, 25
adults, 25, 223
advanced life support (ALS), 142–143
affirmative Action (AA), 189
African Americans, 184, 186
"After the Fire," 13
air, situation, location, intent, and personal accountability (ASLIP), 107–108, 123
American Civil Liberties Union (ACLU), 195
American National Standards Institute (ANSI), 141
American steel companies, 246
Americans with Disabilities Act (ADA), 142
   procedures/regulations, 190
   Title VII of, 189
anger, 57, 103, 208–209
   definition of, 201–202
   management of disappointments, 202–204, 206
   violence in workplace, 199–211
Anglo-Saxons, 185
answers, direct, 37
anticipation, 70–71
appeals, 174
appointments, negligent, 124
arguments, 147
Asians, 186
asking, 43
assignments, negligent, 124
Atlantic Ocean, 335
attendance, 177
attitudes
   documentation and, 178–179
   positive, 237
   wait-and-see, 245
Augustine, Norman, 53
authority, 37, 204–205
auto makers, 246

B

backpacks, 94–95
Blanco, Kathleen Babineaux, 324
base of operations (BOO), 110

basic life support (BLS), 142
Beecher, Henry Ward, 39
benchmarking, 303–307
Best Practices in Emergency Services, 198
blacks, 187
Blackwell, Chris, 111
The Book Of Five Rings for Executives, 256
bosses, 364
  bunker mentality and, 366
  work style of, 365–366
Boulder, Colorado, 186
brain, human, 33
  cortex of, 213–214
  growth of, 33
  information and, 33–34
  memory, 33, 122–123
  men and, 33
  neurological composition of, 33–34
  women and, 33
Braveheart, 55
Brown, Michael, 324
Brown, Mike, 252
Brunacini, Alan, 203
Brunacini, Nick, 44
Buckingham, Marcus, 32, 34, 40
Buffett, Jimmy, 43
Bulova, 246
Bureau of Labor Statistics, 199
Butler, Frank, 222

C

Cain and Abel at Work (Domke & Lange), 47
California, 69, 195–196
canine training facilities, 252
caring, 15
Carter, James, 101
Carter, Jimmy, 376

cell phones, 59
censorships, 195
Census of Fatal Occupational Injuries (CFOI), 199
Central Intelligence Agency (CIA), 326
centralized control, 26
CEOs (chief executive officers), 10, 375
  errors made by, 19
  successful, 32
challenges, 37, 46
change
  as educational tool, 328–339
  it up, 46
  management of, 319–336
  model of placebo, 322–323
  within organizations, 8–9, 20, 42, 46
  people and, 31–32
  perilous journey of, 319–321
  reality of, 321–322
  rule, inverse, 7
  successful, 323–328
character, 72–73, 93
charge, take, 77
Chertoff, Michael, 324
Chesapeake Bay, 335
Chicago Cubs, 79
children, 25
choices
  making, 221–223
  people, 223–228
Chugani, Wayne, 34
Churchill, Winston, 31
circle of influence, 143–144, 155
City Communications and Information Technology Department (COMIT), 250
Civil Rights Act of 1964, Title VII, 189
Cizik, Robert, 329, 330
Clay County, Florida, 186
Clinton, Bill, 195
Clinton, Hillary Rodham, 172

Cobb County Fairgrounds, 109
Code of Federal Regulations, Title 29, 189
Coffman, Curt, 32, 34, 40
Cohen, Alan, 368
collective bargaining, 160, 367
Collins, John, 39–40
Colorado, 186
comfort zones, 46
commands, 262–263, 267
commendations, 177
commitments, 98, 133–134, 239–240
committees, 144
communication(s), 58
   accountability and, 239–240
   attorney/client, 177
   constant/open, 105–106
   in firefighting, 262–263
   lines of, 313
   within organizational leadership, 20
   poor, 330
   responsibility and, 239–240
   in senior leadership, 27
   shaping of vision, 26
   systems, 267
   two-way, 59
   undivided attention in, 59–60
Communications Decency Act (DCA), 195
compensation. *See also* pay
   bar/minimum requirements of, 162–163
   change orders, 177
   merit raises and, 160–165
   pay for performance and, 161
   pay ranges as, 162
   of people, 159–181
   performance, 162
   step raises and, 160
   top out, 161
   workers, 177

competence, 72, 121, 240.
   See also technical competence
   emotional/social, 135
   hiring and, 98
   mental, 134–135
   physical, 134
   spiritual, 135
   technical, 85
competition, 293
compression, 162
computer
   -based training, 128
   use of, 197
conclusions, rush to, 79
Congress, 33
consensus standards, 141–142
consequences, 223
consideration, 84
consistency, 238
control, 26, 37, 262–263
cooperation, 26
coordination, 266–267
Corps Business, 149
Cottrell, David, 92
counseling, 177
courage, 83
courts, 193–194
Cover, Steve, 88, 376
Covey, Stephen, 1, 42, 95, 108, 134, 143, 260
credibility, 120, 285
   establishment of, 136
   expertise and, 121
criticisms, 207
Croker, Edward, 337
Crowe, Russell, 55
culture
   changing, 373–376
   policies/procedure making, 139–144
   team, 288

curiosity, 296–297
customers, 40

D

Dahmer, Jeffrey, 63
debt collection, 190
decentralized execution, 26
decision-making
  choices in, 221–223
  completed staff work and, 219–221
  experiences/training and, 214–215
  fire ground, 223–225
  gut, 218–219
  moving on, 228–229, 368–371
  neurology and, 213–214
  pitfalls of, 214–216
  Powell on, 216–218
  team, 295–297
  values-based, 61–64, 226
dedication, 132
deniability, 50
denials, 246
Department of Homeland Security (DHS), 323, 325
Department of Justice, 184
desperation, 207
Detroit, 246
Diezel, Harry, 24, 80, 85–86, 88
differences, 34
dignity, 84
directions, 59, 125, 164
disappointments, 202–204
DiSC profile, 35–36
Discipline, 119, 177
  of employees, 169–171
  progressive, 169–170
  time frames for, 171
  truths/myths regarding, 169–171
discrimination
  age, 190
  employment, 189
  of national origin, 189
  religious, 189
  sex, 189
  sexual harassment, 190
disorder, 271
Disraeli, Benjamin, 1
dissent, 81
District of Columbia Fire Department, 186
distrust, 83
diversity
  prejudice/sexual harassment and, 183–198
  in workplace, 187–188
documentation
  attitudes and, 178–179
  evaluations and, 175–181
  the Hoover and, 179–181
  personnel files as, 176–178
  VBF&R, 126
doers, 82
Domke, Todd, 47
Downey, Ray, 43, 111, 342
drive, 72
Drucker, Peter, 235
duty
  definition of, 69, 70
  dereliction of, 169–170
  as leader/survival, 69–76

E

ears, 60
Edmund, Oklahoma, 200
education
  appropriate, 65–66
  failures, 354–357
  individual leadership and, 23
  people and, 32
  senior leadership and, 26–27
  technical competence and, 119–120
  tools of, 328–339

Index

effectiveness, 38
effort, 45, 107–108
egos, 72, 224–225
Einstein, Albert, 252–253
Eisenhower, Dwight, 83, 344
Ellis, R.B., 376
Ellis, Teresa, 376
emergency medical technicians (EMTs), 113, 163, 201
Emergency Services Unit (ESU), 112, 141
emotional bank accounts, 108–113
emotional competence, 135
emotions, 38, 40, 79, 226
   decisions and, 226
   false, 94
   judgments and, 213
empathy, 24
employee(s)
   behavior of, 173–174
   discipline of, 169–171
   fringe, 171–174
   intimidation of, 206
   violence, warning signs for, 206–207
employee assistance programs (EAPs), 174
employment
   discrimination, 189
   equal, 189–191
   letters, 177
employment Act, 190
EMS (emergency medical services), 64, 200
   grounds, 224
   skills, 101, 102
   systems, 142–143
EMT's Bill of Rights, 172
Encyclopedia of Ethics, 195
Enron, 246
entrustment, negligent, 124
environment
   hostile work, 191–192

leaders and, 1, 17–18
people and, 37
structuring an, 37
Equal Employment Opportunity (EEO), 189–191
Equal Pay Act, 189, 190
equipment, 285
ethics, 100
ethnicity, 184
Europe, 344
evaluation(s)
   desk/door, 167–168
   documentation and, 175–181
   formation of, 159–160
   HR and, 159–160, 162
   knowledge of, 164–165
   of people, 159–181
   positive influence, 168–169
   process, 165–167
   of teams, 280–283
   written, 167
expectations
   clear, 104
   establishment of, 100–101, 359
   grooming, 147, 148
   high, 87
   of members, 105–106, 149
   within organizations, 6
   test of, 99
experiences, 289
   decision-making and, 214–215
   levels of, 125–126
   past, 226–227
   practical, 37
expertise, 80–81
   credibility and, 121
   of individual leadership, 22–25, 144
eyes, 57, 60

F

facts, 217–218
failure(s), 289
   to educate, 354–357

human, 234
individual, 357–361
leadership, 234, 347–349
organizational/political, 350–371
standards, 234
of supervision, 124–125
survival after, 361–371
training, 124–125, 234
familiarity, 49
Fanning, Jack, 112
favoritism, 105
fears, 38
Federal Bureau of Investigation (FBI), 326
Federal Emergency Management Agency (FEMA), 80, 112, 324–325, 376
feedback, 58–59
continual, 299
instant, 298
The Fifth Discipline (Senge), 219
Filipinos, 187
finger-pointing, 246
fire chiefs, 1, 7
fire department
captains/officers in, 7, 14, 122
critical job aspects of, 121–122
minority, 183–188
multi-company drills of, 129
urban, 186
Fire Department of New York (FDNY), 112
fire service, 7–8, 27–28
battalion chiefs/officers in, 7, 8, 14, 116–117
changing face of, 185–187
definitions used by, 61, 65–66
frontline members in, 7–8, 57–58
global and, 65, 67, 68–69
junior members of, 64
leadership in, 53
middle managers in, 7–9, 58
officers/riverboat captains, 64–69
personnel, 139
senior staff of, 7–9

service within, 6–9
super models and, 66
supervisors, 57–58
Firefighter's Bill of Rights, 157, 172
irefightersCloseCalls.com, 113
firefighting, 7, 28, 109, 150, 340–341
battlefield, 259–272
command/communications/control in, 262–263
dangers of, 342–344
doing it right, 272–273
enemies of, 263–272
essence of, 337–339
final considerations regarding, 276
5% rule of, 339–340
hypercool, 273–276
prioritization and, 260–261
problem identification and, 261–262
ruination of career in, 259
firehouse
/fire ground/accountability, 231–257
pornography, 195–198
showstoppers in, 198
Firehouse (magazine), 87
First Amendment, 195, 196
First, Break All the Rules (Buckingham & Coffman), 32, 34, 40
fiscal considerations, 227–228
Florida, 186, 197
Florida Fire Department, 197
fluidity, 270–271
Forest Service, 109
Fortune 500, 32
Franklin, Benjamin, 144
frankness, 90
freedom, 37
Freedom of Information Act, 150
friction
asymmetrical, 266
external, 266
mental, 265
physical, 265–266

self-induced, 266–267
at work, 265–266
fun, 24, 89

G

Gallup Organization, 32–33
Galton, Francis, 296–297
Garvin, David, 329, 330
General Electric, 167–168, 327
geography, 144–145
Georgia, 110
Germans, 185
Gibson, Mel, 55
Giuliani, Rudolph, 204, 326
Gladiator, 55
Global Positioning System (GPS), 68
goals, 38, 266
Goethe, 99
Goldfeder, Billy, 113, 376
Good to Great (Collins), 39–40
governance model, 2–5
  feedback loops of, 5
  governance process within, 4
  management process within, 4
  poop loop of, 5
  service delivery system of, 4–5, 66
Government Employee Rights Act, 189
Graham, Charlene, 186
Graham, Gordon, 66, 125
Gravin, David, 319
Gray, A.M., 259, 263, 270
grievances, 172–173, 174, 177
Grimm, Charlie, 79
guidance, 45
guidelines, 61, 145
  response, 99
  tactical, 141
Gulf of Mexico, 68
Gurley, Rex, 252

H

hallucination, 55
Hanks, Tom, 148
happiness, 57
Harvard Business Review, 213
Harvard Business School, 32, 218, 241, 331
health-related/medical information, 177
hearing, 20
  individual leadership and, 23
  survival and, 56–60
hiring
  competence and, 98
  for jobs, 32
  practices, 184
Hispanics, 184, 186–187
Hitler, Adolf, 14–15, 63
homicides, 199
honesty, 15, 90
hostility, 204
HR Monthly, 198
Hubbard, Elbert, 73
human behaviors, 35–36, 47–48
human factors, 271–272
human resources (HR)
  evaluations and, 159–160, 162
  files, 173, 175–178
  work, 314
Hurricane Katrina, 310, 324

I

I syndrome, 358–359
identification, 37, 46, 288
immigration, 177
implementation/planning
  at company level, 311–315
  consequences of no, 309–311
  meetings, 312
  paralysis by analysis, 316
  paying dues and, 315–316

strategic, 310
time, 314
tools, 317–318
Incident Support Team (IST), 109
individuals, 55.
See also leadership, individual
influence, 38, 187
  circle of, 143–144, 155
  expansion of, 144–147
  policies, 143–144
  political, 145
  positive, 168–169
information
  flow, 26
  governance model for, 2–5
  health-related/medical, 177
  human brain and, 33–34
  initial, 79
  sharing, 46
initiative, 269–270
innovations, 165
instincts, 218–219, 222
integrity, 15, 71, 93, 96
intelligence, 70
intentions, 107
International Association of Fire Fighter (IAFF), 367
intimidation, 206
intuition, 218
inverse change rule, 7
Irish, 185
Israeli Special Forces, 334
Italians, 185

J

job(s)
  caring for, 85
  hiring for, 32
  maturity/technical competence, 121–129
  supervisors in, 40

Johnson v. County of Los Angeles Fire Department, 196
Johnston, Mark, 132–133
Jones, Bob, 155–156
Jordan, Michael, 44
judgment(s), 38, 70, 268
  emotions and, 213
  errors, 116
  personal, 217
  understanding/use of, 84

K

Kansas City, 200
Kanter, Moss, 331
karma
  operational, 79
  wheel, 53
Keith, Kent M., 90
Kennedy, John F., 279
Kidwell, Doug, 252
King, Martin Luther, 187
kittens in box, 105–106
Klein, Gary, 214, 226, 296, 301, 304
knowledge, 242
  Abel/Cain/workplace and, 47–50
  of A-players, 44
  base, diversification of, 23
  of B-players, 44–47
  of C-players, 45
  of D-players, 45
  of evaluations, 164–165
  of F-players, 45
  is power, 48–49
  of oneself/others, 24, 31–38
  styles, tendencies and, 38–50
Krishnamurti, Jiddu, 199
Krohn, Duane, 22
KSAs (knowledge, skills and abilities), 12, 78, 131, 241, 376
  lack of, 20
  operational, 288, 325
  required, 40

Index

specific, 126
standard set of, 133, 136
teaching of, 215

L

LaFountain, Al, 376
Lange, Gerry, 47
Langone, Tommy, 111
language, body, 57
Lasorda, Tommy, 273
law(s), 141–142
  of Average, 73
  obscenity, 196
leader(s), 242, 250
  environment and, 1, 17–18
  everyday action and, 5, 13
  personnel and, 99–104
  survival/duty as, 69–76
  teams and, 92
  tested as new, 99–117
  theories regarding, 1
leadership, 250, 285, 344–345
  cooperation and, 26
  core values and, 10–11, 14–16, 62–63
  critical commandments of, 377
  evolution of, 53
  failures, 234, 347–349
  in fire service, 53
  forms of, 16–24
  governance model for, 2–5
  humanity in, 89–98
  mission/vision statements and, 10–14, 67–68
  organizational foundation for, 1–30
  principles, 1
  profile, 35
  senior, 25–27
  styles, 63
  survival/ABCs of, 77–82, 282, 361
  survival/best characteristics of, 84–88
  time-management model for, 6–9
  turnover, 291
Leadership (Giuliani), 204
"Leadership in a Combat Zone," 26

leadership, individual
  charisma/presence and, 16, 21–22
  diversification of knowledge base by, 23
  education and, 23
  expertise of, 22–25
  failures of, 357–361
  hearing/listening/respect and, 23
  learning and, 23
leadership, organizational
  chain of command in, 18
  communication within, 20
  conversion of words into action in, 16, 18
  sound executive teams in, 18–19
  staff support operations within, 19
learning, 43, 322
  individual leadership and, 23
  moments of, 113
Learning in Action (Cizik & Garvin), 329, 330
Lee, Robert E., 360
Legal Briefings for Fire Chief, 198
Lesinski, Steve, 111
letters, 177
liability
  avoidance/creation of, 125–129
  vicarious, 124
lies, 49, 240
Ligachev, Yegor, 139
listening, 20, 43
  active, 237
  individual leadership and, 23
  survival and, 56–60
logic, 40
Lombardi, Vince, 294
Los Angeles County, 196
loyalty, 71, 227

M

MacGregor, Douglas, 295
Machiavelli, Niccolo, 81, 319
Madison, Wisconsin, 186

management
  of anger/disappointments, 202–204, 206
  of change, 319–336
  credit for, 49–50
  micro, 9, 78, 83, 305
  unions and, 160–161
Manson, Charles, 15, 63
Marine programs, 335–336
Mark Johnston Training Manual (Johnston), 133
Mars, 73
Maryland, 186, 200
Mauch, Paul, 85, 88
McGill University, 321
McKee, Robert, 197
meaning, 111
medical information, 177
memory, brain, 33, 122–123
Memphis Fire Department, 103
men, 33, 186
mental illness, 33
"A Message to Garcia," 73–76
messages, 58–59
Metro-Dade Task Force (FLTF-1), 109–110
Miami, 188
Military Review, 296
Miller, John Nathan, 60
Miller v. California, 195–196
Mintzberg, Henry, 321
mission/vision statements, 287
  examination of, 12
  examples of, 13
  implementation of, 287–288
  leadership and, 10–14, 67–68
  of VBF&R, 12, 153–154
Mississippi River, 68
mistakes, 97–98, 178–179, 253–255
Monday Morning Leadership (Cottrell), 92

money, 162
monkeys, story of, 146–147
Montgomery County, Maryland, 186, 200
morale, 41–42, 96, 286, 359–360
  high, 287, 288–289, 290–291
  improvement of, 54–55
  issues, 54
Morse, Gardiner, 213
motivation, 35, 41
Mount St. Helens, 39
mouths, 56–57, 60
muscles, 122–123
Myers-Briggs Indicator, 35

N

Nagin, Ray, 324
National Fire Protection Association (NFPA), 124, 136, 141–142
National Incident Management System (NIMS), 126, 267
National Institute of Occupational Health and Safety (NIOSH), 124, 141, 200–201
National Security Agency (NSA), 60
NATO, 60
negligence, 124, 169–170, 178–179
neurons, 33
New Orleans, 68, 310, 324
New York City, 112, 188, 246
New York City Police Department (NYPD), 112
9/11, 83, 340–342
Norfolk FBI SWAT team, 376

O

Oakland Fire Department, 186
obscenity laws, 196

Occupational Safety and Health Administration (OSHA), 124, 136, 141
   Guidelines for Preventing Workplace Violence for Health Care and Social Service Workers, 201
   rules, 272
   standard 148, 201
   on violence, 199–200, 201

Oklahoma, 21, 200

Oklahoma City, 21

Ontario, California, 69

Operation Overload, 344

operations, 164

orders, giving of, 106–108

ordinances, 141–142

organizational leadership
   mistakes of, 334–336
   plan of action within, 20

organizations, 38, 223
   change within, 8–9, 20, 42, 46
   expectations within, 6
   large, 2
   members of, 12
   objectives of, 26–27
   people-based, 91–92
   political system within, 2
   senior staff of, 12
   support within, 286
   training within, 132–133
   values of, 15–16

over inspect, 82–83

The Oz Principle (Connors, Hickman, Smith), 249

P

Page, James O., 43, 87, 159, 351

Pagonis, William, 26

Paradoxical Commandments, 90

Parikh, Jagdish, 219

passions, 57

Past Experiences (Gary), 226

pay. *See also* compensation; salaries
   for performance, 161
   plans, 165
   ranges, 162
   roll, 177

people, 98
   -based organizations, 91–92
   change and, 31–32
   choices, 223–228
   compensation/evaluation of, 159–181
   education and, 32
   environment and, 37
   growth of, 166
   as important assets, 129–133
   motivation of, 35
   before plans, 92–95
   talents of, 32–33
   tendencies of, 35–36
   understanding, 31–35, 86–87

performance, 287
   compensation, 162
   high, 290–291
   pay for, 161
   principles, 164

personal alert locator (PAL), 123

personal protective equipment (PPE), 130

personnel, 285
   files, 176–178
   fire service, 139
   leaders and, 99–104
   mentoring of, 106
   selection, 287
   turnover, 291

persuasiveness, 194

Peter principle behaviors, 20

Phoenix, 44

physical training (PT)
   gears, 104
   policies, 149
   providing, 101–102

plan(s)
   of action, 20
   complicated/unclear, 267
   conception of, 93

contingency, 269
effective, 92, 106
people before, 92–95
simple, flexible, 269
succession, 285–286
treatment, 228
planning/implementation
  at company level, 311–315
  consequences of no, 309–311
  meeting, 312
  paralysis by analysis, 316
  paying dues and, 315–316
  strategic, 310
  time, 314
  tools, 317–318
Plato, 373
Playboy, 196
Plumb, Charles, 28–29, 249
Pol Pot, 63
policies, 61, 164
  adherence to, 101
  conclusions regarding, 158
  creation of, 140, 146
  crisis, 142
  enforcement/understanding, 105, 139–158
  formal, 99
  good sound, 141
  incident management, 141
  influence of, 143–144
  informal, 99
  mayday, 141
  nonenforcement of, 147–149
  operations of emergency vehicles, 141
  political, 142–143
  procedure making/culture, 139–144
  PT, 149
  retooling, 149–154
  ride-along, 141
  vehicle maintenance, 141
  violation of, 154–158, 173
  zero-tolerance pornographic, 196–197
  Policy, procedure, guidelines, and rules (PPGR), 150, 153–154, 215–216
politics, 47–48, 226, 227

pornography, 185
  determination of, 197
  developments in, 195–196
  firehouse, 195–198
  zero-tolerance policies on, 196–197
Powell, Colin, 333
  on decision-making guidelines, 216–218
  views of, 26, 57, 70, 73, 164
power, 37, 48, 141, 364
prejudice
  description of, 183–185
  diversity/sexual harassment and, 183–198
pressure, 38
prestige, 37
Privacy Act, 189
probabilities, 268–269
procedures, 61, 146, 164
promotions, 144
punishment, 169

R

rapid intervention crew (RIC), 272–273
rapid intervention team (RIT), 21, 272–273
Reagan, Ronald, 231
reality, 45, 321–322
reasonable person standard, 194
records retention, 177–178
Red Cross, 13
Reed, Richard, 326
Reich, Gary, 24, 100
religions, 184, 189
Reno v. ACLU, 195
resolve, 99
respect, 120–121
  individual leadership and, 23
  trust and, 352–354
responses, 121, 313

Index

responsibility, 360
  communications and, 239–240
  definition of, 249
  firehouse/fire ground and, 231–257
  lack of, 245–248
  leadership traits for, 255–256
  look-over-the fence mentality and, 243–244
  maintaining, 251–253
  mistakes and, 253–255
  as normal operating procedure, 247–248
  success and, 248–251
  talk the talk, walk the walk and, 238–239, 359–361
  trust and, 234–243
  virus and, 244
results, 95
retention, negligent, 124
retired on active duty (ROAD), 121, 245
retirement, 68
retooling, 149–154
  benefits of, 150
  methods/techniques of, 151–152
  rules/values and, 152–154
rewards, 165
Rickover, Hyman, 89
Ridge, Tom, 324
risk(s)
  encouragement of, 79–80, 272
  gains and, 269
  management, 125
  managers, 123–125
  taking, 87
Rogers, Will, 86, 119
Rommel, Erwin, 94, 230
rookies, 106, 133
Roosevelt, Theodore, 25, 295
rules, 61–62, 164
  definition of, 145–146
  of engagement, 108
  values/retooling and, 152–154
  violation of, 173

S

Safe City working group, 55
St. Petersburg, 197
salaries, baseline, 160.
  See also compensation; pay
Salt Lake City, 185, 188, 197
San Francisco, 186, 188, 197
San Francisco Fire Department, 186, 188
Sargent, Chief, 91, 156
satisfaction, 82
Saving Private Ryan, 147–148
Schoomaker, Peter J., 222
screamers, 273
second chances, 97–98, 116
self-assured, 72
self-awareness, 72
self-contained breathing apparatus (SCBA), 22, 103, 104, 232
self-evaluation, 82
self-improvement, 82
self-interests, 73
self-monitoring, 290, 302, 307
self-starters, 82
Senge, Peter, 219
senior staff block, 330
service
  customer, 57–58, 166
  delivery system, 4–5, 66
  within fire service, 6–9, 57–58
The 7 Habits of Highly Effective People (Covey), 108, 135, 260
sexual harassment
  courts on, 193–194
  description of, 191–195
  discrimination, 190
  forms of, 190
  hostile work environment, 191–192
  prejudice/diversity and, 183–198
  quid pro quo, 191–192

recognition of, 192–193
severity of, 193–194
Sheffield Steel, 246
Sheriff, R.C., 228
Sherrill, Patrick Henry, 199
showmanship, 83–84
simplify, 81–82, 282
Simpson, Liz, 364–365
Sinclair, John, 376
skills
   administrative, 101
   current, 47
   human interaction, 32
   technical, 32
Smith, Sydney, 373
SOP (standard operating procedure), 102, 151, 269
soul, 54, 59
Sources of Power, How People Make Decisions (Klein), 214
Spain, 73
speaking, 84
Spec Rescue, 112
Spore, James, 4
Stalin, Joseph, 15, 63
standards, high, 87
Stanton, Earl, 84–85, 88
statutes, 141–142
success, 86, 90
   accountability and, 248–251
   key decisions for, 331–334
   octagon factors, 287, 292
Sunshine Act, 189
supervision, 37, 124–125
supervisors, 177
   actions of, 99, 194–195
   challenges of, 5
   everyday action of, 5, 13
   fire service, 57–58
   in jobs, 40
   over, 82–83
   rules for, 40–41
Surowiecki, James, 296
survival
   ABCs of leadership and, 77–82, 282, 361
   best leadership characteristics and, 84–88
   duty as leader and, 69–76
   fire service officers/riverboat captains and, 64–69
   hearing/listening and, 56–60
   keeping your cool and, 79, 84
   leadership methods for, 82–84
   moment of truth and, 64–65, 344–345
   requirements of, 122–123
   universal rules for, 53–88
   universal three in, 77
   values-based decision-making and, 61–64
   vision and, 54–55
Swarzkopf, H. Norman, 72
synapses, 33

T

talent
   inclusive, 81
   manage for, 96
   scouts, 40–43
Tarver, Bret, 272
teaching, 131–132
   f KSAs, 215
   with stories, 111–112
team(s)
   conceptual level, 290, 302, 305–307
   culture, 288
   decision training, 295–308
   evaluation of, 280–283
   identity, 290, 301
   leaders and, 92
   life cycles, 279–294
   members, 92, 95–97
   residences, 297–298
   self-monitoring, 290, 302, 307
   small, 361
   special operations, 241

specialty/task, 283
succession, 290
technical competence
  credibility and, 120
  education and, 119–120
  final word on, 135–137
  job maturity and, 121–129
  maintaining, 119–137
  other aspects of, 134–137
  people and, 129–133
  respect/trust/other concepts and, 120–121
technical skills, 32
teenagers, 25
Telecommunications Decency Act (1996), 195
tendencies
  of people, 35–36
  styles and, 39–50
terminations, 177
Thomas, Franklin, 183
Thordsen, Marvin L., 296, 301, 308
2001, lessons from, 325–326
2006, lessons from, 326–328
traditional programs, 233, 314
training, 59, 119, 285, 289–290.
  See also physical training; technical competence
  appropriate, 65–66, 83
  computer-based, 128
  concepts, 128–129
  decision-making and, 214–215
  department, 121
  documented/ongoing/realistic/ verifiable, 126–129
  facilities, canine, 252
  failures, 124–125, 234
  intense, 288
  officer, 116, 133
  within organizations, 132–133
  physical, 101–102, 104, 313
  specialty, 133
  team decision, 295–308
  tempo, 292–293

themes related, 124–129
virtual reality, 128
transition drills, 123
Transportation Security Administration (TSA), 326
trust, 15, 20, 41, 352–354
  relationship-based, 235–237
  subordinates, 77–78
  technical competence and, 120–121
  understanding, 234–243
truth, moment of, 64–65, 344–345
Twin Towers, 83
Tyco, 246

U

uncertainties, 267–270
unions, 160–161
United States (U.S.), 73, 195, 241
U.S. Army Special Operations Command, 222
U.S. Citizenship and Immigration Services, 326
U.S. Coast Guard, 324
U.S. District Court, 195, 196
U.S. Equal Employment Opportunity Commission (EEOC), 189
U.S. Government Printing Office, 189
U.S. Naval Academy, 28
U.S. Navy Medical Corps, 222
U.S. Supreme Court, 195
US&R (Urban Search & Rescue), 252, 324

V

values
  -based decision-making, 61–64, 226–227
  caring as, 15
  definition of, 62
  honesty as, 15
  integrity as, 15
  leadership/core, 10–11, 14–16, 62–63

organizational, 15–16
rules/retooling and, 152–154
trust as, 15

Vancouver, 197

ventilation, 241

verbalization, 37

verification, 126–127

victim cycle, 247–248

Vietnam, 28

violence
anger in workplace, 199–211
conclusions on, 211
definition of, 201
NIOSH on, 200–201
OSHA on, 199–200, 201
potential for, 204–207
random, 204
situations leading to, 209–210
supervisors and, 207–208
threats of, 208–209
warning signs for employee, 206–207

Virginia Beach, 4, 55, 80

Virginia Beach Fire and Rescue (VBF&R), 21, 47, 84
apparatus of, 254
completed staff work and, 219–221
documentation and, 126
Hazmat team, 376
members of, 132–133
minority firefighters at, 184–185, 188
mission/vision statements of, 12, 153–154
programs of, 250
reputation of, 101–102
suspensions, 157
tech team, 376
traditional programs of, 233, 314

Virginia Task Force 2 (VATF-2), 109

vision
communication/shaping of, 26
creation of, 9
development of, 78–79, 282
essence of, 54–55
implementation of, 55

statements of leadership, 10–14
survival and, 54–55

Von Clausewitz, Clause, 259, 265

W

wage garnishments, 177

Wallace, Rusty, 160

war, 263

Warfighting: The U.S. Marine Corps Book of Strategy (Gray), 259, 263

Washington, Booker T., 234

Wayne State University, 34

weapons, 206

Webster's Dictionary, 69

Welch, Jack, 327

White, Keith, 85, 87–88

Wisconsin, 186

The Wisdom of Crowds (Surowiecki), 296

women, 33, 186–187

work
completed staff, 219–221
environment, hostile, 191–192
friction at, 265–266
place, 47–50, 186–188, 199–210
plan, 174
right-to-, 160, 367

World Trade Center, 21, 111–112

World War II, 344

WorldCom, 246

writing, 84

Y

yelling, 84

Z

Zsambok, Caroline E., 296, 301, 308